INTRODUCTION TO ENHANCED OIL RECOVERY (EOR) PROCESSES AND BIOREMEDIATION OF OIL-CONTAMINATED SITES

Edited by **Laura Romero-Zerón**

Introduction to Enhanced Oil Recovery (EOR) Processes and Bioremediation of Oil-Contaminated Sites
Edited by Laura Romero-Zerón

Published by InTech
Janeza Trdine 9, 51000 Rijeka, Croatia

Publishing Process Manager Vana Persen

Technical Editor Teodora Smiljanic

Cover Designer InTech Design Team

Additional hard copies can be obtained from orders@intechopen.com

Introduction to Enhanced Oil Recovery (EOR) Processes
and Bioremediation of Oil-Contaminated Sites, Edited by Laura Romero-Zerón
p. cm.
ISBN 978-953-51-0629-6

For EU product safety concerns:
IN TECH d.o.o., Prolaz Marije Krucifikse Kozulić 3, 51000 Rijeka, Croatia,
info@intechopen.com or visit our website at intechopen.com.

Meet the editor

Laura Romero-Zerón is a chemical engineer and an associate professor in the Chemical Engineering Department, at the University of New Brunswick, Fredericton, Canada. Previously, she worked at the Research Center of "Petróleos the Venezuela" (PDVSA) as a research scientist in the area of enhanced oil recovery (EOR). She holds a PhD degree in Chemical & Petroleum Engineering from the University of Calgary, Alberta, Canada. Her research interests include optimization of EOR chemical flooding, evaluation of the mechanisms of polymer gels as conformance-improvement agents, heavy oil recovery, and bitumen upgrading. Romero-Zerón has published several scientific papers and is the co-author of a book in the area of reservoir conformance-improvement. She is a SPE member and holds the SPE Petroleum Engineering Certification.

Contents

Preface XI

Section 1 Introduction to Enhanced Oil Recovery (EOR) 1

Chapter 1 **Advances in Enhanced Oil Recovery Processes 3**
Laura Romero-Zerón

Chapter 2 **The Application of a New Polymeric Surfactant for Chemical EOR 45**
Khaled Abdalla Elraies and Isa M. Tan

Chapter 3 **Microbial Enhanced Oil Recovery 71**
Hamid Rashedi, Fatemeh Yazdian and Simin Naghizadeh

Chapter 4 **Enhanced Oil Recovery in Fractured Reservoirs 89**
Martin A. Fernø

Section 2 Environmental Management Through Bioremediation 111

Chapter 5 **Microorganisms and Crude Oil 113**
Dorota Wolicka and Andrzej Borkowski

Chapter 6 **Comprehensive Perspectives in Bioremediation of Crude Oil Contaminated Environments 143**
Chukwuma S. Ezeonu, Ikechukwu N.E. Onwurah and Obinna A. Oje

Chapter 7 **Hydrocarbon Pollution: Effects on Living Organisms, Remediation of Contaminated Environments, and Effects of Heavy Metals Co-Contamination on Bioremediation 185**
Shukla Abha and Cameotra Swaranjit Singh

Chapter 8 **Bioremediation of Crude Oil Contaminated Soil by Petroleum-Degrading Active Bacteria 207**
Jinlan Xu

Chapter 9 **Microbial Outlook for the Bioremediation of Crude Oil Contaminated Environments 245**
Rachel Passos Rezende, Bianca Mendes Maciel, João Carlos Teixeira Dias and Felipe Oliveira Souza

Chapter 10 **Crude Oil Metagenomics for Better Bioremediation of Contaminated Environments 261**
Ines Zrafi-Nouira, Dalila Saidane-Mosbahi, Sghir Abdelghani, Amina Bakhrouf and Mahmoud Rouabhia

Chapter 11 **Oil-Spill Bioremediation, Using a Commercial Biopreparation "MicroBak" and a Consortium of Plasmid-Bearing Strains "V&O" with Associated Plants 291**
Andrey Filonov, Anastasia Ovchinnikova, Anna Vetrova, Irina Puntus, Irina Nechaeva, Kirill Petrikov, Elena Vlasova, Lenar Akhmetov, Alexander Shestopalov, Vladimir Zabelin and Alexander Boronin

Preface

Worldwide, crude oil demand is unceasingly increasing. As a response, Enhanced Oil Recovery (EOR) processes have re-gained interest from the research and development phases to the oilfield EOR implementation stage. This renewed interest has been also furthered by the current high oil price environment, the maturation of oilfields worldwide, and few new-well discoveries. Concurrently, environmental concerns and public pressure related to crude oil pollution control and remediation of oil-contaminated sites are becoming greater than ever.

This book provides a concise and accurate introduction of enhanced oil recovery (EOR) processes and bioremediation of oil-contaminated environments. The book is structured in two sections. The first section, which contains 4 chapters, presents a brief and simplified description of the mechanisms of primary and secondary oil recovery processes and clearly explains the need for the implementation of EOR processes to boost the recovery of oil and/or to displace and produce the significant amounts of oil left behind in the reservoir after or during the course of any primary and secondary recovery process. The mechanisms, limitations, and problems encounter during the application of EOR methods are succinctly reviewed. This section also highlights the emerging EOR technological trends and presents key EOR areas that need to evolve through research and development.

The second section of the book focuses on the use of biological action to remediate the inevitable environmental footprint of crude oil production; such is the case of accidental oil spills on aqueous (i.e. marine, river) and land environments. This section is structured in seven chapters and begins introducing specific microorganisms that are capable of living in and/or degrading crude oil. Then, it reviews the concepts of bioremediation, phytoremediation, and biomonitoring. It also provides practical information on the isolation and evaluation of petroleum degrading microorganisms and discusses the main variables affecting bioremediation of aqueous (marine, river) and land sites contaminated with crude oil. Furthermore, it introduces crude oil metagenomics, which is an emerging biotechnology useful to improve the evaluation of the diversity, identification, isolation, and functional properties of microorganisms involved in the degradation of different petroleum compounds. The molecular analysis of the bacterial composition present in polluted sites represents an essential stage to improve the current understanding of the microorganisms involved in the degradation of hydrocarbons. This section closes with insights on the mixture of microorganisms to formulate highly efficient bio-consortia to target the degradation and efficient removal of crude oil pollutants from contaminated environments.

This book is target to a wide range of readers who might have some basic knowledge in these subjects but are looking for more matter-of-fact information. Thus, through this book practical concepts in enhanced oil recovery methods and bioremediation of oil-polluted sites were grouped, structured, and placed all together in context with the ultimate goal of being useful for a non-specialized audience. Thus, this book is useful for university students (undergraduate and graduate students), oilfield operators, reservoir engineers, environmental engineering professionals, oil business managers and/or executives, among others.

The sources used to aid in the writing process consisted mainly on the review of the most recent scientific literature in these fields and on discussion of current research and development conducted directly by the authors and co-authors of some of the chapters. The main objective is to provide state of the art information on the emerging technological trends in the fields of EOR and bioremediation.

Finally, I would like to express my appreciation and thankfulness to the authors and coauthors who contributed writing the different chapters of the book. I also gratefully recognize, the effort and time devoted by the technical reviewers and editors team, who helped and contributed towards the writing process and publishing of this book.

Laura Romero-Zerón
Chemical Engineering Department
University of New Brunswick
Canada

Section 1

Introduction to Enhanced Oil Recovery (EOR)

1

Advances in Enhanced Oil Recovery Processes

Laura Romero-Zerón
University of New Brunswick, Chemical Engineering Department
Canada

1. Introduction

In the last few years, Enhanced Oil Recovery (EOR) processes have re-gained interest from the research and development phases to the oilfield EOR implementation. This renewed interest has been furthered by the current high oil price environment, the increasing worldwide oil demand, the maturation of oilfields worldwide, and few new-well discoveries (Aladasani & Bai, 2010).

Oil recovery mechanisms and processes are concisely reviewed in this chapter. A brief introduction to primary and secondary oil recovery stages is provided; while the main focus of the chapter is given to EOR processes with emphasis on EOR emerging technological trends.

2. Hydrocarbon recovery

Hydrocarbon recovery occurs through two main processes: primary recovery and supplementary recovery. Primary recovery refers to the volume of hydrocarbon produced by the natural energy prevailing in the reservoir and/or artificial lift through a single wellbore; while supplementary or secondary hydrocarbon recovery refers to the volume of hydrocarbon produced as a result of the addition of energy into the reservoir, such as fluid injection, to complement or increase the original energy within the reservoir (Dake, 1978; Lyons & Plisga, 2005).

2.1 Primary oil recovery mechanisms

The natural driving mechanisms of primary recovery are outlined as follows.

- Rock and liquid expansion drive
- Depletion drive
- Gas cap drive
- Water drive
- Gravity drainage drive
- Combination drive

Hydrocarbon reservoirs are unique; each reservoir presents its own geometric form, geological rock properties, fluid characteristics, and primary driving mechanism. Yet, similar reservoirs are categorized based on their natural recovery mechanism. Table 1 summarizes the performance of each of the primary recovery mechanisms in terms of pressure decline rate, gas-oil ratio, water production, well behaviour, and oil recovery as presented by Ahmed & McKinney (2005).

Primary recovery from oil reservoirs is influenced by reservoir rock properties, fluid properties, and geological heterogeneities; so that on a worldwide basis, the most common primary oil recovery factors range from 20% and 40%, with an average around 34%,while the remainder of hydrocabon is left behind in the reservoir (Satter et al., 2008).

Once the natural reservoir energy has been depleted and the well oil production rates decline during primary recovery, it is necessary to provide additional energy to the resevoir-fluid system to boost or maintain the production level through the application of secondary production methods based on fluid injection (Satter et al., 2008).

2.2 Supplementary or secondary hydrocarbon recovery

Secondary hydrocarbon (oil and/or gas) involves the introduction of artificial energy into the reservoir via one wellbore and production of oil and/or gas from another wellbore. Usually secondary recovery include the immiscible processes of waterflooding and gas injection or gas-water combination floods, known as water alternating gas injection (WAG), where slugs of water and gas are injected sequentially. Simultaneous injection of water and gas (SWAG) is also practiced, however the most common fluid injected is water because of its availability, low cost, and high specific gravity which facilitates injection (Dake, 1978; Lyons & Plisga, 2005; Satter et al., 2008).

The optimization of primary oil recovery is generally approached through the implementation of secondary recovery processes at early stages of the primary production phase before reservoir energy has been depleted. This production strategy of combining primary and secondary oil recovery processes commonly renders higher oil recovery if compared to the oil production that would be obtained through the single action of the natural driving mechanisms during primary oil recovery (Lyons & Plisga, 2005).

2.2.1 Waterflood process

Waterflooding is implemented by injecting water into a set of wells while producing from the surrounding wells. Waterflooding projects are generally implemented to accomplish any of the following objectives or a combination of them:

- Reservoir pressure maintenance
- Dispose of brine water and/or produced formation water
- As a water drive to displace oil from the injector wells to the producer wells

Over the years, waterflooding has been the most widely used secondary recovery method worldwide. Some of the reasons for the general acceptance of waterflooding are as follows (Satter et al. 2008). Water is an efficient agent for displacing oil of light to medium gravity, water is relatively easy to inject into oil-bearing formations, water is generally available and inexpensive, and waterflooding involves relatively lower capital investment and operating costs that leads to favourable economics.

Primary Recovery Mechanism	Characteristics	
① **Rock and liquid expansion drive**	Reservoir pressure	Above bubble point: rapid and continuous pressure decline until bubble point is reached.
	Gas-oil ratio (GOR)	Above bubble point: GOR remains low and constant.
	Water production	Little or no water production
	Well behavior	Requires pumping at early stage
	Oil recovery	Least efficient driving mechanism. Oil recovery efficiency typically varies from 1% to 5%, with an average of 3%.
② **Depletion drive:** Solution gas drive Dissolved gas drive Internal gas drive	Reservoir pressure	Declines rapidly and continuously
	Gas-oil ratio (GOR)	Increases to a maximum and then declines
	Water production	Little or no water production
	Well behavior	Requires pumping at early stage
	Oil recovery	Very inefficient driving mechanism. Varies from less than 5% to about 30%, with an average of 16%.
③ **Gas cap drive**	Reservoir pressure	Declines slowly and continuously
	Gas-oil ratio (GOR)	Increases continuously and as the expanding gas cap reaches the producing intervals, the gas-oil ratio increases sharply and finally drops.
	Water production	Absent or negligible
	Well behavior	Tends to flow longer than depletion drive reservoirs
	Oil recovery	Ranges from 20% to 40%, with an average of 25%
④ **Water drive** Usually the most efficient reservoir driving force	Reservoir pressure	Remains high
	Gas-oil ratio (GOR)	Remains low
	Water production	Stars early and increases to appreciable amounts
	Well behavior	Flows until water production gets excessive
	Oil recovery	Ranges from 30% to 80%
⑤ **Gravity drainage drive**	Reservoir pressure	Rapid pressure decline
	Gas-oil ratio (GOR)	Commonly low gas-oil ratios
	Water production	Little or no water production
	Well behavior	Structurally low wells show low GOR Structurally high wells show increasing GOR
	Oil recovery	Varies broadly but usually high oil recoveries are observed. Recoveries up to 80% have been reported.
⑥ **Combination drive** Usual combinations ②+④ ②+③+④	Reservoir pressure	Relatively rapid pressure decline
	Gas-oil ratio (GOR)	Manage to maintain low GOR
	Water production	Slow increase of water production
	Well behavior	Structurally low wells show low GORs Structurally high wells show increasing GORs
	Oil recovery	Usually higher than depletion drive reservoirs but less than recovery from water drive or gas cap drive reservoirs. Ultimate recovery depends on the degree to which it is possible to reduce the magnitude of recovery by depletion drive.

Table 1. Primary Recovery Mechanisms Performance (Adapted from Ahmed & McKinney, 2005; Satter et al. 2008).

Waterflooding is generally implemented by following various types of well flooding arrangements such as pattern flooding, peripheral flooding, and crestal flooding, among others. Pattern flooding is used in reservoirs having a small dip (not flat-lying reservoirs) and a large surface area. Figure 1 presents the geometry of common pattern floods. Economic factors are the main criteria for the selection of a specific pattern geometry; these factors include the cost of drilling new wells, the cost of switching existing wells to a different type (i.e., a producer to an injector), and the loss of revenue from the production when making a switch from a producer to an injector. For instance, the direct-line-drive and staggered-line-drive patterns are frequently used because they require the lowest investment. However, if the reservoir characteristics yield lower injection rates than those desired, the operator should consider using either a seven- or a nine-spot pattern where there are more injection wells per pattern than producing wells as suggested by Craft & Hawkins, (1991).

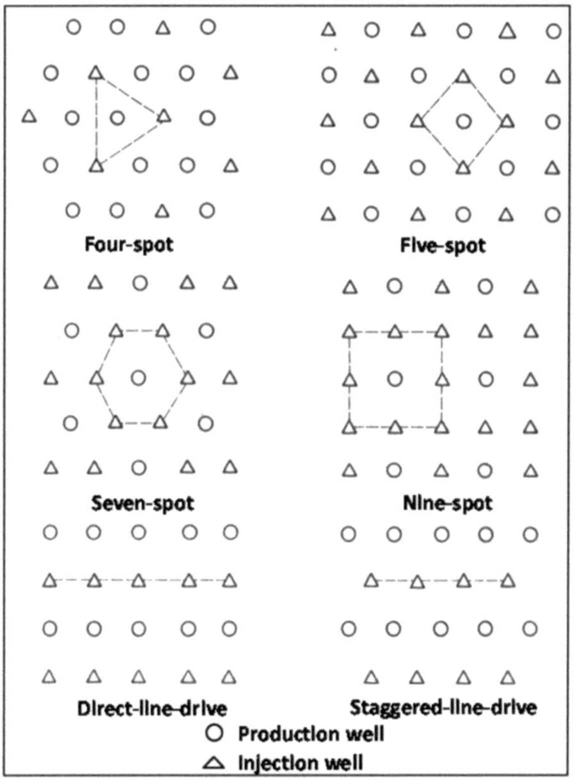

Fig. 1. Geometry of common regular pattern floods (Craft & Hawkins, 1991)

In the regular patterns shown in Fig. 1, the producer wells are always located in the centre of the pattern, surrounded by the injector wells; while the opposite is true for the inverted pattern floods, where the injectors are drilled in the middle of the pattern, and producers are at the corners.

Optimization of oil recovery during the life of a waterflood project is approached by changing over time the injector/producer pattern and well spacing. Thus, based on simulation studies and economic analyses, producers are converted to injectors, infill wells are drilled, and a relatively dense well spacing is implemented at certain stages of recovery.

However, the implementation of any pattern flood modification is conditioned to the expected increase in oil recovery and whether the incremental oil justifies the capital expenditure and operating costs. Figure 2 shows an example in which a waterflood operation was initiated using an inverted 9-spot pattern that was gradually transformed to a regular 5-spot pattern at later stages of waterflooding through well conversion and infill drilling (Satter et al., 2008).

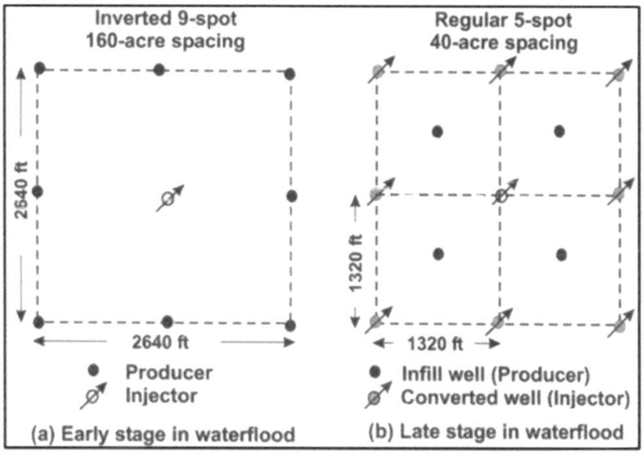

Fig. 2. Modifications of the injector/producer pattern and well spacing over the life of a waterflooding project to optimize the recovery of oil: (a) Early stage and (b) Late stage (Satter et al., 2008).

In peripheral flooding, the injection wells are positioned around the periphery of a reservoir. In Figure 3, two cases of peripheral floods involving reservoirs with underlying aquifers are shown. In the anticlinal reservoir of Fig. 3a, the injector wells are placed in such a manner that the injected water either enters the aquifer or is near the aquifer-reservoir interface displacing oil towards the producer wells located at the upper part of the reservoir, thus in this case the geometrical well configuration is similar to a ring of injectors surrounding the producers. For the monoclinal (dipping or not flat lying) reservoir illustrated in Fig. 3b, the injector wells are placed down dip to take advantage of gravity segregation, thus the injected water either enters the aquifer or enters near the aquifer-reservoir interface. In this situation, the well configuration renders the grouping of all the injector wells on the structurally lower side of the reservoir (Craft & Hawkins, 1991).

In reservoirs having sharp structural features, the water injection wells can be located at the crest of the structure to efficiently displace oil located at the top of the reservoir; this is known as crestal injection. In any case, injection well configuration and well spacing depend on several factors that include rock and fluid characteristics, reservoir heterogeneities, optimum injection pressure, time frame for recovery, and economics (Satter et al., 2008).

Under favorable fluid and rock properties, current technology, and economics, waterflooding oil recovery ranges from 10% to 30% of the original oil in place (OOIP).

2.2.1.1 Low-salinity waterflooding

Low-salinity waterflooding is an emerging process that has demonstrated to increase oil recovery. The mechanisms associated to this processes are still unclear, however the favorable oil recovery are attributed to fine migration or permeability reduction and wettability alteration in sandstones when the salinity or total solids dissolved (TSD) in the injected water is reduced. In the case of carbonate formations, the active mechanisms are credited to wettability alteration and to interfacial tension reductions between the low salinity injected water and the oil in the carbonate formation (Okasha & Al-Shiwaisk, 2009; Sheng 2011).This waterflooding process requires more research in order to clearly establish the mechanisms involved and an understanding of the application boundaries based on the type of reservoir formation to avoid adverse effects on reservoir permeability caused by the injection of water that could negatively interact with the formation water and the formation rock (Aladasani & Bai, 2010).

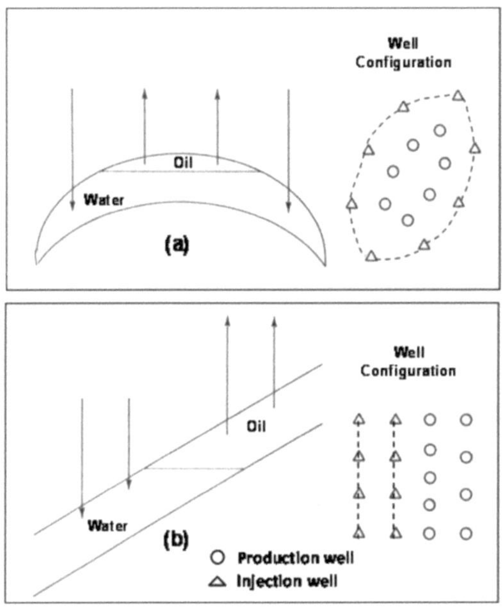

Fig. 3. Well Configuration for peripherial waterflooding of reservoirs with underlying aquifers: (a) Anticlinal reservoir and (b) Monoclinal reservoir (Craft & Hawkins, 1991).

2.2.2 Gas injection

Immiscible gas (one that will not mix with oil) is injected to maintain formation pressure, to slow the rate of decline of natural reservoir drives, and sometimes to enhance gravity drainage. Immiscible gas is commonly injected in alternating steps with water to improve recovery. Immiscible gases include natural gas produced with the oil, nitrogen, or flue gas. Immiscible gas injected into the well behaves in a manner similar to that in a gas-cap drive: the gas expands to force additional quantities of oil to the surface. Gas injection requires the use of compressors to raise the pressure of the gas so that it will enter the formation pores (Van Dyke, 1997).

Immiscible gas injection projects on average render lower oil recovery if compared to waterflooding projects, however in some situations the only practicable secondary recovery process is immiscible gas injection. Those situations include very low permeability oil formations (i.e. shales), reservoir rock containing swelling clays, and thin formations in which the primary driving mechanism is solution-gas drive, among others (Lyons & Plisga, 2005).

Table 2 summarizes the oil recovery efficiencies from primary and secondary recovery processes obtained from production data from several reservoirs in the United States.

	Recovery Efficiency			
Reservoir Location	**Primary %OOIP**	**Type of Secondary Recovery**	**Secondary % OOIP**	**Oil Remaining % OOIP**
California Sandstones	26.5		8.8	64.7
Louisiana Sandstones	36.5		14.7	48.8
Oklahoma Sandstones	17.0	**Pattern Waterfloods**	10.6	72.4
Texas Sandstones	25.6		12.8	61.6
Wyoming Sandstones	23.6		21.1	55.3
Texas carbonates	15.5		16.3	68.2
Louisiana Sandstones	41.3	**Edge Water Injection**	13.8	44.9
Texas carbonates	34		21.6	44.4
California Sandstones	29.4	**Gas Injection Into Cap**	14.2	56.4
Texas Sandstones	35.3		8.0	56.7

Table 2. Oil Recovery Efficiencies as % of OOIP from Primary and Secondary Recovery (Adapted from Lyons & Plisga, 2005).

As Table 2 shows after primary and secondary oil recovery, a significant amount of oil is left behind in the reservoir. Average recovery efficiency data on a worldwide basis indicates that approximatelly one-third of the original oil in place, or less, is recovered by conventional primary and secondary methods (Hirasaki et al., 2011). The efficiency of conventional primary and secondary oil recovery methods can be improved through the implementation of oilfield operations such as infill drilling and the use of horizontal wells, among other improved oil recovery techniques. Figure 4 presents a mind mapping of conventional oil recovery processes.

Tertiary recovery processes refer to the application of methods that aim to recover oil beyond primary and secondary recovery. During tertiary oil recovery, fluids different than just conventional water and immiscible gas are injected into the formation to effectively boost oil production. Enhanced oil recovery (EOR) is a broader idea that refers to the injection of fluids or energy not normally present in an oil reservoir to improve oil recovery that can be applied at any phase of oil recovery including primary, secondary, and tertiary recovery. Thus EOR can be implemented as a tertiary process if it follows a waterflooding or an immiscible gas injection, or it may be a secondary process if it follows primary recovery directly. Nevertheless, many EOR recovery applications are implemented after waterflooding (Lake, 1989; Lyons & Plisga, 2005; Satter et al., 2008; Sydansk & Romero-Zerón, 2011). At this point is important to establish the difference between EOR and Improved Oil Recovery (IOR) to avoid misunderstandings. The term Improved Oil Recovery (IOR) techniques refers to the application of any EOR operation or any other advanced oil-recovery technique that is implemented during any type of ongoing oil-recovery process. Examples of IOR applications are any conformance improvement technique that is applied during primary, secondary, or tertiary oil recovery operations. Other examples of IOR applications are: hydraulic fracturing, scale-inhibition treatments, acid-stimulation procedures, infill drilling, and the use of horizontal wells (Sydansk & Romero-Zerón, 2011).

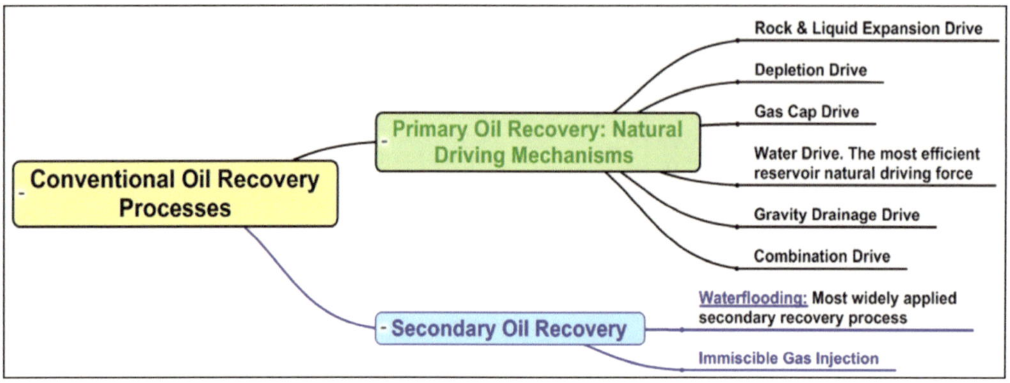

Fig. 4. Summary of Conventional Oil Recovery Processes

2.3 Enhanced Oil Recovery (EOR) processes

EOR refers to the recovery of oil through the injection of fluids and energy not normally present in the reservoir (Lake, 1989). The injected fluids must accomplish several objectives as follows (Green & Willhite, 1998).

- Boost the natural energy in the reservoir
- Interact with the reservoir rock/oil system to create conditions favorable for residual oil recovery that include among others:
  - Reduction of the interfacial tension between the displacing fluid and oil
  - Increase the capillary number
  - Reduce capillary forces
  - Increase the drive water viscosity
  - Provide mobility-control
  - Oil swelling
  - Oil viscosity reduction
  - Alteration of the reservoir rock wettability

The ultimate goal of EOR processes is to increase the overall oil displacement efficiency, which is a function of microscopic and macroscopic displacement efficiency. Microscopic efficiency refers to the displacement or mobilization of oil at the pore scale and measures the effectiveness of the displacing fluid in moving the oil at those places in the rock where the displacing fluid contacts the oil (Green & Willhite, 1998). For instance, microscopic efficiency can be increased by reducing capillary forces or interfacial tension between the displacing fluid and oil or by decreasing the oil viscosity (Satter et al., 2008).

Macroscopic or volumetric displacement efficiency refers to the effectiveness of the displacing fluid(s) in contacting the reservoir in a volumetric sense. Volumetric displacement efficiency also known as conformance indicates the effectiveness of the displacing fluid in sweeping out the volume of a reservoir, both areally and vertically, as well as how effectively the displacing fluid moves the displaced oil toward production wells (Green & Willhite, 1998). Figure 5 presents a schematic of sweep efficiencies: microscopic and macroscopic (areal sweep and vertical sweep).

The overall displacement efficiency of any oil recovery displacement process can be increased by improving the mobility ratio or by increasing the capillary number or both (Satter et al., 2008). Mobility ratio is defined as the mobility of the displacing fluid (i.e. water) divided by the mobility of the displaced fluid (i.e. oil).

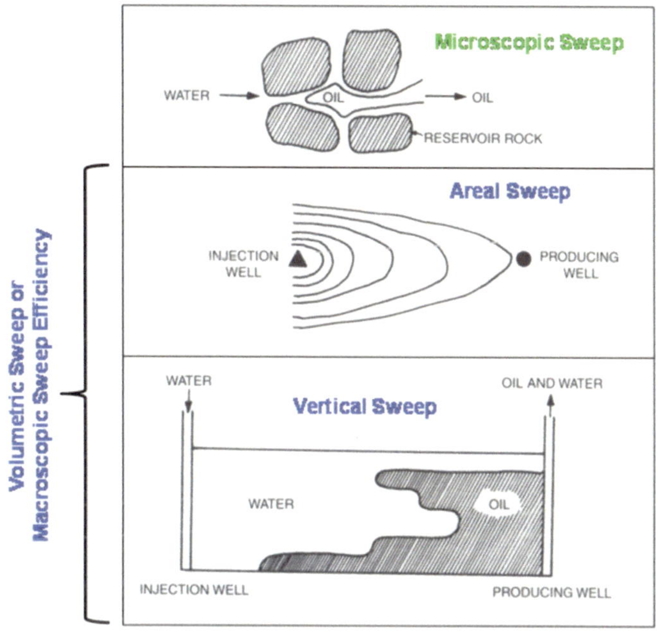

Fig. 5. Schematics of microscopic and macroscopic sweep efficiencies (Lyons & Plisga, 2005)

For waterfloods, this is the ratio of water to oil mobilities. The mobility ratio, M, for a waterflood is given by the following expression:

$$M = \frac{\text{Mobility}_{\text{Water}}}{\text{Mobility}_{\text{Oil}}} = \frac{\lambda_w}{\lambda_o} = \frac{k_{rw}/\mu_w}{k_{ro}/\mu_o} = \frac{k_{rw}\mu_o}{k_{ro}\mu_w} \tag{1}$$

where λ_w and λ_o are water and oil mobilities, respectively, in md/cp; k_{rw} and k_{ro} are relative permeabilities to water and oil, respectively, is μ_o oil viscosity and μ_w is water viscosity (Lyons & Plisga, 2005).

Volumetric sweep efficiency increases as M decreases, therefore mobility ratio is an indication of the stability of a displacement process, with flow becoming unstable (nonuniform displacement front or viscous fingering) when M> 1.0. Thus, a large viscosity contrast between the displacing fluid (i.e. water) and the displaced fluid (i.e. oil) causes a large mobility ratio (unfavorable M) which promotes the fingering of water through the more viscous oil (Fig. 6) and reduces the oil recovery efficiency. As such mobility ratio can be improved by increasing the drive water viscosity using polymers.

The capillary number, N_c, is a dimensional group expressing the ratio of viscous to capillary (interfacial) forces as follows:

$$N_c = \frac{\text{viscous forces}}{\text{capillary forces}} = \frac{\nu\mu_w}{\sigma_{ow}} \tag{2}$$

where ν is the interstitial velocity of the displacing fluid (i.e. water), μ_w is the viscosity of the displacing fluid (i.e. water), and σ_{ow} is the interfacial tension between the oil and the displacing fluid. Capillary numbers for a mature waterflooding process are commonly in the order of 10^{-7} to 10^{-6} (Green & Willhite, 1998). At the end of the waterflooding process, experience has shown that at these low capillary numbers an important amount of oil is left behind in the reservoir trapped by capillary forces at the pore scale. Thus, if the capillary number is increased through the application of EOR processes, residual oil will be mobilized and recovered. The most practical alternative to significantly increase the capillary number is through the application of surfactants or alkaline flooding (chemical flooding) (Sydansk & Romero-Zerón, 2011).

EOR processes are classified in five general categories: mobility-control, chemical, miscible, thermal, and other processes, such as microbial EOR (Green & Willhite, 1998). Figure 7 shows this EOR classification in more detail.

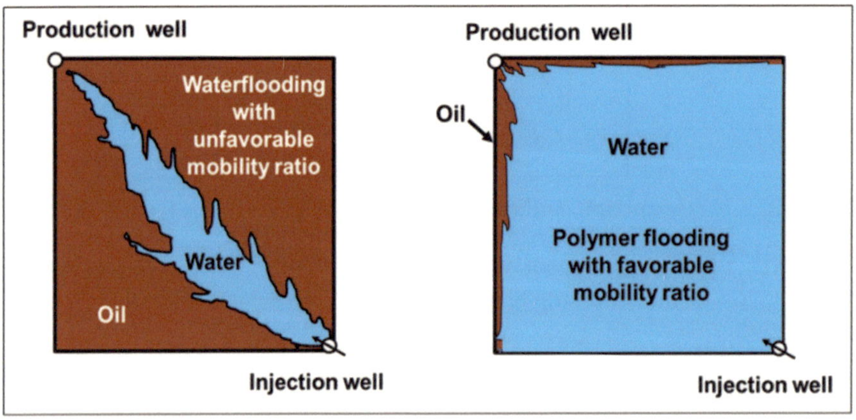

Fig. 6. (a) Waterflooding with unfavorable mobility ratio (M> 1), (b) Polymer augmented waterflooding with favorable mobility ratio (M ≤ 1) (Sydansk & Romero-Zerón, 2011).

A typical EOR fluid injection sequence is presented in Fig. 8. Some of the requirements for the ideal EOR flooding include among others (Singhal, 2011):

- Appropriate propagation of fluids and/or chemicals (i.e. polymers or surfactants) deep inside the reservoir rock.
- Low or minimum chemical adsorption, mechanical entrapment, and chemical consumption onto the formation rock.
- Fluids and/or chemicals tolerance to formation brine salinity and hardness.
- Fluids and/or chemicals stability to high reservoir temperatures.
- Polymers stability to mechanical degradation.
- Advanced polymer mobility-control to improve sweep efficiency.
- Efficient reductions of interfacial tension between oil and water.

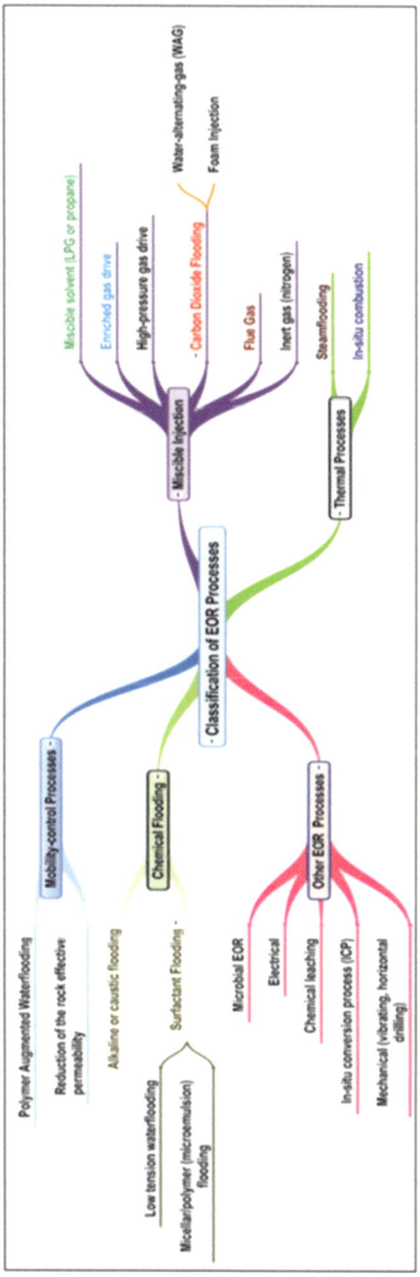

Fig. 7. Classification of Enhanced Oil Recovery Processes (Lake, 1989; Lyons & Plisga, 2005)

Although, EOR has been practiced for decades, and the petroleum industry has actively cooperated towards the advancement of EOR technology, there are still several challenges to the implementation of EOR projects that must be overcome. In the subsequent paragraphs a brief description of each EOR process is given with emphasis on the emerging technological trends.

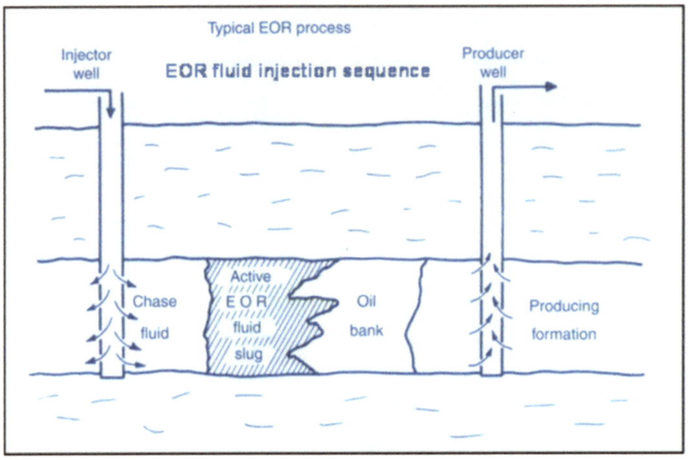

Fig. 8. Common EOR fluid injection sequence (Source: Lyons & Plisga, 2005)

2.3.1 Mobility-control processes

High mobility ratios cause poor displacement and sweep efficiencies, which can be caused by a large viscosity contrast between the displacing fluid (i.e. water) and oil or by the presence of high permeability flow channels that result in early breakthrough of the displacing fluid (i.e. water) at the producer well (Lyons & Plisga, 2005). Large volumes of produced water significantly increase operational costs due to water handling and the disposal of water according to the environmental regulations in place. According to Okeke & Lane, (2012), on average, it is estimated that seven barrels of water are produced per barrel of oil in the U.S.A., and the associated treatment and disposal cost is estimated around $ 5-10 billion annually. There are two techniques that can be applied within the oil-reservoir rock to successfully control volumetric sweep and/or conformance problems as follows (Sydansk & Romero-Zerón, 2011).

- **Polymer flooding:** consist in increasing the viscosity of the oil-recovery drive fluid.
- **Gels or crosslinked polymers:** consist in the placement of permeability-reducing material in the offending reservoir high-permeability flow channels.

2.3.1.1 Polymer flooding

Polymer flooding or polymer augmented waterflooding consist of adding water-soluble polymers to the water before it is injected into the reservoir. Polymer flooding is the simplest and most widely used chemical EOR process for mobility control (Pope, 2011). The most extensively used polymers are hydrolyzed polyacrylamides (HPAM) and the biopolymer Xanthan. Normally low concentrations of polymer are used often ranging from 250 to 2,000 mg/L and the polymer solution slug size injected is usually between 15% to 25% of the reservoir pore volume (PV). For very large field projects, polymer solutions may be injected over a 1-2 year period of time; after wich the project reverts to a normal waterflood. Incremental oil recovery is on the order of 12% of the original oil in place (OOIP) when polymer solution is injected for about one pore volume and values as high as 30% OOIP have been reported for some field projects (Pope, 2011). Furthermore, the displacement is more efficient in that less injection water is required to produce a given amount of oil (Lyons & Plisga, 2005; Sydansk, 2007). To produce an incremental barrel of oil, about 1 to 2 lbs of polymer are required, which means that currently the polymer cost is approximately USD 1.5/bbl to USD 3/bbl. The affordable price of polymer compared to the price of oil, explains why presently, the number of polymer flooding projects is increasing exponentially; for instance, in the U.S.A approximatelly 1 billion lbs of polymer was used in 2011 for mobility-control EOR (Pope, 2011). Mobility-control performance of any polymer flood within the porous media is commonly measured by the resistance factor, (RF), which compares the polymer solution resistance to flow (mobility) through the porous media as compared to the flow resistance of plain water. As exemplified by Lyons & Plisga (2005), if a RF of 10 is observed, it is 10 times more difficult for the polymer solution to flow through the system, or the mobility of water is reduced 10-fold. As water has a viscosity around 1cP, the polymer solution, in this case, would flow through the porous system as though it had an apparent or effective viscosity of 10 cP even though a viscosity measured in a viscometer could be considerably lower. Figure 9 presents a mind map of the mechanisms, limitations, and problems linked to polymer flooding applications.

2.3.1.1.1 Polymer flooding: emerging trends

The emerging technological trends in polymer flooding include the development of temperature and salinity resistant polymers (i.e. Associative polymers or hydrophobically modified polymers), high-molecular weight polymers, the injection in the reservoir of larger polymer concentrations, and the injection of larger slugs of polymer solutions (Aladasani & Bai, 2010; Dupuis et al. 2010, 2011; Lake, 2010; Pope, 2011; Reichenbach-Klinke et al., 2011; Seright et al. 2011; Sheng, 2011; Singhal, 2011; Sydansk & Romero-Zerón, 2011; Zaitoun et al., 2011) among others. Figure 10 summarizes the emerging trends in polymer flooding.

2.3.1.2 Polymers or polymer gel systems

Permeability-reducing materials can be applied from both the injection-well and the production-well side. There are a variety of materials that can be used for this purpose including polymer gels, resins, rigid or semi-rigid solid particles, microfine cement, etc. In this chapter the attention is focused on polymers or polymer gel systems.

Relative-Permeability-Modification (RPM) water-shutoff treatments, also termed Disproportionate Permebility Reduction (DPR) treatments are water-soluble polymer systems and weak gels that reduce the permeability to water flow to a greater degree than to oil and gas flow (Sydansk & Seright, 2007); particularly in wells where water and oil are produced from the same zone and the water-bearing cannot easily be isolated. These systems perform due to adsorption onto the pore walls of the formation flow paths (Chung et al., 2011). Several mechanisms for RPM have been proposed, including changes in porous media wettability, lubrication effects, segregation of flow pathways, gel dehydration, and gel displacement, among others (Sydansk & Romero-Zerón, 2011). However, there is no agreement upon a definite mechanism or mechanisms responsible for RPM, the most accepted mechacnism is gel dehydration (Seright et al., 2006). Therefore, this topic remains a subject under active investigation. The outcome of RPM oilfield applications has been mixed and the magnitude of the effect of RPM has been unpredictable from one application to another (Chung et al., 2011, Sydansk & Romero-Zerón, 2011).

Polymer gel systems are formed when low concentrations of a water-soluble high-molecular-weight polymer reacts with a chemical crosslinking agent to form a 3D crosslinked-polymer network that shows solid-like properties with rigidities up to and exceeding that of Buna rubber. In oilfield applications, polymer gels contain polymer concentrations ranging from 1,500 ppm to 12,000 ppm and gels are injected into the reservoir as a "watery" gelant (pre-gel) solution, or as a partially formed gel; after the gelation-onset time for the particular gel at reservoir conditions, the gelant solution (or partially formed gel) matures and sets up in the reservoir acquiring solid-like properties. Gelant solution must be pumped at very low rates to preferentially flow into the water channels, reducing the invation of gelant into the matrix rock containing oil (Chung et al., 2011).

Chromium (III)-Carboxylate/Acrylamide-Polymer (CC/AP) gels are the most popular and widely applied polymer-gel technology as mobility control treatments and as water and gas-shutoff treatments. These CC/AP gels are produced by croslinlinking aqueous soluble acrylamide-polymers with chromium (III) carboxylate or chromic triacetate ($CrAc_3$). Depending on the particular oilfield application, the gelation reaction rate can be accelerated or retarded by adding the proper chemical agents or combination of chemicals. For instance, chromic trichloride can be used as an additive to accelarate the CC/AP gelation reaction. If the CC/AP are applied to high-temperature reservoirs, it may be necessary to delay the gelation-rate. In this situation, the gelation-rate is slowed down by using gelation-rate retardation agents such as carboxylate ligands (i.e. lactate), the use of ultra-low-hydrolysis polyacrylamides within the gel formulation, and the use of low molecular weight (MW) acrylamide polymers. The proper application of gel treatments can generate in a profitable manner large volumes of incremental oil production and/or substantial reductions in oil-production operating costs via the shutting off of the production of excessive non-oil fluids, such as water and gas (Norman, et al. 2006; Sydansk & Romero-Zerón, 2011). A summary of the mechanisms, benefits, limitations, and problems eoncountered during oilfield applications of CC/AP gels is presented in Fig. 11.

2.3.1.2.1 Gels or crosslinked polymers: emerging trends

Some of the emerging trends in polymer gels or similar permeability-reducing materials that are under development include: thermally expandable particulate material, pH sensitive polymers, colloidal dispersion gels (CDGs), nano-size microgels, organically crosslinked polymer (OCP) systems, and preformed particle gel technologies, among others.

- **Thermally expandable particulate material**. This technology, which is still under development and field testing, is based on the expansion of thermally sensitive microparticles which can be used to cause blocking effect at the temperature transition in an oil reservoir. The polymeric material is a highly crosslinked, sulfonate-containing polyacrylamide microparticles in which the conformation is constrained by both unstable and stable internal crosslinks. As the particles reach areas of elevated temperatures within the reservoir, decomposition of the unstable crosslinker takes place releasing the constrains on the polymer molecule(s) in the particle allowing absorption of water and re-equilibration to render a larger particle size that provides resistance to fluid flow in porous media. The fluid flow blocking activity of the expanded polymeric particles in the porous media include particle/wall interactions, particle/particle

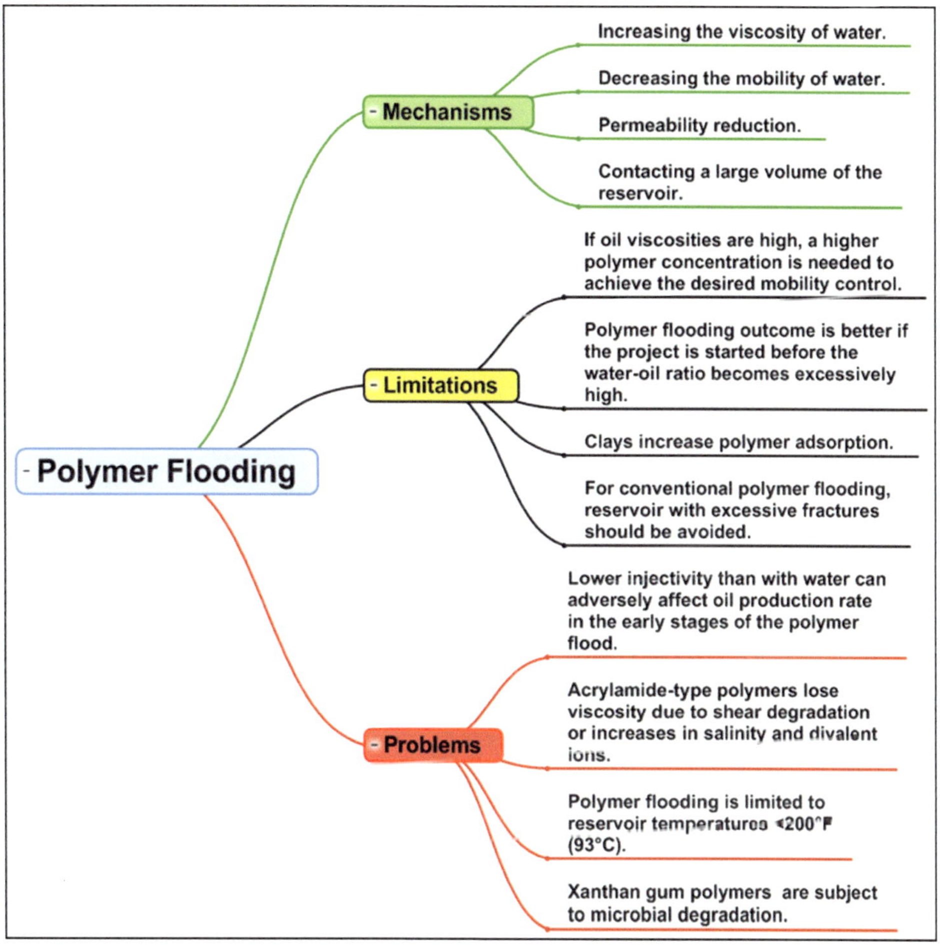

Fig. 9. Polymer Flooding: mechanisms, limitations, and problems (Adapted from Lyons & Plisga, 2005).

interactions, and bridging (Garmeh, et al. 2011; Frampton, et al,. 2004). Critical parameters that affect the application of this technology in the field were recently evaluated by Izgec & Shook (2012). The oilfield application of this technology is still at the evaluation stage. Some oilfield trial applications are presented by Mustoni et al. (2010); Ohms et al. (2010); and Roussennac & Toschi (2010). Independent technical and economic comparative analyses between this thermally activated polymeric material for in-depth profile modification and conventional polymer flooding conducted by Okeke & Lane, (2012) and Seright et al. (2011) concluded that in the long-term a properly design polymer flood program is advantageous over the application of thermally activated deep diverting materials.

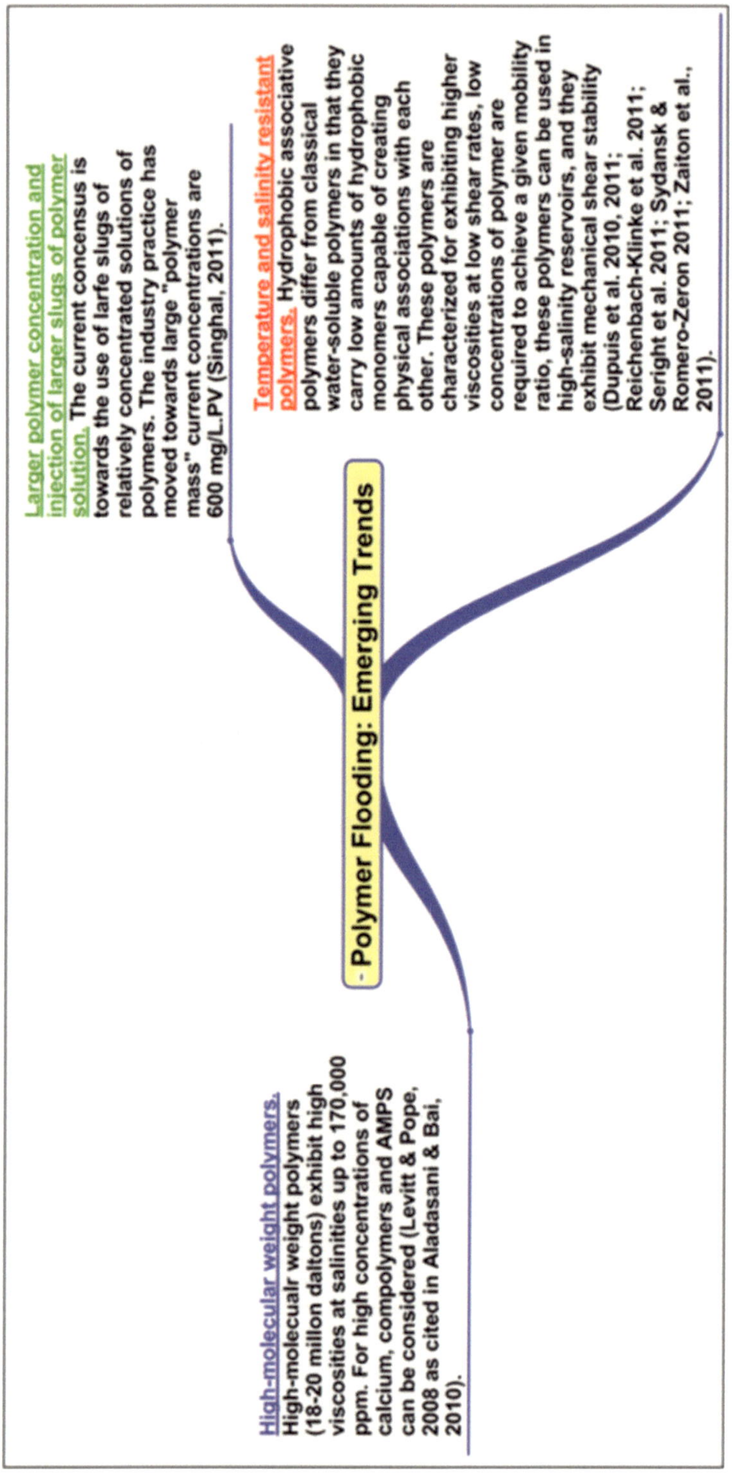

Fig. 10. Polymer Flooding: Emerging Trends

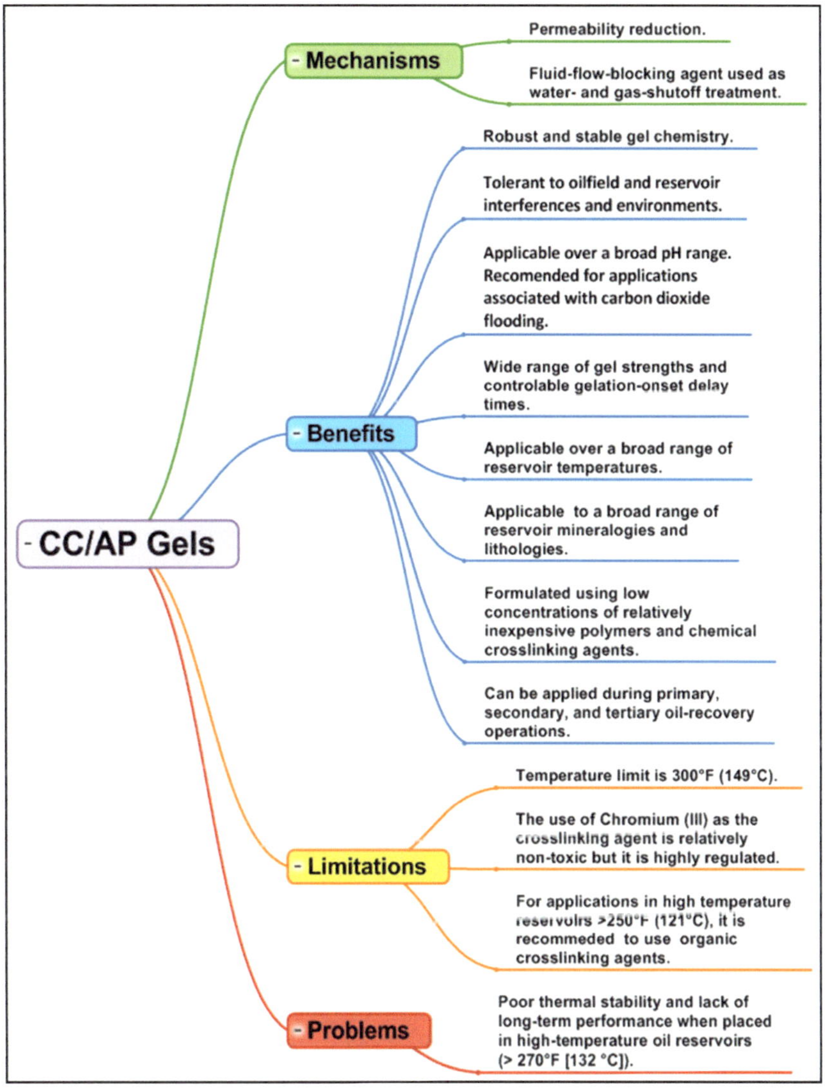

Fig. 11. CC/AP Gels: mind map of the mechanisms, benefits, limitations, and problems encountered during oilfield applications of CC/AP gels (Lyons & Plisga, 2005; Sydansk & Romero-Zerón, 2011).

- **pH Sensitive Polymers.** This technology aims the use of low-cost, pH-triggered polymer to improve reservoir sweep efficiency and reservoir conformance in chemical flooding. The idea is to use polymers or microgels containing carboxyl functional groups to make the polymer viscosity pH dependent. Thus, the polymer or microgel solution is injected into the reservoir at low pH conditions, in which the polymer molecules are in an unswelled tightly-coiled state and the viscosity is low, so that the polymer and/or microgel solution flows through the near wellbore region with a relatively low pressure drop avoiding the generation of unwanted fractures near the wellbore. Away from the near wellbore region, at a neutral pH the polymer and/or microgel solution swells and becomes thickened by a spontaneous reaction between the injected polymer acid solution and the resident rock mineral components thus, lowering the brine mobility and increasing oil displacement efficiency (Choi, 2005; Choi et al., 2006; Huh et al. 2005; Lalehrokh, 2009; Sharma et al., 2008).

- **Colloidal Dispersion Gels (CDGs).** As defined by Fielding et al. (1994), CDG is a solution containing low concentrations of high molecular weight polymer and a crosslinker that has a slow rate of formation and is considered semi-fluid. The CDGs microgels are injected from the injection-well side for the purpose of improving vertical and areal conformance deep in heterogeneous "matrix-rock" of sandstone reservoirs, while maintaining high temperature stability. CDG gels flow a high-pressure differentials and resist flow at low-pressure differentials (Fielding et al. 1994). Although numerous publications (Chang et al. 2004; Diaz et al. 2008; Fielding et al. 1994; Lu et al. 2000; Muruaga et al. 2008; Norman et al. 1999; Shi et al. 2011a, 2011b; Smith et al. 1996; Smith et al. 2000; Spildo et al. 2008) have discussed over the years the effectiveness of CDGs through laboratory studies and field trials, this technology has proven to be controversial (Al-Assi et al,. 2006; Ranganathan et al., 1998; Seright, 2006; Sydansk & Romero-Zerón, 2011; Wang et al., 2008).
- **Nano-size Microgels.** Recently Wang et al. (2010) proposed the use of crosslinked polyacrylamide (PAM) nanospheres for in-depth profile control to improve sweep efficiency. It is speculated that owing to the nano size, water absorbing selectivity, brine tolerance, high water absorption, good dispersion in water, low aqueous solution viscosity, nanospheres can easily migrate into the high-permeability zones (channels of low-resistance to flow) where the nanospheres would swell due to their high water absorption capacity blocking off the thief zones.
- **Organically crosslinked polymer (OCP) systems**. These systems are based on PEI (polyethyleneimine) crosslinker with a copolymer of acrylamide and *t*-butyl acrylate (PAtBA) for in-depth profile control. The main advantage of this crosslinker is its lower toxicity while the low molecular weight copolymer PAtBA enhances the biodegradability of the material and facilitates the formation of thermally stable, rigid gels that are insensitive to formation fluids, lithology, and/or heavy metals. Numerous field trials have been conducted worldwide (Chung et al., 2011).
- **Preformed Particle Gels (PPGs).** PPGs consist on crushed dry gels that are sieved to obtain different cuts of gel particles, which swell in water and form a stable suspension that flows within the porous media. These diverting agents aim to provide in-depth fluid diversion (Coste et al., 2000). PPGs are currently gaining attention and popularity for use in conformance-improvement treatments. Bai et al. (2009) reported an extensive review of PPGs for conformance control that covers from PPGs mechanisms to field applications. Recently a new PPG enhanced surfactant-polymer system has been proposed by Cui, et al. (2011).

2.3.2 Chemical flooding

Chemical flooding is a generic term for injection processes that use special chemicals (i.e. surfactants) dissolved in the injection water that lower the interfacial tension (IFT) between the oil and water from an original value of around 30 dynes/cm to 10^{-3} dynes/cm; at this low IFT value is possible to break up the oil into tiny droplets that can be drawn from the rock pores by water (Van Dyke, 1997). There are two common chemical flooding: micellar-polymer flooding and alkaline or caustic flooding.

2.3.2.1 Micellar-polymer flooding

Micellar-polymer flooding is based on the injection of a chemical mixture that contains the following components: water, surfactant, cosurfactant (which may be an alcohol or another surfactant), electrolytes (salts), and possible a hydrocarbon (oil). Micellar-polymer flooding is also known as micellar, microemulsion, surfactant, low-tension, soluble-oil, and chemical flooding. The differences are in the chemical composition and the volume of the primary slug injected. For instance, for a high surfactant concentration system, the size of the slug is often 5%-15% pore volumes (PV), and for low surfactant concentrations, the slug size ranges from 15%-50% PV. The surfactant slug is followed by polymer-thickened water. The concentration of polymer ranges from 500 mg/L to 2,000 mg/L. The volume of the polymer solution injected may be 50% PV, depending on the process design (Green & Willhite, 1998; Satter et al., 2008). Flaaten et al. (2008) reported a systematic laboratory approach for chemical flood design and applications. Some of the main surfactant requirements for a successful displacement process are as follows (Hirasaki et al., 2011).

- The injected surfactant slug must achieve ultralow IFT (IFT in the range of 0.001 to 0.01 mN/m) to mobilize residual oil and create an oil bank where both oil and water flows as continuous phases.
- It must maintain ultralow IFT at the moving displacement front to prevent mobilized oil from being trapped by capillary forces.
- Long-term surfactant stability at reservoir conditions (temperature, brine salinity and hardness).

Figure 12 summarizes the mechanisms, limitations, and problems eoncountered during micellar-polymer flooding projects.

2.3.2.2 Alkaline or Caustic flooding

In this type of chemical flooding, alkaline or caustic solutions are injected into the reserovoir. Common caustic chemicals are sodium hydroxide, sodium silicate, or sodium carbonate. These caustic chemicals react with the natural acids (naphtenic acids) present in crude oils to form surfactants in-situ (sodium naphthenate) that work in the same way as injected synthetic surfactants (reduction of interfacial tension, IFT, between oil/water) to move additional amounts of oil to the producing well. These chemicals also react with reservoir rocks to change wettability. Alkaline flooding can be applied to oils in the API gravity range of 13° to 35°, particulary in oils having high content of organic acids. The preferred oil formations for alkaline flooding are sandstone reservoirs rather than carbonate formations that contain anhydride or gypsum, which can consume large amounts of alkaline chemicals. These chemicals are also consumed by clays, minerals, or silica, and the higher the temperature of the reservoir the higher the alkali consumption. Another common problem during caustic flooding is scale formation in the producing wells. During alkaline flooding, the injection sequence usually includes: (1) a preflush to condition the reservoir before injection of the primary slug, (2) primary slug (alkaline chemicals), (3) polymer as a mobility buffer to displace the primary slug. Modifications of alkaline flooding are the alkali-polymer (AP), alkali-surfactant (AS), and alkali-surfactant-polymer (ASP) processes (Green & Willhite, 1998; Satter et al., 2008; Van Dyke, 1997). In addition to the beneficial formation of natural

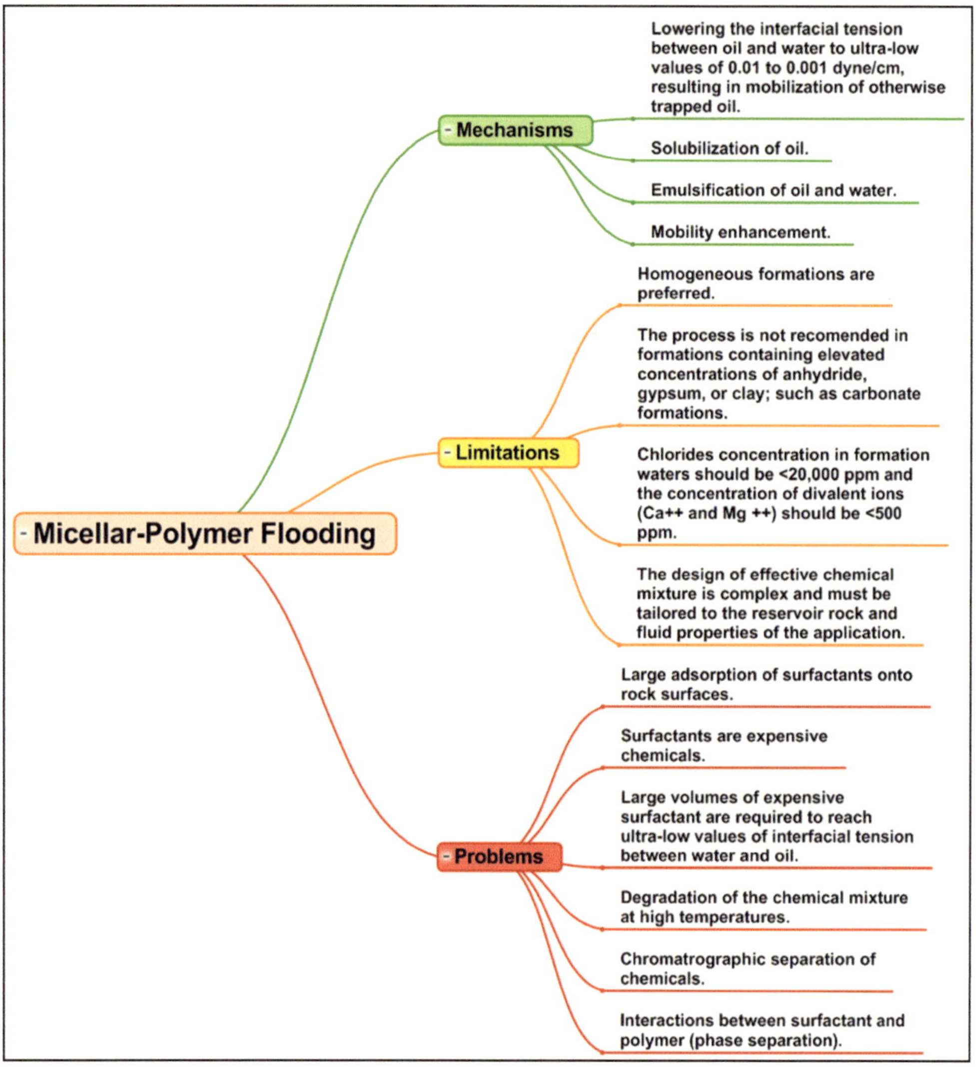

Fig. 12. Micellar-Polymer Flooding: mechanisms, limitations, and problems (Adapted from Feitler, 2009; Lyons & Plisga, 2005; Satter et al., 2008).

surfactants (surfactants in-situ) driven by the reaction of alkali with naphtenic acids in the crude oil, the role of the alkali in the AS and ASP processes is to reduce the adsorption of the surfactant during displacement through the formation and sequestering of divalents ions. The presence of alkali can also alter formation wettability to reach either more water-wet or more oil-wet states. For instance, in fractured oil-wet reservoirs, the combined effect of alkali and surfactant in making the matrix preferentially water-wet is essential for an effective process. These benefits of alkali will occur only when alkali is present (Hirasaki et al., 2011). Jackson (2006) reported a detailed experimental study of the benefits of sodium carbonate on surfactants for EOR. Surfactants are also used to change the wettability of the porous media to boost oil recovery. Wu et al., (2006) reviewed the mechanisms responsible for wettability changes in fractured carbonate reservoirs by surfactant solutions.

Micellar-polymer flooding was considered a promising EOR process during the 1970s, however the high surfactant concentrations required in the process and the cost of surfactants and cosurfactants, combined with the low oil prices during the mid 1980s limited its applications. Worldwide, oilfield applications of chemical flooding have been insignificant since the 1990s (Hirasaki et al., 2011; Manrique, et al. 2010) and although practiced, is rarely reported by operators (Enick & Olsen, 2012). The main reason is again the dependance of chemical flooding processes on the volatility of the oil markets because these processes are capital intensive and carry a high degree of risk (Bou-Mikael et al. 2000). Nevertheless, the development of both SP and ASP EOR technology and advances on surfactant chemistry during the last 5 years have brought a renewed attention for chemical floods (Pope, 2011), specially to boost oil production in mature and waterflooded fields. Currently, there are numerous active ASP flooding projects worldwide, with the ASP flooding implemented at the Daqing field in China considered one of the largest ASP ongoing projects (Manrique et al. 2010). Several current ASP oilfield applications are reported in the literature (Buijse et al., 2010; Manrique et al. 2010).

2.3.2.3 Chemical flooding: emerging trends

In the area of micellar-polymer flooding, the emerging trends are related to the development of surfactant systems having the following capabilities (Adkins et al., 2010; Azira et al., 2008; Banat et al., 2000; Barnes et al., 2010; Berger & Lee, 2002; Cao & Li, 2002; Elraies et al., 2010; Feitler, 2009; Flaaten et al., 2008; Hirasaki et al., 2011; Iglauer et al., 2004; Levitt, 2006; Levitt et al., 2009; Ovalles et al., 2001; Puerto et al., 2010; Sheng, 2011; Yang et al., 2010; Wang et al., 2010; Wu et al., 2005):

- Reduce the oil-water interfacial tension to ultra-low values (0.01-0.001 dyne/cm) at low surfactant concentrations (<0.1 wt%) to significantly reduce the amount of expensive surfactant used in oil recovery.
- Production of low cost surfactants.
- Development of surfactants active only upon contact with hydrocarbon fluids.
- The reduction or minimization of surfactant loss due to adsorption on rock surface.
- Development of single-phase injection composition systems (no phase separation when polymer is present).
- Development of mechanistic chemical flood simulators useful for the designing and performance prediction of EOR processes.
- The evaluation and development of novel alkalis.
- Advances of protocols for surfactant selection, laboratory testing, and scaling up of laboratory data to the field (Pope, 2011).

Some of the best surfactants presently available have molecular weights 10 times larger than the surfactants previously used and the surfactant molecule is highly branched, which minimizes the surfactant adsorption onto both sandstones and carbonates. Furthermore, these new surfactants are up to three times more efficient in terms of oil recovery per pound of surfactant (Pope, 2011). A brief review on new polymeric surfactants for EOR is presented in Chapter 2. Figure 13 summarizes recent advances in chemical flooding technology.

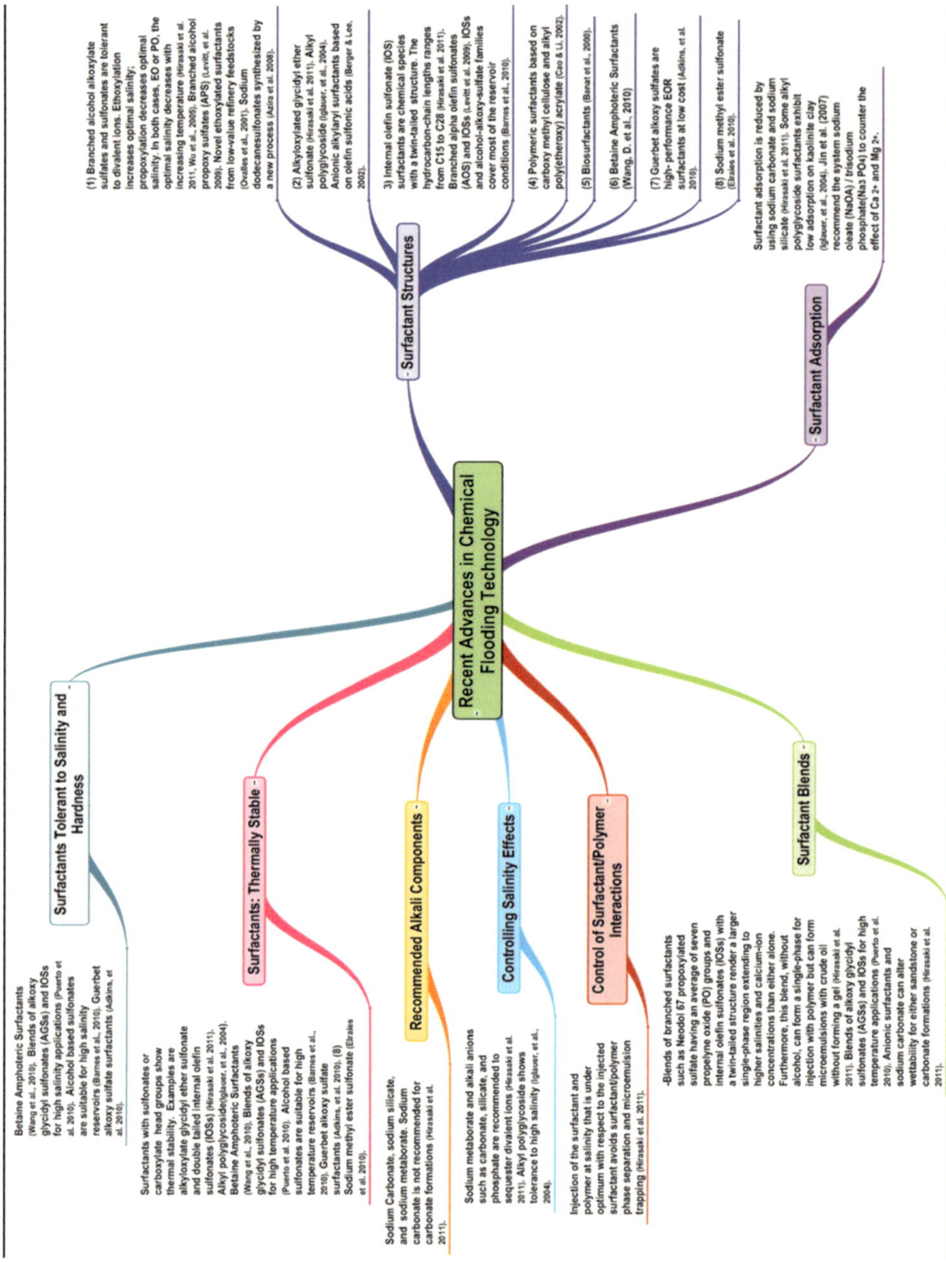

Fig. 13. Recent Advances in Chemical Flooding Technology.

2.3.3 Miscible injection

Miscible injection uses a gas that is miscible (mixable) with oil and as the gas injection continues, the gas displaces part ot the oil to the producing well. Injection gases include liquefied petroleum gases (LPGs) such as propane, methane under high pressure, methane enriched with light hydrocarbons, nitrogen under high pressure, flue gas, and carbon dioxide used alone or followed by water. LPGs are appropriate for use in many reservoirs because they are miscible with crude oil on first contact. However, LPGs are in such demand as marketable commodity that their use in EOR is limited (Van Dyke, 1997). As presented by Lyons & Plisga (2005), the mechanisms of oil recovery by hydrocarbon miscible flooding are:

- Miscibility generation
- Increasing the oil volume (swelling)
- Decreasing the viscosity of the oil

Limitations of hydrocarbon miscible flooding are the reservoir pressure needed to maintain the generated miscibility. Pressures required range from 1,200 psi for the LPG process to 5,000 psi for the high pressure methane or lean gas drive. For these applications a steeply dipping (not flat-lying) formation is preferred to allow gravity stabilization of the displacement, which normally has an unfavorable mobility ratio. The main problems associated to hydrocarbon miscible flooding are poor vertical and horizontal sweep efficiency due to viscous fingering, large quantities of expensive gases are required, and solvent may be trapped and not recovered (Lyons & Plisga, 2005). Hydrocarbon gas injection projects have made a relatively marginal contribution in terms of total oil recovered in Canada and U.S.A. other than on the North Slope of Alaska where large natural gas resources are available for use that do not have a transportation system to markets. Generally, hydrocarbon miscible flooding is applied in water-alternating-gas (WAG) injection schemes (Manrique et al. 2010). Injection of flue gas (N_2, CO_2) and N_2 had essentially been replaced by CO_2 since the 1980s. While, EOR from miscible CO_2 floods has been steadily increasing over the last two decades, as has the number of projects. Currently, CO_2 EOR provides about 280,000 barrels of oil per day, just over 5 % of the total U.S.A. crude oil production. Recently CO_2 flooding has become so technically and economically attractive that CO_2 supply, rather than CO_2 price, has been the constraining developmental factor. CO_2 EOR is likely to expand in the U.S.A. in the near future due to "high" crude oil prices, natural CO_2 source availability, and possible large anthropogenic CO_2 sources through carbon capture and storage (CCS) technology advances. It is estimated that in the U. S. A., the "next generation" CO_2 EOR can provide 137 billion barrels of additional technically recoverable domestic oil (Enick & Olsen, 2012).

CO_2 flooding is carried out by injecting large quantities of CO_2 (15% or more of the hydrocarbon pore volume, PV) into the reservoir. Typically it takes about 10 Mcf of CO_2 to recover an incremental barrel of oil and about half of this gas will be left in the reservoir at the economic limit (Pope, 2001). Although CO_2 is not truly miscible with the crude oil, CO_2 extracts the light-to-intermediate components from the oil, and, if the pressure is high enough, develops miscibility to displace the crude oil from the reservoir (Lyons & Plisga, 2005). Basically, during CO_2 displacements miscibility takes place through in-situ composition changes resulting from mutliple-contacts and mass transfer between reservoir oil and the injected CO_2 (Green & Willhite, 1998). CO_2 has a greater viscosity under pressure than many other gases and displaces oil at low pressures (Van Dyke, 1997). Figure 14 summarizes the mechanisms, limitations, and main downsides of CO_2 flooding.

To overcome mobility control during CO_2 flooding, the current technologies of choice are the use of water-alternating-gas (WAG) injection schemes and mechanical techniques including cement, packers, well control, infield drilling, and horizontal wells (Enick & Olsen, 2012).

CO_2 flooding has been the most widely used EOR recovery method of medium and light oil production in sandstone reservoirs during recent decades, especially in the U.S.A., due to the availability of cheap and readily available CO_2 from natural resources such as the Permian Basin (Pope, 2011). It is expected that CO_2 floods will continue to grow globally in sandstone and carbonate reservoirs (Manrique et al. 2010). Furthermore, the pressing issue of sequestering considerable volumes of industrial CO_2 due to environmental concerns would play an importan role in the grow of miscible CO_2 oilfield applications. Several CO_2-EOR field projects are described by Enick & Olsen, (2012) and Manrique et al. (2010).

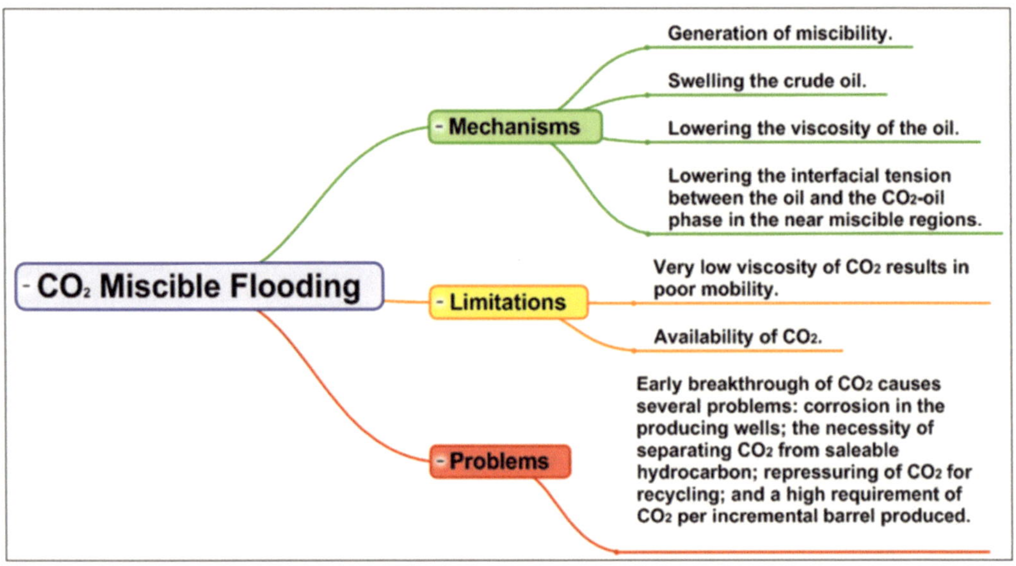

Fig. 14. CO_2 Flooding: mechanisms, limitations, and main problems

2.3.3.1 Miscible CO_2 flooding: emerging trends

According to a recent and comprehensive literature review on the history and development of CO_2 mobility control and profile modification technologies prepared by Enick & Olsen (2012) for the U.S.A. National Energy Technology Laboratory (NETL), the emerging trends in CO_2 flooding include the following: increasing CO_2 injection volumes by 50% or more, horizontal wells for injection or production, improving mobility ratio and flood conformance, extending the conditions under which miscibility between the oil and CO_2 can be achieved, innovative flood design and well placement, and the application of advanced methods for monitoring flood performance. However, the short term emphasis of the emerging technologies will be on the development and identification of techniques that allow to overcome the most urgent CO_2 miscible process limitations and operational problems such as poor sweep eficiency, unfavorable injectivity profiles, gravity override, high ratios of CO_2 to oil produced, early breakthrough, and viscous fingering. Thus it is necessary to identify effective and affordable CO_2 thickeners that would allow increase the viscosity of CO_2 in a carefully controlled manner simply by changing the thickener concentrations. Ideally, the CO_2-soluble thickener would be brine- and crude oil-insoluble, which would inhibit its partitioning into these other fluid phases and its adsorption onto the reservoir rock. Other technologies that will be carefully looked at are the effective design of CO_2 foams for mobility reduction, nanoparticle-stabilized foams, the alternating or simultaneous injection of aqueous nano-silica dispersions and CO_2, especially for high temperature reservoirs where chemical degradation of surfactants might be a concern.

2.3.4 Thermal recovery processes

Thermal recovery processes use thermal energy in some form both to increase the reservoir temperature and therefore to decrease the viscosity of the oil, which makes possible the displacement of oil towards the producing wells. Thermal recovery processes are globally the most advanced EOR processes, which are classified in three main techniques as follows (Green & Willhite, 1998).

- Steam Drive
- Cyclic Steam Injection
- In-situ Combustion or Fire flooding

2.3.4.1 Steam drive

Steam drive, also known as steam injection or continuous steam injection, involves generating steam of about 80% quality on the surface and forcing this steam down the injection wells and into the reservoir. When the steam enters the reservoir, it heats up the oil and reduces its viscosity. As the steam flows through the reservoir, it cools down and condenses. The heat from the steam and hot water vaporizes lighter hydrocarbons, or turn them into gases. These gases move ahead of the steam, cool down, and condense back into liquids that dissolve in the oil. In this way, the gases and steam provide additional gas drive. The hot water also moves the thinned oil to production wells, where oil and water are produced (Van Dyke, 1997).

2.3.4.2 Cyclic steam injection

Cyclic steam injection is also termed huff and puff; this operation involves only one well that functions either as injection and production well. In this process steam is injected into the reservoir for several days or weeks to heat up the oil. Then, steam injection is stopped and the well is shut in to allow the reservoir soak for several days. In the reservoir, the steam condenses, and a zone of hot water and less viscous oil forms. Later on, the well is brought into production and the hot water and thinned oil flow out. This cyclic process of steam injection, soaking, and production can be repeated until oil recovery stops (Van Dyke, 1997). Figure 15 summarizes the mechanisms, limitations, and problems encountered during steam drive and cyclic steam injection.

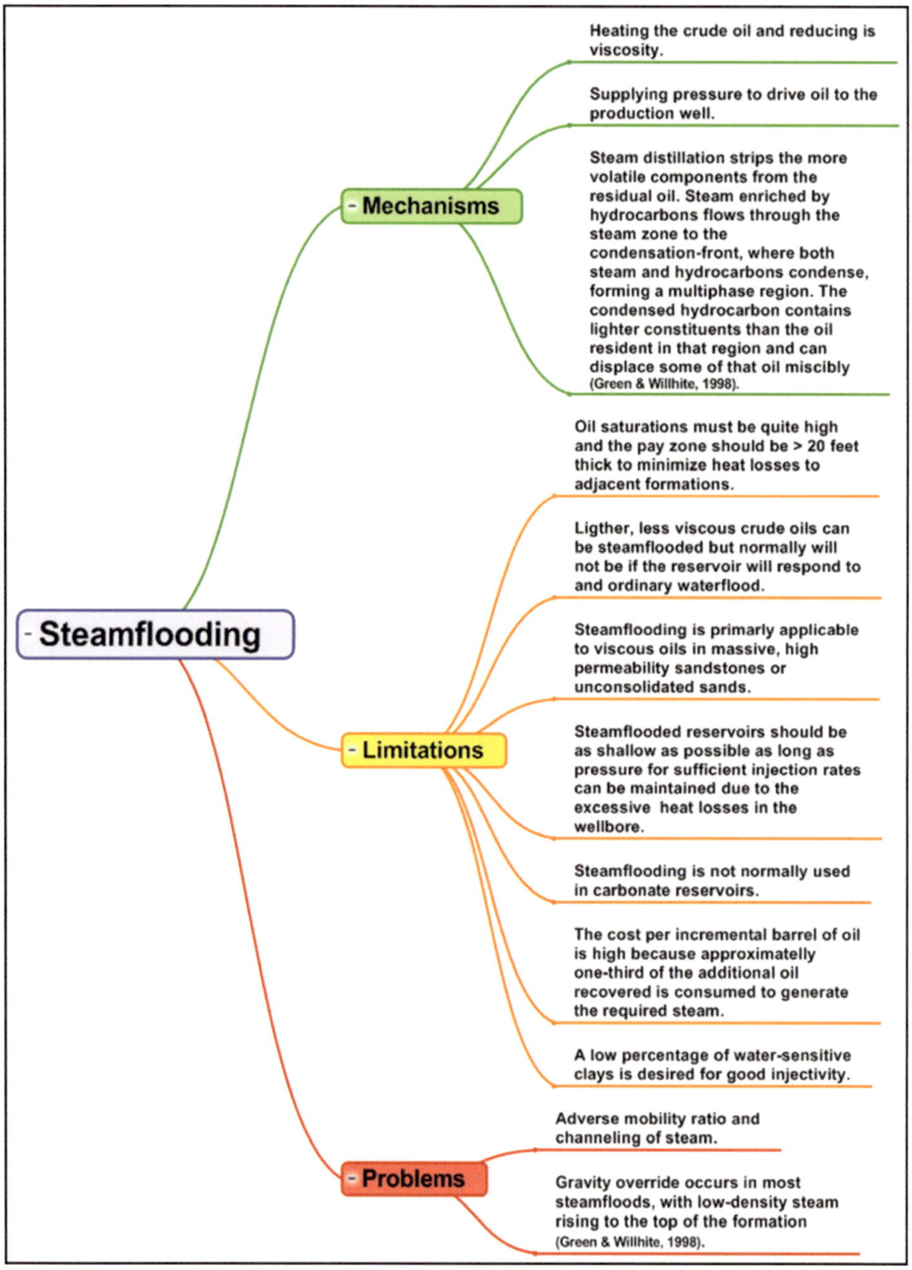

Fig. 15. Steamflooding: mechanisms, limitations, and problems (Adapted from Lyons & Plisga, 2005).

2.3.4.3 In-situ combustion

In-situ combustion or fire flooding is a process in which an oxygen containing gas is injected into a reservoir where it reacts with the oil contained within the pore space to create a high-temperature self-sustaining combustion front that is propagated through the reservoir.

Ignition may be induced through electrical or gas igniters of may be spontaneous if the crude oil has sufficient reactivity. In most cases the injected gas is air and the fuel consumed by the combustion is a residuum produced by a complex process of cracking, coking, and steam distillation that occurs ahead of the combustion front. The heat from the combustion thins out the oil around it, causes gas to vaporize from it, and vaporizes the water in the reservoir to steam. Steam, hot water, and gas all act to drive oil in front of the fire to production wells. In-situ combustion is possible if the crude-oil/rock combination produces enough fuel to sustain the combustion front. In-situ combustion field tests have been carried out in reservoir containing API gravities from 9 to 40°API (Green & Willhite, 1998; Van Dyke, 1997). The mechanisms, limitations, and problems occurring during in-situ combustion applications are briefly presented in Figure 16.

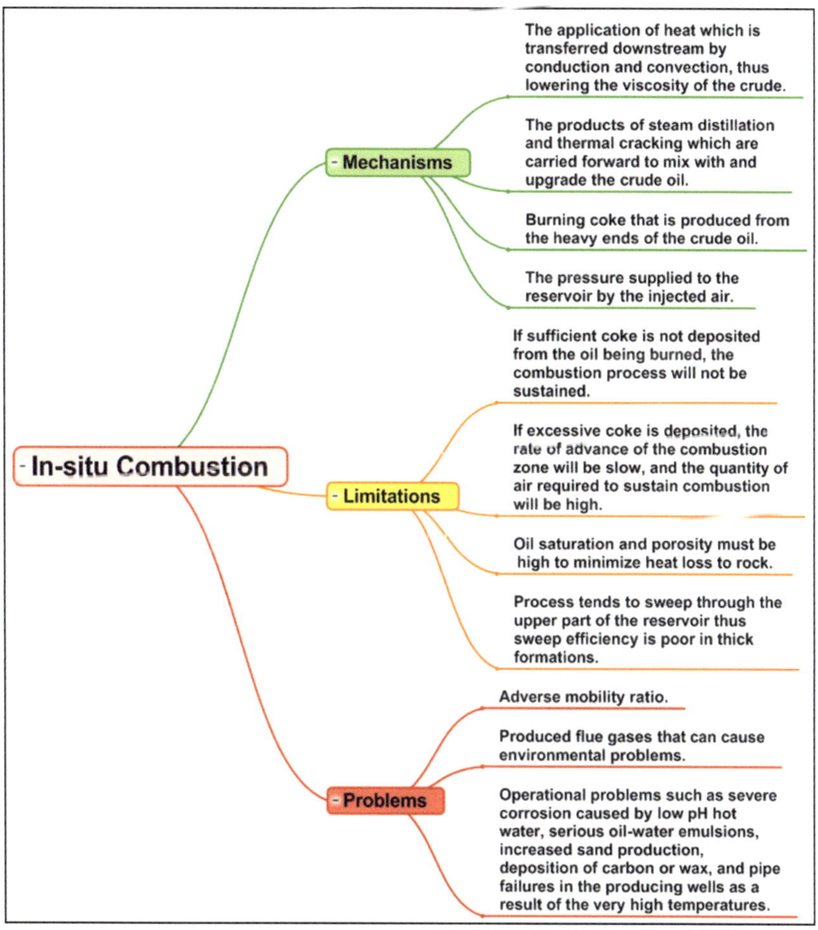

Fig. 16. In-situ Combustion: mechanisms, limitations, and problems (Adapted from Lyons & Plisga, 2005).

As previously indicated, thermal recovery processes are the most advanced EOR processes and contribute significant amounts of oil to daily production. Most thermal oil production is the result of cyclic steam and steamdrive. New thermal processes have derived from steamflooding and in-situ combustion, however several of these newly proposed methods at early stages of evaluation and are not expected to have an impact on oil production in the near future (Green & Willhite, 1998; Manrique et al., 2010). Figure 17 outlines several of the recent advances in thermal oil recovery processes (AOSC, 2008; OSDG, 2009; Shah et al., 2009, 2010; Meridian, 2006).

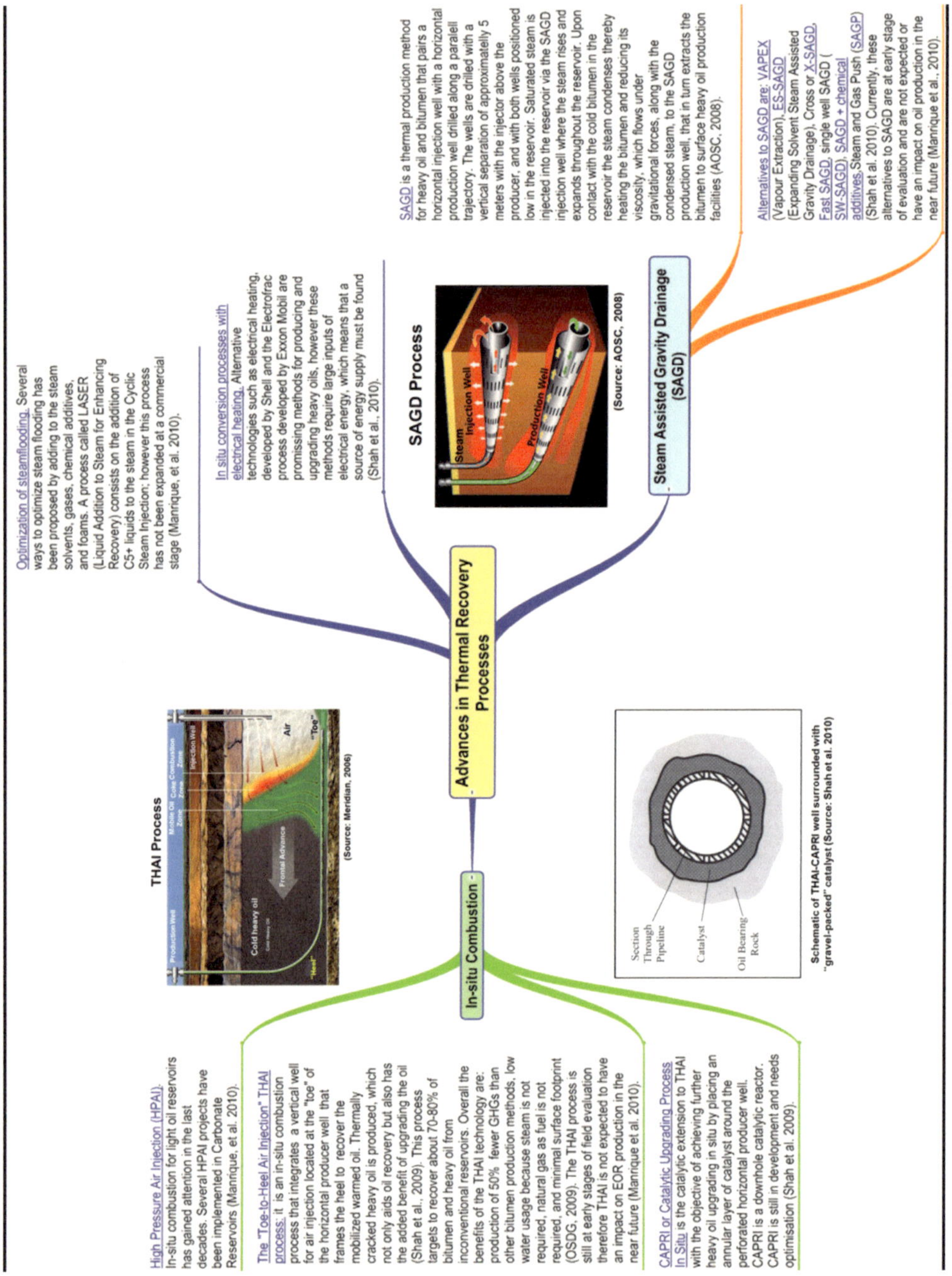

Fig. 17. Advances in Thermal Oil Recovery

As reported in a comprehensive review of EOR projects prepared by Manrique et al., (2010), thermal EOR projects have been concentrated mostly in Canada, the Former Sovietic Union, U.S.A., and Venezuela. Several EOR thermal projects have been also reported in Brazil and China but in a lesser extend. For the specific case of bitumen production, it is expected that the SAGD process will continue to expand for the production of bitumen from the Alberta's oil sands. Numerous thermal oil recovery projects are reported in the literature; for instance Manrique et al. (2010) presents several examples of recent thermal projects conducted worldwide.

2.3.5 Other EOR processes

Other important EOR processes include foam flooding and microbial enhanced oil recovery (MEOR), among others.

2.3.5.1 Foam flooding

Foam is a metastable dispersion of a relatively large volume of gas in a continuous liquid phase that constitutes a relatively small volume of the foam. The gas content in classical foam is quite high (often 60 to 97 vol%). Bulk foams are formed when gas contacts a liquid containing a surfactant in the presence of mechanical agitation (Sydansk & Romero-Zerón, 2011).

In oilfield applications, the use of CO_2 foams has been considered a promising technique for CO_2 mobility control (Enick & Olsen, 2012) and steamflooding mobility control (Hirasaki et al., 2011). The use of foams for mobility control in surfactant flooding, specifically at high temperatures (due to polymer degradation), in alkaline-surfactant flooding, surfactant/polymer projects, and in alkaline/surfactant/polymer flooding have been reported (Hirasaki et al., 2011).

The reduced mobility of CO_2 foams in porous media is attributed to the flow of dispersed high-pressure CO_2 droplets separated by surfactant-stabilized lamellae within the porous of the formation (Enick & Olsen, 2012). The largest pores are occupied by cells/droplets of the non-wetting CO_2 phase, which enables the gas to be transported through the pores along with the lamellae that separate them. These "trains" of flowing CO_2 bubbles encounter drag forces related to the pore surfaces and constrictions that lead to alteration of the gas-liquid interface by viscous and capillary forces. Further, the transport of surfactant from the front to the rear of moving bubbles establishes a surface-tension gradient that impedes bubble flow. These phenomena give the flowing foam a non-Newtonian character and an apparently high viscosity, or low mobility, compared to pure CO_2 and water flowing through the pores in the absence of surfactant and lamellae (Tang & Kovscek, 2004, as cited in Enick & Olsen, 2012). The intermediate size pores can become filled with immobile, trapped bubbles of the gas phase, which reduces the pore volume available for the flow of CO_2 foam through the rock. The majority of the gas within foam in sandstone at steady state can be trapped in these intermediate size pores (Chen et al., 2008, as cited in Enick & Olsen, 2012). The gas trapping leads to gas blocking which, in turn, reduces the gas mobility even further (Enick & Olsen, 2012). In oilfield operations, foams can be applied as a viscosity-enhancement agent or as a permeability-reducing treatment (Sydansk & Romero-Zerón, 2011).

- A mobility control foam is one in which the mobility of the foam is reduced approximately to a level that is comparable to the oil being displaced in an attempt to supress fingering and channeling. Typically aternated slugs of surfactant solution and CO_2 are injected into the reservoir. Once the foam is formed, it is intended to propagate throughout the formation as an in-depth mobility control to improve sweep efficiency through the CO_2 flood (Enick & Olsen, 2012).
- Conformance control foam is intended to selectively generate strong, very low mobility foams in highly permeable, watered-out thief zones. These foams are also referred to as blocking/diverting foams, or injection profile improvement foams. This is achieved by employing higher concentrations of surfactant in the aqueous solution that is injected alternately with CO_2 (Enick & Olsen, 2012).

Although, there are numerous reports on oilfield application of foams and their performace (Enick & Olsen, 2012; and Marinque et al. 2011); foams for conformance improvement to date have not been applied widely in a succesful, commercially atractive, and profitable manner; on the contrary the application of foams is considered to be an advanced and nonroutine conformance-improvement technology (Pope, 2011; Sydansk & Romero-Zerón 2011). Nonetheless, with the further development of technlogies such is the case of new surfactants for CO_2 and nanoparticles as foam stabilizers, the oilfield application of foams might be revitalized (Pope, 2011; Sydansk & Romero-Zerón, 2011).

2.3.5.2 Microbial Enhanced Oil Recovery (MEOR)

Microbial Enhanced Oil Recovery (MEOR) relies on microbes to ferment hydrocarbons and produce by-products such as biosurfactants and carbon dioxide that help to displace oil in a similar way than in conventional EOR processes. Bacterial growth occurs at exponential rates, therefore biosurfactants are rapidly produced. The activity of biosurfactants compare favourably with the activity of chemically synthesized surfactants. The injection of nutrients such as sugars, nitrates or phosphates stimulates the growth of the microbes and aid their performance. MEOR applications are limited to moderate reservoir temperatures, because high temperatures limit microbial life and the availability of suitable nutrients (Shah et al., 2010). In the 1980's, researches and industry focused considerable efforts toward EOR processes including MEOR, which has always been an attractive EOR method due to its low cost and potential to improve oil recovery efficiencies (Aladasani & Bai, 2010). At present, researchears are still evaluating and advancing MEOR processes (Soudmand-asli, et al., 2007). For instance, a chemical flooding simulator the "UTCHEM" developed at the University of Texas at Austin by Delshad et al., (2002) has recently incorporated a model that is capable of qualitatively mimic the oil recovery mechanisms occuring during MEOR processes. Presently, few commercial MEOR operators continue to offer customized microbial process applications.

MEOR oilfield applications have shown mixed results. On average, MEOR field trials results have been poor and published studies offer little insights into the potential field viability of MEOR. The main reason is that the current state of knowledge of bioreaction kinetics such as nutrient reaction kinetics, selectivity, and level of conversion are still lacking. For instance, microbial gas production, CO_2 and CH_4, are commoly cited as contributing to oil recovery, however is unlikely that these gases could be produced in-situ in the quantities needed for effective oil displacement. Similarly, the in-situ generation of viscosifying agents is intrinsically unstable, which indicates that ex-situ polymer injection should be required for mobility control of MEOR (Bryant & Lockhart, 2002) . Therefore, MEOR is potentially a "high risk, high reward" process as concluded by Bryant & Lockhart, (2002) in "Reservoir Engineering Analysis of Microbial Enhanced Oil Recovery"; in which the reward refers to the fact that the difficulty and the logistical costs of implementing the process would approach those of implementing a waterflood rather than an EOR process. The risk is associated with the many and severe performance constraints that a microbial system would have to satisfy to take advantage of an in-situ carbon source. Therefore MEOR feasibility still requires considerable research and development.

Conversely, published information of MEOR oilfield applications have demonstrated the benefits of MEOR as is the case of the application of microbial enhanced oil recovery technique in the Daqing Oilfield reported by Li et al., (2002), which claims and incremental oil recovery by 11.2% and a reduction of oil viscosity of 38.5% in the pilot tests. Another report by Portwood J. T. (1995), presented the analysis of the effectiveness and economics of 322 MEOR projects carried out in more than 2,000 producing oil wells in the United States that applied the same single MEOR technology. This analysis indicated that the MEOR process applied effectively mobilized residual crude oil. 78% of the MEOR projects demonstrated on average 36% of incremental oil production. The MEOR process implemented worked effectively under reservoir conditions and is environmentally friendly. Common operational problems associated with paraffin, emulsion, scale, and corrosion were significantly reduced and oil production decrease was not observed during the MEOR projects time frame. This MEOR technology demonstrated to be economically and technologically feasible. The producers average return on investment from MEOR was 5:1 within the first 24 months of MEOR and the average time to project payback was six months. In Chapter 3, a more detailed review of the MEOR process is presented.

2.3.6 EOR processes and technical maturity

According to Regtien (2010), mature EOR processes that are well established and therefore can be implemented without significant adaptations are: vertical well steam drive, cyclic steam stimulation, miscible gas injection, and polymer flooding, as shown in Figure 18. The second group of EOR processes in the middle of the curve (Fig. 18) presents the technologies that require a significant amount of optimization and field trials to de-risk the concept or get the correct full field design. These EOR processes are alkaline-surfactant-polymer (ASP) flooding, in-situ combustion such as High Pressure Air Injection (HPAI), steam assisted gravity drainage (SAGD), low-salinity waterflooding, and high pressure steam injection. The EOR technologies at the bottom-left of the curve (Fig. 18) outlines processes that are relatively immature but very promising that still require extensive research and development. These processes are in-situ upgrading, foams, and hybrid processes (Regtien, 2010).

2.3.7 Selection of EOR processes

According to Pope (2011) the selection of a suitable EOR process for a specific oil formation requires and integrated study of the reservoir and its characteristics. Some basic questions that the operator must address during the decision making process, rather than follow conventional wisdom or simplified screening criteria that may be out of date are outlined in Figure 19.

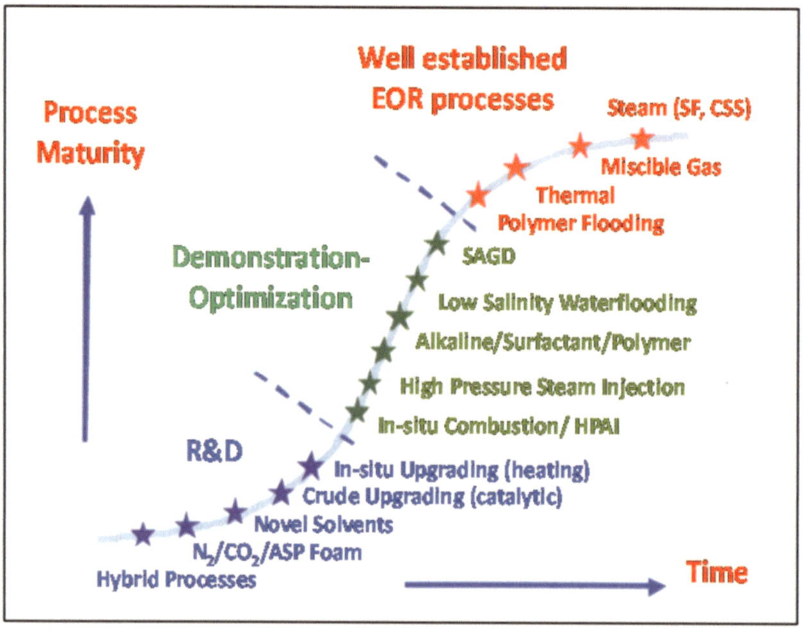

Fig. 18. Maturation Curve for Enhanced Oil Recovery (Adapted from Regtien, 2010).

As Pope (2011) continues…once these questions are carefully addresed based on sound technical analysis the ideal strategy is to use both simplified models and detailed reservoir simulation models to evaluate the options assuming the process might be economic. If initial calculations indicate the process may be profitable, then there will be a need for additional laboratory and field measurements followed by more modeling. In many cases, a single well test will be justified to evaluate injectivity, reduction in oil saturation, and other performance indicators that can only be assessed with field tests. When comparing the economics of different processes, many factors must also be taken into account. Chemical methods have the advantage of lower capital costs than miscible gas and thermal methods, and commercial projects can start small and be expanded if successful without the need for expensive infrastructure such as pipelines (Pope, 2011).

Detailed outlines of EOR screening criteria are available in the literature that are based on matching particular EOR processes to reservoir properties such as °API gravity, reservoir depth, oil saturation, rock permeability and porosity distribution, oil viscosity distribution, type of rock formation, and reservoir temperature distribution, among others. Aladasani & Bai, (2010) recently published an updated paper on EOR screening criteria. Practical information on EOR field planning and development strategies are provided by Alvarado and Manrique (2010). A comprehensive screening criteria for EOR based on oilfield data obtained from successful EOR projects worldwide and on oil recovery mechanisms was proposed by Taber et al., (1997a, 1997b).

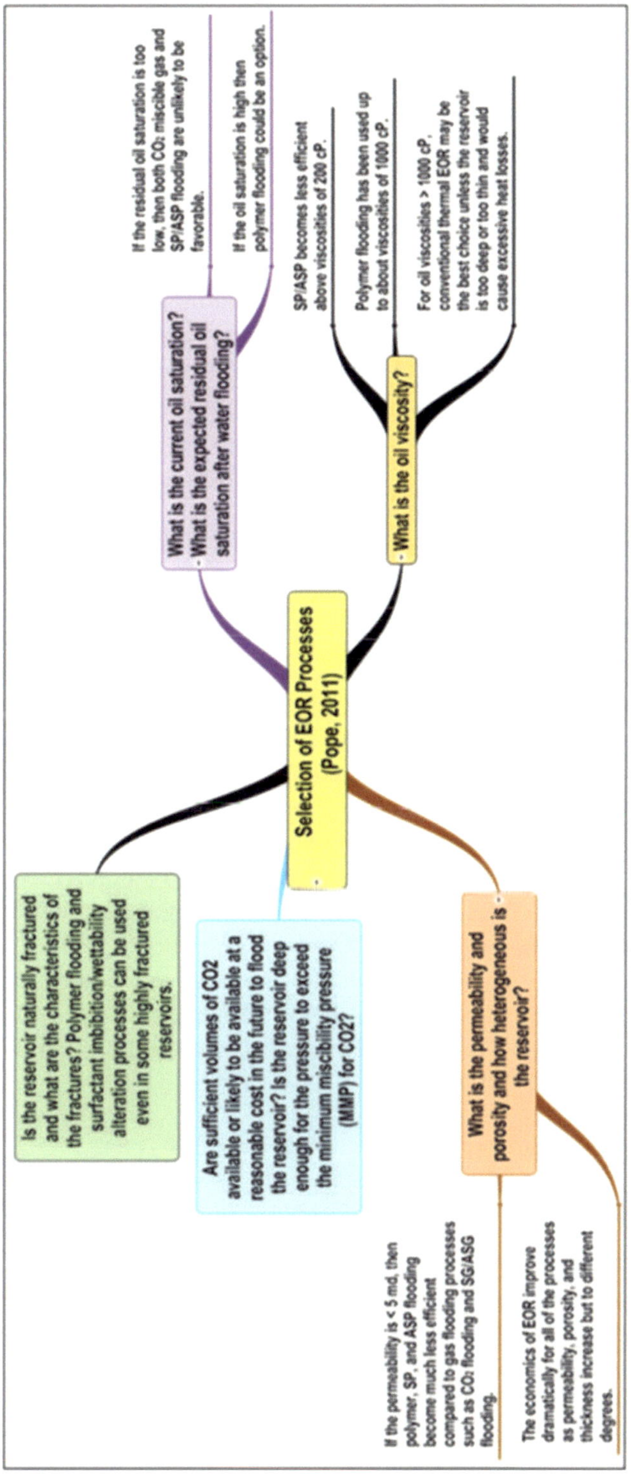

Fig. 19. Selection of EOR Processes: basic questions (Source: Pope, 2011).

Lately, publications have been focused on reservoir selection for anthropogenic CO_2 sequestration and CO_2-flood EOR. CO_2 sequestration in oil reservoirs is not a straightforward application of existing oil field technology and operating practices (Kovscek 2002). The key issue in this process is to maximize the volume of CO_2 that can be retained in a reservoir by physical trapping or by maximizing the CO_2 solubility in the reservoir fluids (Aladasani & Bai, 2010). For instance, Shaw & Bachu, (2002) developed a method for the rapid screening and ranking of oil reservoirs suited for CO_2-flood EOR that is useful for preliminary analyses and ranking of large number of oil pools. To determine reservoir suitability for CO_2 flooding, oil reservoirs are screened on the basis of oil gravity, reservoir temperature and pressure, minimum miscibility pressure and remaining oil saturation. Kovscek (2002) proposed screening criteria for CO_2 storage and in a subsequent paper Kovscek & Cakici (2005) reported strategies to cooptimize oil recovery and CO_2 storage via compositional reservoir simulation. This study proposed a form of production well control that limits the fraction of gas relative to oil produced as an effective practice for the cooptimization of CO_2 sequestration and oil recovery.

3. Summary

The current renewed interest on research and development of EOR processes and their oilfield implementation would allow targeting significant volumes of oil accumulations that have been left behind in mature reservoirs after primary and secondary oil recovery operations. The potential for EOR is real and achivable. However, improvements of the operational performance and the economical optimization of EOR projects in the future would require the application of a synergistic approach among EOR processes, improved reservoir characterization, formation evaluation, reservoir modeling and simulation, reservoir management, well technology, new and advanced surveillance methods, production methods, and surface facilities as stated by Pope (2011). This synergistic approach is in line with the Smart Fields Concept, also known as Intelligent Field, Digital Field, i-Field or e-Field, developed by Shell International Exploration and Production that involves an integrated approach, which consists of data acquisition, modeling, integrated decision making, and operational field management, each with a high level of integration and automation (Regtien, 2010).

4. References

Adkins S., Liyanage P., Pinnawala Arachchilage G., Mudiyanselage T., Weerasooriya U. & Pope, G. "A New Process for Manufacturing and Stabilizing High-Performance EOR Surfactants at Low Cost for High-Temperature, High-Salinity Oil Reservoirs." *SPE Paper 129923 presented at the 2010 SPE Improved Oil Recovery Symposium.* Tulsa, Oklahoma, U.S.A., 24-28 April: Society of Petroleum Engineers, 2010. 1-9.

Ahmed T. & McKinney P. *Advanced Reservoir Engineering.* Burlington, MA: Elsevier, 2005.

Aladasani A. & Bai B. "Recent Developments and Updated Screening Criteria of Enhanced Oil Recovery Techniques." *SPE 130726 presented at the CPS/SPE International Oil & Gas Conference and Exhibition.* Beijing, China, 8-10 June: Society of Petroleum Engineers, 2010. 1-24.

Al-Assi A. A., Willhite G. P., Green D. W. & McCool C. S. "Formation and Propagation of Gel Agregates Using Partially Hydrolyzed Polyacrylamide and Aluminun Citrate." *SPE Paper 100049 presented at the 2006 SPE/DOE Symposium on Improved Oil Recovery.* Tulsa, Oklahoma, U.S.A., 22-26 April: Society of Petroleum Engineers, 2006. 1-13.

Alvarado V. & Manrique E. *Enhanced Oil Recovery: Field Planning and Development Strategies.* Burlington, MA, USA.: Gulf Professional Publishing, 2010.

AOSC -Athabasca Oil Sands Corporation. *Technology. Steam Assisted Gravity Drainage (SAGD) vs. Mining.* 2008. http://www.aosc.com/corporate-overview/technology.html (accessed April 5, 2012).

Azira H., Tazerouti A. & Canselier J. P. "Phase Behaviour of Pseudoternary Brine/Alkane/Alcohol-secondary Alkanesulfonates Systems. Surfactant ratio effects of salinity and alcohol." *Journal of Thermal Analysis and Calorimetry, 92 (3),* 2008: 759-763.

Bai, B., Shuler P., Qu Q. & Wu Y. *Preformed Particle Gel For Conformance Control.* Progress Report. RPSEA Subcontract # 07123-2, Missouri: Missouri University of Science and Technology (MS & T), 2009.

Banat I. M., Makkar R. S. & Cameotra S. S. "Potential commercial applications of microbial surfactants." *Appl Microbiol Biotechnol, 53,* 2000: 495-508.

Barnes J., Dirkzwager H., Smit J.R., Smit J. P., On A., Navarrete R. C., Ellison B. H., Buijse M. A. & Rijswijk B. V. "Application of Internal Olefin Sulfonates and Other Surfactants to EOR. Part 1: Structure-Performance Relationships for Selection at Different Reservoir Conditions." *SPE Paper 129766 presented at the 2010 SPE Improved Oil Recovery Symposium.* Tulsa, Oklahoma, U.S.A., 24-28 April: Society of Petroleum Engineers, 2010. 1-16.

Berger P. D. & Lee H. C. "New Anionic Alkylaryl Surfactants Based on Olefin Sulfonic Acids." *Journal of Surfactants and Detergents,* 5 (1) January, 2002: 39-43.

Bou-Mikael S., Asmadi F., Marwoto D. & Cease C. "Minas Surfactant Field Trial Tests Two Newly Designed Surfactants with High EOR Potential." *SPE Paper 64288 presented at the SPE Asia Pacific Oil and Gas Conference and Exhibition.* Brisbane, Australia October: Society of Petroleum Engineers Inc., 2000. 1-12.

Bryant S. & Lockhart T. "Reservoir Engineering Analysis of Microbial Enhanced Oil Recovery." *SPE Reservoir Evaluation & Engineering,* 5 (5), 2002: 365-374.

Buijse M. A., Prelicz R. M., Barnes J. R. & Cosmo C. "Application of Internal Olefin Sulfonates and Other Surfactants to EOR. Part 2: The Design and Execution of an ASP Field Test." *SPE Paper 129769 presented at the 2010 SPE Improved Oil Recovery Symposium.* Tulsa, Oklahoma, U.S.A., 24-28 April: Society of Petroleum Engineers, 2010. 1-12.

Cao Y. & Li H. "Interfacial activity of a novel family of polymeric surfactants." *European Polymer Journal 38,* 2002: 1457-1463.

Chang H., Sui X., Xiao L., Liu H., Guo S., Yao Y., Xiao Y., Chen G., Song K. & Mack J. C. "Successful Field Pilot of In-Depth Colloidad Dispersion Gel (CDG) Technology in Daqing Oil Field." *SPE Paper 89460 presented at the 2004 SPE/DOE Fourteenth Symposium on Improved Oil Recovery.* Tulsa, Oklahoma, U.S.A., 17-21 April: Society of Petroleum Engineers Inc., 2004. 1-15.

Choi, S. K. *A Study of a pH-sensitive Polymer for Novel Conformance Control Applications.* Master Science Thesis, Austin, Texas: The University of Texas , 2005.

Choi S. K., Ermel Y. M., Bryant S. L., Huh C. & Sharma M. M. "Transport of a pH-Sensitive Polymer in Porous Media for Novel Mobility-Control Applications." *SPE Paper 99656 presented a the 2006 SPE/DOE Symposium on Improved Oil Recovery.* Tulsa, Oklahoma, U.S.A., 22-26 April 2006: Society of Petroleum Engineers Inc., 2006. 1-15.

Chung T., Bae W., Nguyen N. T. B., Dang C. T. Q., Lee W. & Jung B. "A Review of Polymer Conformance Treatment: A Successful Guideline for Water Control in Mature Fields." *Energy Sources, Part A: Recovery, Utilization, and Environmental Effects* 34(2), 2011: 122-133.

Coste J. P., Liu Y., Bai B., Li Y., Shen P., Wang Z. & Zhu G. "In-Depth Fluid Diversion by Pre-Gelled Particles. Laboratory Study and Pilot Testing." *SPE Paper 59362 presented at the 2000 SPE/DOE Improved Oil Recovery Symposium.* Tulsa, Oklahoma, 3-5 April: Society of Petroleum Engineers Inc., 2000. 1-8.

Craft B. C. & Hawkins, M. F. *Applied Petroleum Reservoir Engineering (Second Edition).* New Yersey, ISBN0-13-039884-5: Prentice Hall PTR, 1991.

Cui X., Li Z., Cao X., Song X. & Zhang X. "A Novel PPG Enhanced Surfactant-Polymer System for EOR." *SPE Paper 143506 presented at the 2011 SPE Enhanced Oil Recovery Conference.* Kuaka Lumpur, Malaysia, 19-21 July: Society of Petroleum Engineers, 2011. 1-8.

Dake L. (1978). *Fundamentals of Reservoir Engineering.* Elsevier Inc. 0-444-41830-X, San Diego, CA, 1978.

Delshad M., Asakawa K., Pope G. & Sepehrnoori K. "Simulations of Chemical and Microbial Enhanced Oil Recovery Methods." *SPE Paper 75237 presented at the SPE/DOE Improved Oil Recovery Symposium.* Tulsa, Oklahoma, 13-17 April: Society of Petroleum Engineers Inc., 2002. 1-13.

Diaz D., Somaruga C., Norman C. & Romero J. "Colloidal Dispersion Gels Improve Oil Recovery in a Heterogeneous Argentina Waterflood." *SPE Paper 113320 presented at the 2008 SPE/DOE Improved Oil Recovery Symposium.* Tulsa, Oklahoma, U.S.A. 19-23 April: Society of Petroleum Engineers, 2008. 1-10.

Dupuis G., Rousseau D., Tabary R. & Grassl B. "How to get the Best Out of Hydrophobically Associative Polymers for IOR? New Experimental Insights." *SPE paper 129884 presented at the 2010 SPE Improved Oil Recovery Symposium.* Tulsa, 24-28 April: Society of Petroleum Engineers, 2010. 1-12.

Dupuis G., Rousseau D., Tabary R., & Grassl B. "Injectivity of Hydrophobically Modified Water Soluble Polymers for IOR: Controlled Resistance Factors vs. Flow-Induced Gelation." *SPE Paper 140779 presented at the 2011 SPE International Symposium on Oilfield Chemistry.* The Woodlands, Texas, U.S.A. 11-13 April: Society of Petroleum Engineers, 2011. 1-13.

Elraies K. A., Tan I. M., Awang M. & Saaid I. "The Synthesis and Performance of Sodium Methyl Ester Sulfonate for Enhanced Oil Recovery." *Petroleum Science and Technology,* 28(17), 2010: 1799-1806.

Enick R. & Olsen, D. *Mobility and Conformance Control for Carbon Dioxide Enhanced Oil Recovery (CO_2-EOR) via Thickeners, Foams, and Gels-A Detailed Literature Review of 40 Years of Research.* Contract DE-FE0004003. Activity 4003.200.01, Pittsburgh: National Energy Technology Laboratory (NETL), 2012.

Feitler D. *The Herculean Surfactant for Enhanced Oil Recovery, Request #60243.* Cleveland, April 17, 2009.

Fielding Jr. R. C., Gibbons D. H. & Legrand F. P. "In-Depth Drive Fluid Diversion Using an Evolution of Colloidal Dispersion Gels and New Bulk Gels: An operational Case History of North Rainbow Ranch Unit." *SPE/DOE Paper 27773 presented at the 1994 SPE/DOE Ninth Symposium on Improved Oil Recovery.* Tulsa, Oklahoma, U.S.A., 17-20 April: Society of Petroleum Engineers, Inc., 1994. 1-12.

Flaaten A. K., Nguyen Q. P., Pope G. & Zhang J. "A Systematic Laboratory Approach to Low-Cost, High-Performance Chemical Flooding." *SPE Paper 113469 presented at the 2008 SPE/DOE Improved Oil Recovery Symposium.* Tulsa, Oklahoma, U.S.A. 19-23 April: Society of Petroleum Engineers, 2008. 1-20.

Frampton H., Morgan J. C., Cheung S. K., Munson L., Chang K. T. & Williams D. "Development of a novel waterflood conformance control system." *SPE Paper 89391 presented at the 2004 SPE/DOE Fourteenth Symposium on Improved Oil Recovery.* Tulsa, Oklahoma, U.S.A., 17-21 April: Society of Petroleum Engineers Inc., 2004. 1-9.

Garmeh R., Izadi M., Salehi M., Romero J. L., Thomas C. P. & Manrique, E. J. "Thermally Active Polymer to Improve Sweep Efficiency of Water floods: Simulation and Pilot Design Approaches." *SPE Paper 144234 presented at the 2011 SPE Enhanced Oil Recovery Conference.* Kuala Lumpur, Malaysia, 19-21 July: Society of Petroleum Engineers, 2011. 1-15.

Green D. W. & Willhite G. P. *Enhanced Oil Recovery.* Richardson, Texas: Society of Petroleum Engineers, 1998.

Hirasaki G. J., Miller C. A., & Puerto M. "Recent Advances in Surfactant EOR." *SPE Journal,* 16 (4), 2011: 889-907.

Huh C., Choi S. K. & Sharma M. M. "A Rheological Model for pH-Sensitive Ionic Polymer Solutions for Optimal Mobility-Control Applications." *SPE Paper 96914 presented at the 2005 SPE Annual Technical Conference and Exhibition.* Dallas, Texas, U.S.A. 9-12 October: Society of Petroleum Engineers Inc., 2005. 1-13.

Iglauer S., Wu Y., Shuler P. J., Blanco M., Tang Y. & Goddard III W. A. "Alkyl Polyglycoside Surfactants for Improved Oil Recovery." *SPE Paper 89472 presented at the 2004 SPE/DOE Fourteenth Symposium on Improved Oil Recovery.* Tulsa, Oklahoma, U.S.A. 17-21 April: Society of Petroleum Engineers, 2004. 1-9.

Izgec O. & Shook, G. M. "Design Considerations of Waterflood Conformance Control with Temperature-Triggered Low Viscosity Sub-micron Polymer." *SPE Paper 153898 presented at the 2012 SPE Western Regional Meeting.* Bakersfield, California, U.S.A., 19-23 March: Society of Petroleum Engineers, 2012. 1-12.

Jackson A. C. *Experimental study of the Benefits of Sodium Carbonate on Surfactants for Enhanced Oil Recovery.* Master of Science in Engineering, Austin, Texas: The University of Texas at Austin, 2006.

Kovscek A. R. "Screening Criteria for CO_2 Storage in Oil Reservoirs." *Petroleum Science and Technology,* 20 (7 – 8), 2002: 841-866.

Kovscek A. R. & Cakici, M.D. "Geological storage of carbon dioxide and enhanced oil recovery. II. Cooptimization of storage and recovery." *Energy Conversion and Management,* 46 (11-12), July 2005, 1941-1956.

Lake L. W. "Enhanced Oil Recovery." *SPE ATCE. Training Courses.* Florence, Italy: Society of Petroleum Engineers, September 23, 2010.

Lake L. W. *Enhanced Oil Recovery.* Englewood Cliffs, New Jersey: Prentice Hall, 1989.

Lalehrokh, F. & Bryant, S. L. "Application of pH-Triggered Polymers for Deep Conformance Control in Fractured Reservoirs. "*SPE Paper 124773 presented at the 2009 SPE Annual Technical Conference and Exhibition.* New Orleans, Louisiana, U.S.A. 4-7 October: Society of Petroleum Engineers Inc., 2009. 1-11.

Levitt D. B. "*Experimental Evaluation of High Performance EOR Surfactants for a Dolomite Oil Reservoir.*" Master of Science in Engineering, Austin: The University of Texas at Austin, 2006.

Levitt D.B., Jackson A. C., Heinson C., Britton L. N., Malik T., Dwarakanath V., & Pope G. A. "Identification and Evaluation of High-Performance EOR Surfactants." *SPE Reservoir Evaluation & Engineering,* 12 (2), 2009: 243-253.

Li, Q., Kang C., Wang H., Liu C. & Zhang C. "Application of microbial enhanced oil recovery technique to Daqing Oilfield." *Biochemical Engineering Journal,* 11, 2002: 197-199.

Lu, X., Song K., Niu J. & Chen F. "Performance and Evaluation Methods of Colloidal Dispersion Gels in the Daqing Oil Field." *SPE Paper 59466 presented at the 2000 SPE Asia Pacific Conference on Integrated Modelling for Asset Management.* Yokohama, Japan, 25-26 April: Society of Petroleum Engineers Inc., 2000. 1-11.

Lyons W. & Plisga, B. S. (Eds). *Standard Handbook of Petroleum & Natural Gas Engineering (Second edition).* Burlington, MA: Elsevier Inc. ISBN-13:978-0-7506-7785-1, 2005.

Manrique E., Thomas C., Ravikiran R., Izadi M., Lantz M., Romero J. & Alvarado V. "EOR: Current Status and Opportunities." *SPE Paper 130113 presented at the 2010 SPE Improved Oil Recovery Symposium.* Tulsa, Oklahoma, U. S.A., 24-28 April: Society of Petroleum Engineers, 2010. 1-21.

Meridian. *Energy Insights.* August 1, 2006. http://www.safehaven.com/article/5639/energy-insights (accessed April 6, 2012).

Muruaga E., Flores M., Norman C. & Romero J. "Combining Bulk Gels and Colloidal Dispersion Gels for Improved Volumetric Sweep Efficiency in a Mature Waterflood." *SPE Paper 113334 presented at the 2008 SPE/DOE Improved Oil Recovery Symposium.* Tulsa, Oklahoma, U.S.A., 19-23 April: Society of Petroleum Engineers, 2008. 1-12.

Mustoni J. L., Norman C. A., & Denyer P. "Deep Conformance Control by a Novel Thermally Activated Particle System to Improve Sweep Efficiency in Mature Waterfloods on the San Jorge Basin." *SPE Paper 129732 presented at the 2010 SPE Improved Oil Recovery Symposium.* Tulsa, Oklahoma, U.S.A. 24-28 April: Society Petroleum Engineers, 2010. 1-10.

Norman C. A., Smith J. E. & Thompson, R. S. "Economics of In-Depth Polymer Gel Processes." *SPE Paper 55632 presented at the 1999 SPE Rocky Mountain Regional Meeting.* Gillette, Wyoming, 15-18 May: Society of Petroleum Engineers Inc., 1999. 1 - 8.

Norman C., Turner B., Romero J. L., Centeno G. & Muruaga, E. "A Review of Over 100 Polymer Gel Injection Well Conformance Treatments in Argentina and Venezuela: Design, Field Implementation, and Evaluation." *SPE Paper 101781 presented at the First International Oil Conference and Exhibition in Mexico.* Cancun, Mexico, 31 August - 2 September: Society of Petroleum Engineers, 2006. 1 -16.

Ohms, D., McLeod J., Graff C., Frampton H., Morgam J. C., Cheung S., & Chang, K.T. "Incremental-Oil Success From Waterflood Sweep Improvement in Alaska." *SPE Production & Operations*, 25 (3), August 2010: 1-8.

Okasha T. & Al-Shiwaish, A. "Effect of Brine Salinity on Interfacial Tension in Arab-D Carbonate Reservoir, Saudi Arabia." *SPE paper 119600 presented at the 2009 SPE Middle East Oil & Gas Show and Conference.* Bahrain, Kingdom of Bahrain: Society of Petroleum Engineers, 2009. 1-9.

Okeke T. & Lane, R. "Simulation and Economic Screening of Improved Oil Recovery Methods with Emphasis on Injection Profile Control Including Waterflooding, Polymer Flooding and a Thermally Activated Deep Diverting Gel." *SPE Paper 153740 presented at the 2012 SPE Western Regional Meeting.* Bakersfield, California, U.S.A. 19-23 March: Society of Petroleum Engineers, 2012. 1-14.

OSDG - The Oil Sands Developers Group . *Toe-to-Heel Air Injection (THAI).* 2009. http://www.oilsandsdevelopers.ca/index.php/oil-sands-technologies/in-situ/the-process-2/toe-to-heel-air-injection-thai. (accessed April 6, 2012).

Ovalles C., Bolivar R., Cotte E., Aular W., Carrasquel J. & Lujano E. "Novel ethoxylated surfactants from low-value refinery feedstocks." *Fuel* 80 (4), 2001: 575-582.

Pope G. "Recent Developments and Remaining Challenges of Enhanced Oil Recovery." *JPT,* 2011: 65-68.

Portwood J. T. "A Commercial Microbial Enhanced Oil Recovery Technology: Evaluation of 322 Projects." *SPE Paper 29518 presented at the 1995 Production Operations Symposium.* Oklahoma City, OK, U.S.A., 2-4 April: Society of Petroleum Engineers, Inc., 1995. 1-16.

Puerto M., Hirasaki G. J., Miller C. & Barnes J. R. "Surfactant Systems for EOR in High-Temperature, High-salinity Environments." *SPE Paper 129675 presented at the 2010 SPE Improved Oil Recovery Symposium.* Tulsa, Oklahoma, U.S.A., 24-28 April: Society of Petroleum Engineers, 2010. 1-20.

Ranganathan R., Lewis R., McCool C. S., Green D. W. & Willhite G. P. "Experimental Study of the Gelation Behavior of a Polyacrylamide/Aluminum Citrate Colloidal-Dispersion Gel System." *SPE Journal*, 3 (4), December 1998: 337-343.

Regtien J. M. M. "Extending The Smart Fields Concept To Enhanced Oil Recovery." *SPE Paper 136034 presented at the 2010 SPE Russian Oil & Gas Technical Conference and Exhibition.* Moscow, Russia, 26-28 October: Society of Petroleum Engineers, 2010. 1-11.

Reichenbach-Klinke R., Langlotz B., Wenzke B., Spindler C. & Brodt G. "Hydrophobic Associative Copolymer with Favorable Properties for the Application in Polymer Flooding." *SPE Paper 141107 presented at the SPE International Symposium on Oilfield Chemistry.* The Woodlands, Texas, U.S.A., 11-13 April: Society of Petroleum Engineers, 2011. 1-11.

Roussennac B. & Toschi, C. "Brightwater Trial in Salema Field (Campos Basin, Brazil)." *SPE Paper 131299 presented at the 2010 SPE EUROPEC/EAGE Annual Conference and Exhibition.* Barcelona, Spain, 14-17 June: Society of Petroleum Enginners, 2010. 1-13.

Satter A., Iqbal, G. & Buchwalter, J. *Practical Enhanced Reservoir Engineering.* Tulsa, Oklahoma: PennWell , 2008.

Seright R. S. "Discussion of SPE 89175, Advances in Polymer Flooding and Alkaline/Surfactant/Polymer Processes as Developed and Applied in the People's Republic of China." *J Pet Technol 58 (2):80*, 2006: 84-89.

Seright R. S., Zhang G., Akanni O. O. & Wang D. "A Comparison of Polymer Flooding With In-Depth Profile Modification." *SPE Paper 146087 presented at the 2011 Canadian Unconventional Resources Conference.* Calgary, Alberta, Canada, 15-17 November: Society of Petroleum Engineers, 2011. 1-13.

Seright R. S., Prodanovic M. & & Lindquist B. "X-Ray Computed Microtomography Studies of Fluid Partitioning in Drainage and Imbibition Before and After Gel Placement." *SPE J*, 11(2): June 2006, 159-170.

Seright R. S., Fan T., Wavrik K., Wan H., Gaillard N. & Favero, C. "Rheology of a New Sulfonic Associative Polymer in Porous Media." *SPE Reservoir Evaluation & Engineering*, 14 (6), December 2011: 726-134.

Shah A., Wood J., Greaves M., Rigby, S. & Fishwick R. "In-situ up-grading of heavy oil/natural bitumen: Capri Process Optimisation." *8th World Congress of Chemical Engineering: Incorporating the 59th Canadian Chemical Engineering Conference and the 24th Interamerican Congress of Chemical Engineering.* Montreal: Elsevier B. V., 2009. 520 e.

Shah A., Fishwick R., Wood J., Leeke G., Rigby S. &. Greaves M. "A review of novel techniques for heavy oil and bitumen extraction and upgrading." *Energy & Environmental Science*, 3, 2010: 700-714.

Sharma M., Bryant S. & Huh C. *pH Sensitive Polymers for Improving Reservoir Sweep and Conformance Control in Chemical Flooding.* DOE Final Report, Austin, Texas: The University of Texas at Austin, 2008.

Shaw J. & Bachu, S. "Screening, Evaluation, and Ranking of Oil Reservoirs Suitable for CO_2-Flood EOR and Carbon Dioxide Sequestration." *JCPT*, 41 (9), 2002, 51-61.

Sheng J. *Modern Chemical Enhanced Oil Recovery. Theory and Practice.* Burlington, MA, USA: Gulf Professional Publishing, 2011.

Shi J., Varavei A., Huh C., Delshad M., Sepehrnoori K. & Li, X. "Viscosity Model of Preformed Microgels for Conformance and Mobility Control." *Energy Fuels*, 2011: 25, 5033-5037.

Shi J., Varavei A., Huh C., Sepehrnoori, K., Delshad M. & Li X. "Transport Model Implementation and Simulation of Microgel Processes for Conformance and Mobility Control Purposes." *Energy Fuels*, 2011: 25, 5063-5075.

Singhal A. *Preliminary Review of IETP Projects Using Polymers.* Engineering Report, Calgary, Alberta, Canada: Premier Reservoir Engineering Services LTD, 2011.

Smith J. E., Liu H. & Guo, Z. D. "Laboratory Studies of In-Depth Colloidal Dispersion Gel Technology for Daqing Oil Field." *SPE Paper 62610 presented at the 2000 SPE/AAPG Western Regional Meeting.* Long Beach, California, 19-23 June: Society of Petroleum Engineers Inc., 2000. 1-13.

Smith J. E., Mack J. C. & Nicol, A. B. "The Adon Road-An In-Depth Gel Case History." *SPE/DOE Paper 35352 presented at the 1996 SPE/DOE Tenth Symposium on Improved Oil Recovery.* Tulsa, Oklahoma, 21-24 April: Society of Petroleum Engineers, 1996. 1-11.

Soudmand-asli A., Ayatollahi S. , Mohabatkar H., Zareie. M. & Shariatpanahi F. "The in situ microbial enhanced oil recovery in fractured porous media." *Journal of Petroelum Science and Engineering*, 58, 2007. 161-172.

Spildo K., Skauge A., Aarra, M. G. & Tweheyo M. T. "A New Polymer Application for North Sea Reservoirs." *SPE Paper 113460 presented at the 2008 SPE/DOE Improved Oil Recovery Symposium.* Tulsa, Oklahoma, U.S.A. 19-23 April: Society of Petroleum Engineers, 2008. 1 - 9.

Sydansk R. D. & Romero-Zerón, L. *Reservoir Conformance Improvement.* Richardson, Texas: Society of Petroleum Engineers, 2011.

Sydansk R. D. & Seright, R. S. "When and Where Relative Permeability Modification Water-Shutoff Treatments Can Be Successfully Applied." *SPE Prod & Oper*, 22(2), May 2007: 236-247.

Sydansk R. D. "Polymers, Gels, Foams, and Resins." In *Petroleum Engineering Handbook Vol. V (B), Chap. 13*, by Lake L. W. (Ed.), 1149-1260. Richardson, Texas: Society of Petrleum Engineers, 2007.

Taber J. J., Martin F. D. & Seright, R. S. "EOR Screening Criteria Revisited-Part 1: Introduction to Screening Criteria and Enhanced Recovery Field Projects." *SPE Resevoir Engineering*, 12 (3) August 1997: 189-198.

Taber J. J., Martin F. D. & Seright, R. S. "EOR Screening Criteria Revisited-Part 2: Applications and Impact of Oil Prices." *SPE Reservoir Engineering*, 12 (3) August 1997: 199-205.

Van Dyke K. *Fundamentals of Petroleum (Fourth Edition).* Austin, Texas: The University of Texas at Austin, 1997.

Wang D., Liu C., Wu W. & Wang G. "Novel Surfactants that Attain Ultra-Low Interfacial Tension between Oil and High Salinity Formation Water without adding Alkali, Salts, Co-surfactants, Alcohol, and Solvents." *SPE Paper 127452 at the SPE EOR Conference at Oil & Gas West Asia.* Muscat, Oman, 11-13 April: Society of Petroleum Engineers, 2010. 1-11.

Wang, D., Han P., Shao Z., Hou W. & Seright, R. S. "Sweep-Improvement Options for the Daqing Oil Field." *SPE Reservoir Evaluation & Engineering*, 11 (1) February 2008: 18 - 26.

Wang L., Zhang G. C., Ge J.J., Li G. H., Zhang J. Q. & Ding B. D. "Preparation of Microgel Nanospheres and Their Application in EOR." *SPE Paper 130357 presented at the 2010 CPS/SPE International Oil & Gas Conference and Exhibition.* Beijing, China, 8-10 June: Society of Petroleum Engineers, 2010. 1 - 7.

Wu Y., Shuler P. J., Blanco M., Tang Y & Goddard III W. A. "A Study of Wetting Behavior and Surfactant EOR in Carbonates With Model Compounds." *SPE Paper 99612 presented at the 2006 SPE/DOE Symposium on Improved Oil Recovery.* Tulsa, Oklahoma, U.S.A.: Society of Petroleum Engineers, 2006. 1-11.

Wu, Y., Shuler P., Blanco M., Tang Y. & Goddard, W. A. "A Study of Branched Alcohol Propoxylate Sulfate Surfactants for Improved Oil Recovery." *SPE Paper 95404 presented at the 2005 SPE Annual Technical Conference and Exhibition.* Dallas, Texas, U.S.A., 9-12 October: Society of Petroleum Engineers Inc., 2005. 1-10.

Yang H., Britton C., Liyanage P.J., Solairaj S., Kim D.H., Nguyen Q., Weerasooriya U. & Pope G. "Low-cost, High -Performance Chemicals for Enhanced Oil Recovery." *SPE paper 129978 presented at the 2010 SPE Improved Oil Recovery Symposium.* Tulsa, Oklahoma, U.S.A., 24-28 April: Society of Petroleum Engineers, 2010. 1-24.

Zaitoun A., Makakou P., Blin N., Al-Maamari R.S., Al-Hashmi A. R. & Abdel-Goad M. "Shear Stability of EOR Polymers." *SPE Paper 141113 presented at the 2011 SPE International Symposium on Oilfield Chemistry.* The Woodlands, Texas, U.S.A., 11-13 April: Society of Petroleum Engineers, 2011. 1-7.

2

The Application of a New Polymeric Surfactant for Chemical EOR

Khaled Abdalla Elraies and Isa M. Tan
Universiti Teknologi PETRONAS, Ipoh
Malaysia

1. Introduction

Crude oil makes a major contribution to the world economy today. The provision of heat, light, and transportation depends on oil and there has not been yet a single energy source to replace crude oil that is widely integrated. Moreover, the global economy currently depends on the ability to acquire the energy required and it is indisputable that oil is the main contributor to this demand. Currently, there is no an energy source available that could compete with oil, making the world, and mainly the high energy consumers to rely on countries with large reserves (Energy Information Administration, 2003).

Traditionally oil production strategies have followed primary depletion, secondary recovery, and tertiary recovery processes. Primary depletion uses the natural reservoir energy to accomplish the displacement of oil from the porous rocks to the producing wells (Craft et al., 1991). An average of 10 to 20 percent of original oil in place (OOIP) can be recovered through primary recovery. Secondary recovery methods are processes in which the oil is subjected to immiscible displacement with injected fluids such as water or gas. It is estimated that about thirty to fifty percent of the OOIP can be produced through the entire life of a mature reservoir that has been developed under primary and secondary recovery methods (Green & Willhite, 1998). The remaining oil is still trapped in the porous media. This is attributed to surface and interfacial forces (capillary forces), viscosity forces, and reservoir heterogeneities which results in poor displacement efficiency (Green & Willhite, 1998). Recognition of these facts has led to the development and use of many enhanced oil recovery (EOR) methods. EOR methods hold promise for recovering a significant portion of the remaining oil after conventional methods.

Planning for improving or enhancing oil production strategies through EOR methods is one of the most critical challenges facing the industry today. EOR not only will extend the life of this important non-renewable resource, but it will also delay a world production decline and shortage in the energy supply. Realizing the significant potential of EOR, most of oil companies embarked on a massive journey to advance EOR processes.

Various modifications of EOR methods have been developed to recover at least a portion of the remaining oil. Thermal processes are the most common type of EOR, where a hot invading phase, such as steam, hot water or a combustible gas, is injected in order to increase the temperature of oil and gas in the reservoir and facilitate their flow to the production wells (Green & Willhite, 1998). Another type of EOR process consists of injecting a miscible phase with the oil and gas into the reservoir to eliminate the interfacial tension effects. The miscible phase can be a hydrocarbon solvent, CO_2, or an inert gas (N_2). Another common EOR technique is chemical flooding which includes alkalis, surfactants, and polymers, or combinations thereof. The injected alkali and surfactant agents can lower interfacial tension (IFT) between oil and water, thereby mobilize the residual oil. Polymers are used to viscosify the aqueous solution for mobility control (Green & Willhite, 1998).

2. Chemical flooding for EOR

Chemical flooding, which has been developed since the early 1950s, is an important method for enhanced oil recovery that includes alkaline flooding, alkali-surfactant flooding, and alkali-surfactant-polymer flooding. Surfactant flooding and its variants are EOR processes that have been employed to recover the residual oil after primary and secondary recovery process.

The efficiency of the chemical EOR is a function of liquid viscosities, relative permeabilities, interfacial tensions, wettabilities, and capillary pressures (Liu, 2008). Even if all the oil is contacted by the injected chemicals, some oil would still remain in the reservoir. This is due to the trapping of oil droplets by capillary forces due to high interfacial tension (IFT) between water and oil (Liu, 2008). The capillary number (*Nc*) is used to express the forces acting on an entrapped droplet of oil within a porous media. *Nc* is a function of the Darcy velocity (v), the viscosity (μ) of the mobile phase, and the IFT (σ) between the mobile and the trapped oil phase (Berger & Lee, 2006). Equation 1 shows the relationship of Darcy velocity, viscosity and IFT to the capillary number.

$$N_C = v\mu / \sigma \tag{1}$$

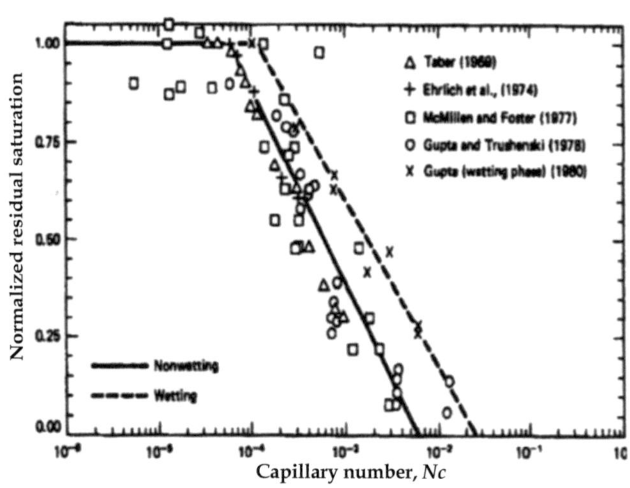

Fig. 1. Capillary pressure curves for sandstone cores (Liu, 2008)

Figure 1 shows capillary desaturation curves that plot residual saturation of oil versus a capillary number on a logarithmic x-axis. From this figure, increasing capillary number reduces the residual oil saturation. The residual oil saturations for both the nonwetting and the wetting cases are roughly constant at low capillary numbers. Above a certain capillary number, the residual saturation begins to decrease. This phenomenon indicates that large capillary number is beneficial to high recovery efficiency because the residual oil fraction becomes smaller. Capillary number must be increased in order to reduce the residual oil saturation. The most logical way to increase the capillary number is to reduce the IFT (Berger & Lee, 2006; Liu, 2008). Therefore, the principal objective of the chemical process is to lower the interfacial tension so that the recovery performance will be improved.

3. Surfactant in enhanced oil recovery

The concept of recovering oil by surfactant flooding dates back to 1929 when De Groot was granted a patent claiming water-soluble surfactants as an aid to improve oil recovery. The surfactant could reduce the interfacial tension between the brine and residual oil. The use of proper surfactant can effectively lower the IFT resulting in a corresponding increase in the capillary number (Berger & Lee, 2006). The success of surfactant flooding depends on many factors such as formulation, cost of surfactants, availability of chemicals, and oil prices in the market. In enhanced oil recovery, surfactants could be used in several formulations to enhance oil production. Some of these formulations are surfactant-alkali flooding, surfactant-polymer flooding, and alkali-surfactant-polymer flooding.

The idea of combining surfactants and alkalis was first proposed by Reisberg and Doscher in 1956. They added non-ionic surfactants to the alkali solution to improve oil recovery at laboratory scale. Recent work has shown that the addition of alkali to the surfactant solution would not only decrease the IFT, but also reduces the surfactant adsorption on the negatively charged sand surface (Touhami et al., 2001). An inexpensive alkali could be used with expensive surfactants to achieve both a technically successful and economically feasible flood.

In order to design an effective surfactant-alkali flooding formulation, it is important to utilize the synergistic effect between the surfactant and alkali. Surfactants tend to accumulate at the oil and water interface where the hydrophilic and hydrophobic ends of the molecules can be in a minimal energy state. This increases the surface pressure and decreases both the interfacial energy and the IFT. Rudin & Wasan (1992) concluded that the dominant mechanism of the synergistic effect is the formation of mixed micelles of the surfactants and the generated in situ surfactant. The mixed micelles cause the IFT to drop significantly (Nelson et al., 1984). At the same time, surfactant adsorption on sand is reduced by the presence of alkali. The sand surface will become increasingly negatively charged with an increase in pH and will thereby retard the adsorption of the anionic surfactant.

A number of alkali-surfactant flooding field tests have been described in the literature (Mayer et al., 1983; McCafferty and McClaflin, 1992). Success of these processes in an actual reservoir will depend on how well and for how long the internally-generated surfactant and the externally-added surfactant work together as intended. Mayer et al., (1983) summarized based on known field tests the amount of alkali injected and the performance results for early alkaline flooding processes. Most of the projects were not as profitable as expected. Falls et al., (1992) reported successful field tests using alkaline-surfactant flooding in recovering waterflood residual oil from sandstone reservoirs in the White Castle Field, USA. The process recovered at least 38% of the residual oil after waterflooding.

Alkali-surfactant flooding is a promising method for enhanced oil recovery. With the combination of alkali and a small amount of surfactant, oil-water IFT can be reduced much more than with either alkali or surfactant alone. However, the recovery factor of this process is usually insufficient due to the unfavourable mobility ratio. Hence, a polymer is added to the surfactant solution to improve the sweep efficiency. The application of alkali-surfactant polymer in the Daqing oil field in China is an example of successful field trials. However, because of the high cost of surfactants, this process has not been expanded (Wang et al., 1997). In order to reduce the cost of the surfactant and to enlarge the swept volume, this technology was upgraded to alkali-surfactant-polymer flooding. The combination of alkali-surfactant-polymer is expected to cause the residual oil to be economically recovered from the reservoir.

Alkali-surfactant-polymer (ASP) is considered to be one of the major EOR techniques that can be successfully used in producing light and medium oils. The advantage of ASP flooding over conventional alkaline flooding is that ASP can be used for low acid number crudes while alkaline flooding can only be applied for medium to high acid number crudes. In the ASP process, alkali reacts with acidic oils to form in-situ surfactant and increases the pH to lower surfactant adsorption on the porous media. Surfactants are used to lower the IFT between oil and water while polymer is used to improve the sweep efficiency by providing mobility control (Elraies et al., 2010a). ASP flooding has been extensively evaluated in the laboratory and widely used in field applications with great success. In recent years, many ASP field pilot tests have been conducted in USA, India, Venezuela, and China (Pitts et al 2006; Pratap & Gauma, 2004; Clara et al 2001; Wang et al., 1999). The ASP process uses the benefits of the three flooding methods, and oil recovery is greatly enhanced by decreasing interfacial tension (IFT), increasing the capillary number, enhancing microscopic displacing efficiency, and improving the mobility ratio (Pingping et al., 2009). However, even with these advantages and the success of ASP projects, the process is not without some disadvantages.

An earlier paper written by Hou and co-workers (2001) addresses the corrosion and scale problems that occurred during the ASP flood in Daqing field. The strong alkali had detrimental effects on polymer performance and in many cases additional polymer was required to achieve the desired viscosity (Wang et al., 2006). Nasr-El-Din et al., (1992) conducted an experimental study to examine the effectiveness of alkali concentration in ASP performance. Their experiments confirmed an exponential decrease in viscosity of the combined ASP slug with the increase in alkali concentration.

The selection of proper surfactants is one of the key factors for chemical EOR application. Surfactant should be stable under reservoir conditions resulting in an ultra low interfacial tension. Wangqi & Dave, (2004) conducted screening studied by interfacial tension experiments using different types of surfactants and validated by core flood tests. The IFT results showed wide range of IFT reduction, depends on the surfactant concentration and type. Core flood results indicated that 11.2% OOIP could be recovered when the selected surfactant concentration and type are combined with alkali and polymer. Flaaten et al., 2008 performed the screening and optimization of surfactant formulations by microemulsion phase behavior using various combinations of surfactants, co-solvents, and alkalis. Branched alcohol propoxy sulfates and internal olefin sulfonates demonstrated a superior performance when mixed with conventional alkali. The recovery performance indicated that nearly 100% of residual oil was recovered with very low surfactant adsorption.

4. Chemical EOR challenges

Most pilot tests reported in the 2000s accomplished a higher oil recovery than those in the 1970s and 1980s. Improvements in the functionality of the chemicals and a better understanding of the process mechanisms are the causes for these successes. These field tests indicate that surfactant flooding and its variants can be technically successful. However, the main downside for these chemical EOR applications was still the high manufacture cost of surfactants and the cost of raw materials. The recovered oil by this process was not economical or the economical and technical risk was too high compared with the oil price (Austad & Milter, 2000). Therefore, a lot of work has been recently conducted to develop an economical surfactant when the crude oil prices remained high. To reduce the cost of surfactant production, much attention is focused toward agriculturally derived oleochemicals as alternative feedstocks (Gregorio, 2005). Many surfactants have been produced from natural oils to satisfy EOR requirements (Wuest et al., 1994; Li et al., 2000). Soybean and coconut oils are the most popular raw materials used to derive oleochemical feedstocks such as fatty alcohol and esters (Hill, 2000).

Paradoxically, these surfactants use edible vegetable oils for its synthesis and it will compete with the food supply in the long-term. As the demand and cost of edible vegetable oils has increased annually in recent years, then their derivative surfactant becomes more expensive (Gregorio, 2005). According to the United States Department of Agriculture Oilseeds 2009, the average cost of soybean oil was approximately $ 395 per tonne at the time. While, the cost of non-edible oils such as Jatropha oil was about $ 250 per tonne and the typical cost of the major petrochemical feedstock such as ethylene was $ 595 per tonne. Therefore, the evaluation and production of Jatropha oil based surfactant was an attractive pursuit for chemical EOR.

5. Development of a new polymeric surfactant

This section of the chapter is focused on the description of the development of a new polymeric surfactant with the aim to overcome some of the existing problems associated with conventional ASP flooding. The goal is to produce a new surfactant that will be both economical and effective for interfacial tension reduction and viscosity control. The basic idea was to attach the sulfonate group to a hydrophobic group of an associative polymer chain. A hydrophobically modified polymer is one class of water soluble associative polymer that contains a small number of hydrophobic groups (Abdala, 2002). Hydrophobically modified polymers have either a telechelic structure in which the chains are end-capped with the hydrophobic groups, or more complicated comb-like structures in which the hydrophobic groups are randomly grafted to the polymer backbone. The backbone has a polyelectrolyte feature and is composed of a polymer of acrylamide or acrylic acid, and ethylacrylate. Upon neutralization, the polymer backbone adopts a more extended conformation allowing the hydrophobic groups to associate forming a transient network structure (Abdala, 2002).

Herein, a single step route that is similar to the method reported by Ye et al., (2004) was used to produce a new polymeric surfactant via the polymerization process. This surfactant was designed to graft the sulfonated group to the polymer backbone as one component system for interfacial tension reduction and viscosity control. Therefore, the polymerization was conducted with an excess of different surfactant to acrylamide ratios. Sodium methyl ester sulfonate (SMES) was used as a surfactant feedstock in the polymerization process. Because the goal was to design a cost effective surfactant, the SMES was synthesized from Jatropha oil as the raw material. Jatropha oil was selected because it is a non-edible oil so it will not compete with food supply and it is not a petroleum derivative. Finally, it is a drought resistant perennial tree that grows in marginal lands and can live over 50 years. Under these conditions, it is expected that the supply and availability of Jatropha oil will not be a major concern.

6. Jatropha oil

Jatropha curcas L. is a plant belonging to the Euphorbiaceae family that produces a significant amount of oil from its seeds. This is a non-edible oil-bearing plant widespread in arid, semi-arid, and tropical regions of the world. Jatropha is a drought resistant perennial tree that grows in marginal lands and can live over 50 years (Bosswell, 2003). Jatropha oil content varies depending on the types of species and climatic conditions, but mainly on the altitude where it is grown (Pant et al., 2006). The oil content in Jatropha seed is reported to be in the ranges from 30 to 50% by weight of the seed and ranges from 45% to 60% weight of the kernel itself (Pramanik, 2003). The Jatropha tree has several beneficial properties such as its stem is being used as a natural tooth paste and brush, the latex from the stem is used as natural pesticides and wound healing, its leaf is used as feed for silkworms among other uses. It is a rapid growing tree that propagates easily.

Density, g/cm^3	0.92
Flash point, ^oC	236
Cloud point, ^oC	8
Iodine value	95-107
Acid value (mgKOH/g)	0.92-10
Sulfur content, ppm	0.13
Phosphate content ppm	290

Table 1. Jatropha oil properties

In this study, non-edible Jatropha oil was used as a starting raw material to produce different types of surfactants for EOR applications. The crude Jatropha oil was purchased from a local oil industry (Bionas) in Kuala Lumpur, Malaysia, and it was used as received. Table 1 summarizes the properties of the Jatropha oil.

7. Experimental design and procedure

The purpose of this work was to develop new polymeric surfactants for enhanced oil recovery applications. Several experiments have been conducted to synthesize different surfactants based on fatty acid methyl ester derived from Jatropha oil. The experimental work started with the production of the methyl ester, followed by the synthesis of the surfactants and their characterizations. Figure 2 shows the experimental steps followed in this study.

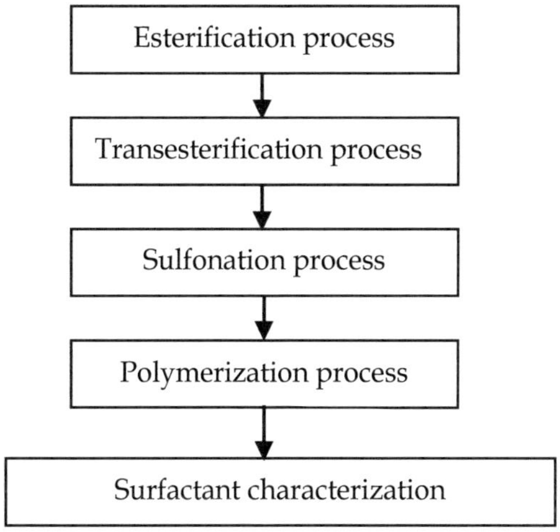

Fig. 2. Flow chart of the surfactant production stages.

7.1 Fatty acid methyl ester production

Fatty acid methyl ester is a renewable and environmentally friendly energy source. The most commonly used technique to produce methyl esters involves transesterification reaction in which triglycerides are reacted with methanol in the presence of a catalyst. However, this process is greatly affected by the free fatty acid (FFA) content of the raw material. The presence of high FFA (i.e. high acid value) in the raw material results in soap formation that could decrease the methyl ester yield and complicate the separation and purification of the product of interest (Vicente et al., 2004). This problem can be avoided by pre-treating the oil with an acid catalyst esterification to convert the FFAs into esters before the alkali catalyst is used. Hence, fatty acid methyl ester is produced via a two-step transesterification as described below.

Step 1. Acid-catalysed esterification:

The main purpose of acid-catalyzed esterification is to reduce the acid value of crude Jatropha oil. This oil has an initial acid value of 10.54 mg KOH/ g-oil corresponding to a free fatty acid of 3.75%. Therefore, the effect of different methanol to oil volume ratios of 0.17 v/v, 0.25 v/v, and 0.30 v/v on the reduction of acid value was investigated using 1.14% v/w of sulfuric acid as a catalyst. In this step, the reaction was carried out at $60^{o}C$ for 120 minutes using 250 ml round bottom flask. After the reaction, the mixture was allowed to settle for three hours and the methanol-water fraction at the top (upper phase) was removed by a separating funnel. The effectiveness of this step was then evaluated by determining the acid value of the product (lower phase in the separating funnel) using the American Oil Chemists' Society method, 2003. The product having an acid value of less than 1 mg KOH/g was subsequently used for the main transesterification reaction in the next step.

The acid value of the crude Jatropha oil was greatly influenced by the methanol-oil ratio. The pre-treatment of the Jatropha oil with a methanol to oil ratio of 0.17v/v reduced the acid value from 10.5 to 0.221 mg KOH/g-oil. With methanol to oil ratios of 0.25 and 0.30, the acid value further decreased to 0.156 and 0.056 mg KOH / g oil, respectively. Because the recommended acid value is 1 mg KOH/g-oil, then, the methanol-oil ratio of 0.17 was selected as the optimum ratio for the acid-catalyzed esterification reaction at 60°C and 120 minutes of reaction time. Based on the weight of oil used in this step, the average product yield was about 90%, which is in agreement with the product yield obtained by Tiwari et al., (2007) who conducted the pre-treatment of Jatropha oil that contained 28 mg KOH/g-oil using a methanol-oil volume ratio of 0.28 over 88 minutes of reaction time.

Step 2. Alkaline-catalysed transesterification:

The transesterification reaction was conducted to produce methyl esters from the pre-treated Jatropha oil. Different methanol- oil ratios were evaluated at a constant ratio of potassium hydroxide (KOH) to oil ratio of 0.5%w/w. In this study, the volume ratios of methanol to oil volume were 0.16 v/v, 0.22 v/v, and 0.26 v/v. The reaction was carried out at 60°C for 35 minutes.

At the reaction time of 35 minutes, the yield of methyl ester obtained was similar for methanol to oil ratios of 0.22 and 0.26. For instance, a maximum yield of 99.8% and 99.3% were obtained for methanol to oil ratios of 0.22 and 0.26 respectively; while a yield of 96.4% was obtained when the lowest methanol to oil ratio of 0.16 was used. Therefore, the optimum methanol to oil ratio was chosen as 0.22. According to Tiwari et al. (2007) a maximum yield of 99% was obtained with a methanol to oil ratio of 0.16 v/v and 24 min of reaction time. As compared to other oils, a maximum yield of 95% was obtained from soybean oil with a methanol to oil molar ratio of 12:1 and 3 hours of reaction time (Xuejun et al., 2008). Therefore, Jatropha oil seems to be a promising source for methyl ester production.

	Jatropha oil (wt %)	Soybean oil (wt %) (Sarin et al., 2007)
Palmitic acid methyl ester	17.24	11.0
Stearic acid methyl ester	9.79	4.0
Margaric acid methyl ester	0.11	-
Myristate methyl ester	0.09	0.1
Palmitoleic acid methyl ester	1.28	0.1
Linoleic acid methyl ester	35.21	53.2
Oleic acid methyl ester	36.28	23.4

Table 2. Analysis of the fatty acid methyl ester

The methyl ester produced at the optimum methanol to oil ratio was characterized by Gas Chromatography-Mass Spectrometry (GC-MS) to confirm the presence of fatty acid methyl esters. Table 2 summarizes the composition of the fatty acid methyl esters produced from Jatropha oil. The presence of methyl esters were assessed using the GC-MS library that is provided with the equipment. Table 2 indicates that the Jatropha oil methyl ester contains 27.23% of saturated fatty acid and 72.77% of unsaturated fatty acid. It was also found that the Jatropha oil has a high quantity of linoleic acid methyl ester (35.21 %) and oleic acid methyl ester (36.28%). If compared to soybean oil (Table 2), Jatropha oil has also great potential as a fatty acid source.

7.2 Surfactant synthesis

7.2.1 Sulfonation process

The fatty acid methyl ester produced from Jatropha oil was then sulfonated according to Chonlin et al., (1990). The purpose of the sulfonation process was to synthesize a sodium methyl ester sulfonate (SMES) based on fatty acid methyl esters as feedstocks. The sulfonation reaction was carried out at laboratory scale using a 250 ml round bottom flask. However, since n-butanol and sodium carbonate are already used in chemical EOR as cosolvent and alkali respectively, the SMES obtained was used in the polymerization reaction without any further purification so as to minimize the cost of surfactant manufacturing (Elraies et al., 2010).

7.2.2 Polymerization process

A single step route similar to Ye et al., (2004) was used to produce polymeric methyl ester sulfonate (PMES) via polymerization process. The principle of this process was to attach the sulfonate group of SMES to the polymer backbone (polyacrylamide) as a one component system for ITF reduction and viscosity control.

The polymerization process was performed using a 250 ml-three necked flask. In a typical run, the reaction was conducted using the methyl ester sulfonate (SMES) as the surfactant and potassium persulfate as the initiator. The initiator solution was prepared by dissolving 0.123 g of potassium persulfate in 10 ml of deionized water and then the pH of the solution was adjusted to 9-10 with sodium hydroxide. The surfactant solution was prepared by dissolving the appropriate amount of SMES in 100 ml deionized water. The appropriate amount of acrylamide monomer was dissolved in 70 ml of deionized water and purged with nitrogen to remove residual oxygen. Afterward, the surfactant solution was added to the acrylamide solution and stirred under a nitrogen atmosphere until a clear solution was observed. This mixture was then heated to 60°C and the initiator was added. The polymerization reaction was conducted at 60°C for 1.5 hours using and auto shaker water bath. The crude product was then extracted with acetone and dried in an oven for 12 hours (Elraies et al., 2011).

The previous experimental procedure was followed for the production of several polymeric methyl ester sulfonates using different SMES to acrylamide weight ratios. Table 3 summarizes the experimental runs conducted.

Experiment No.	Surfactant to acrylamide(v/v) ratio	Polymeric surfactant name
1	1:0.50	SURF 1
2	1:0.60	SURF 2
3	1:0.80	SURF 3
4	1:1.16	SURF 4
5	1:1.33	SURF 5

Table 3. Experiment details for the polymerization reaction

7.3 Surfactant characterization

7.3.1 FTIR spectroscopy analyses

A FTIR spectrophotometer was used to determine the chemical functional groups present in the surfactant. Different functional groups are susceptible to absorb characteristic frequencies of IR radiation. Figure 3 shows the FTIR spectrum of sodium methyl ester sulfonate. All the IR absorption bands are analyzed with reference to the Spectrometric identification of organic compounds by Silverstein et al., (2005). The broad absorbance peaks between 3300-2500 cm^{-1} represent the O-H stretching of carboxylic acid. The presence of esters is indicated by the absorbance peak of C=O stretching vibration between 1730-1715 cm^{-1}. The presence of the significant peak at 1450 cm^{-1} corresponds to the asymmetrical bending vibration band of methyl group (C-H). Peaks between 1160 - 1120 cm^{-1} indicate the presence of sulfonate groups due to S=O stretching (Silverstein et al., 2005; Awang & Goh, 2008). The peaks at 1410 and 1068 cm^{-1} are another indication of the presence of sulfonate groups due to the S=O stretching vibration. These results indicate that this compound must be sodium methyl ester sulfonate.

The polymeric surfactants produced based on sodium methyl ester sulfonate were also characterized by FTIR. The IR spectrums recorded of the five produced surfactants showed similar pattern but the percentage of transmission is different due to the variation in their molecular weights. The IR spectrums indicate that the chemical structures for these five surfactants are the same. Figure 4 presents the FTIR spectrum for surfactant SURF 1. Figure 5 shows the IR spectra of the other four surfactants (SURF 2, SURF 3, SURF 4, and SURF 5).

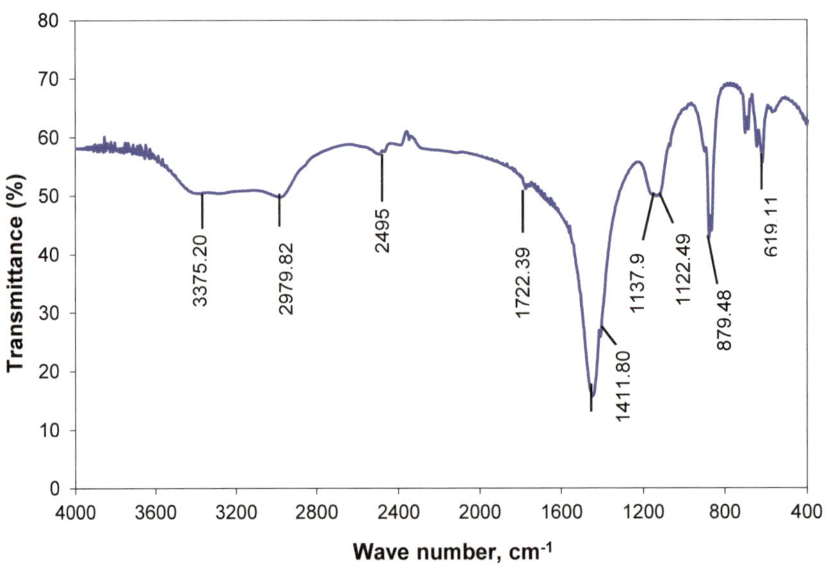

Fig. 3. FTIR spectrum of sodium methyl ester sulfonate

In Figure 4, the peaks between 1160 - 1120 cm^{-1} and 1409 and 1068 cm^{-1} indicate the presence of sulfonate groups due to C=O stretching. The absorbance peaks between 1730-1715 cm^{-1} represent the S=O stretching vibration indicating the presence of esters. The presence of the significant peak at 1450 cm^{-1} corresponds to the asymmetrical bending vibration band of methyl group (C-H). Changes in the absorbance peaks from 2975 to 3352 cm^{-1} are due to the introduction of acrylamide to the surfactant. The tiny peaks from 3350 to 3180 cm^{-1} are an indication of the presence of primary and secondary amides due to N-H stretching. The peaks between 1680 and 1630 cm^{-1} are another indication of the presence of amide groups due to the C=O stretching vibration (Silverstein et al., 2005).

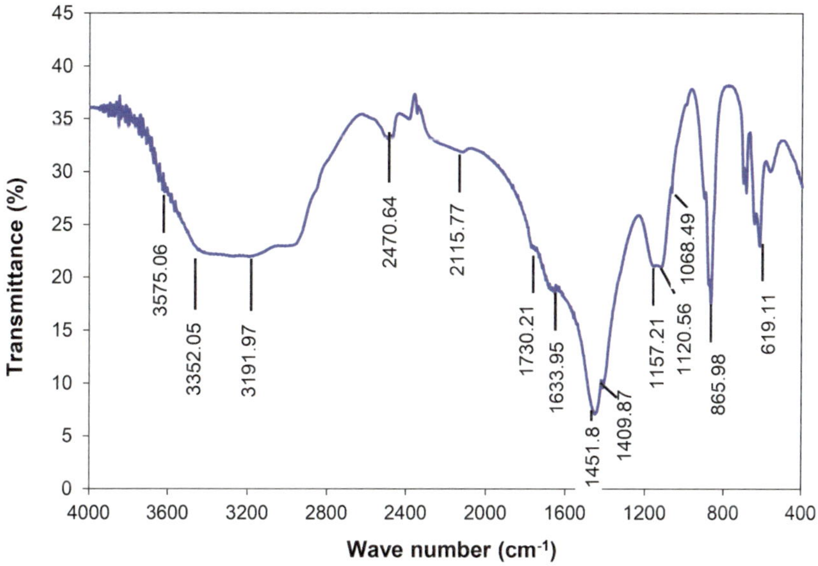

Fig. 4. FTIR spectrum of polymeric SURF 1

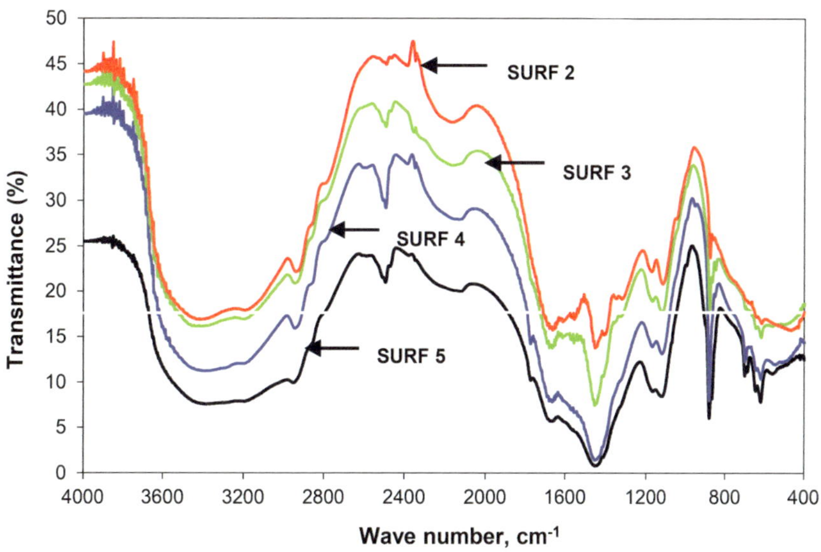

Fig. 5. FTIR spectrum of polymeric surfactants (SURF 2 to SURF 5)

7.3.2 Thermal stability analyses

The thermal degradation of the sodium methyl ester sulfonate (SMES) and the polymeric surfactants were examined by thermogravimetric analysis (TGA) between 30°C and 500°C. Figure 6 illustrates the thermal behavior of the SMES and the polymeric methyl ester sulfonates.

In Figure 6, the TGA profile of SMES shows that 3.4% weight loss occurred at 100°C due to the loss of bound water. Then, 45% weight loss occurred sharply from 100°C to 180°C, revealing that SMES molecules start to decompose at temperatures exceeding 100°C. Beyond 180°C, residual components of the SMES are thermally stable up to 500°C. The TGA curves in Figure 6 indicate that the thermal degradation behavior of the polymeric surfactants is different from the thermal degradation behavior of SMES. The polymeric surfactants show similar thermal degradation trends with three distinctive degradation regions. The first thermal degradation that occurred near 100 °C is attributed to the loss of water bound with an average of 6% weight loss. The second thermal degradation region from 100 to 300°C corresponds to the degradation of amide groups. The third degradation region from 300 to 500°C represents a complex thermal degradation process which may result from the condensation of the residual amide groups and cyclic amide rings (Laishun, 2000).

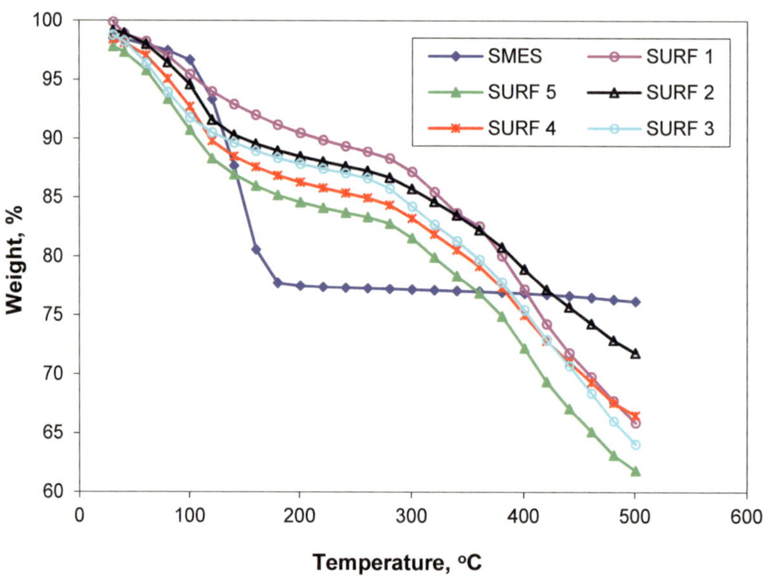

Fig. 6. TGA curves for SMES and polymeric surfactants

From all the TGA curves presented in Figure 6, the SMES shows lower mass loss as compared to the polymeric surfactants at about 100°C, while the polymeric surfactants demonstrate much less mass loss when the temperature exceeded 100°C. It is also shown that the degradation increases as the surfactant to acrylamide ratio decreases. For instance, in case of SURF 1 where the surfactant to acrylamide ratio is 1:0.5, the TGA showed 4% weight loss at 100°C, while about 9% weight loss is recorded for the lowest ratio of 1:1.33 which corresponds to SURF 5. As the reservoir temperature used in this study is 90°C, all the polymeric surfactants retain an average of 95% of their original structure and mass. It could be concluded that these polymeric surfactants are thermally stable under the desired reservoir temperature.

7.3.3 Interfacial tension measurements

Interfacial tension (IFT) measurements between crude oil and aqueous solutions of sodium methyl ester sulfonate (SMES) and polymeric methyl ester sulfonates were performed at several surfactant concentrations. All the measurements were conducted at 29°C using the spinning drop method. Angsi I-68 crude oil (Malaysia) was used throughout this study. The total acid number was 0.478 mg KOH/g. The API gravity was 40.1° and live oil viscosity was 0.3 cP at reservoir temperature.

Figure 7 shows the interfacial tension as a function of SMES concentration and time. At 0.2wt% loading, SMES reduces the interfacial tension between softened water and crude oil from about 13.6 mN/m to 0.82 mN/m. This demonstrates the surface adsorption and aggregative properties of the new surface-active compound. The interfacial tension of the system crude oil-SMES solution reduces drastically as surfactant concentration increases. For instance, when the surfactant concentration is increased from 0.2 wt% to 0.4 wt% and 0.6 wt%, the IFT drops continuously to values of 0.56 mN/m and 0.45 mN/m respectively.

The surface activity of the SMES was also compared with the surface activity of a commercial surfactant (sodium dodecyl sulphate, SDS). Figure 7 shows that at a concentration of 0.2wt%, the SMES and SDS reduced the interfacial tension of the system crude oil-aqueous solution to 0.82 mN/m and 0.63 mN/m respectively. While, at a concentration of 0.4 wt%, the reduction of IFT in the system crude oil-aqueous solution is similar to the IFT value obtained with 0.2wt% of sodium dodecyl sulphate (SDS). These results indicate that there is no much difference in the interfacial tension reduction provided by the SMES compared to SDS, especially considering the fact that the manufacture cost of the SMES is lower than the cost of the commercial SDS.

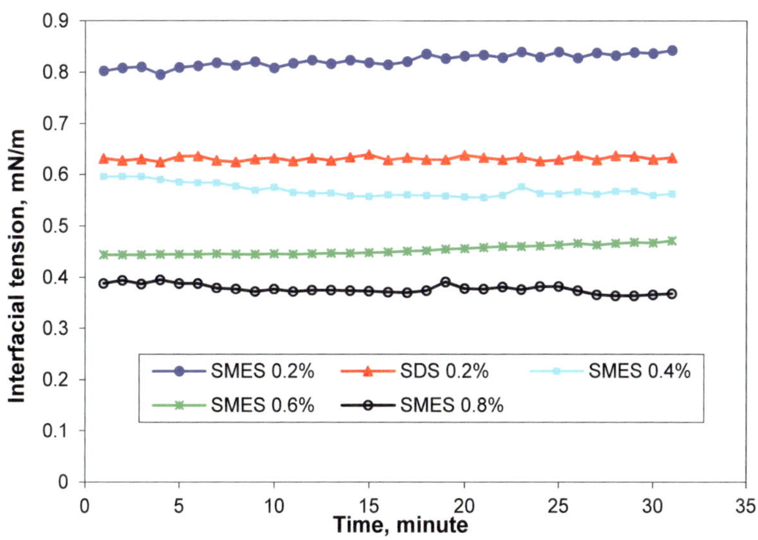

Fig. 7. IFT for the system crude oil- aqueous solution as a function of surfactant concentration and measuring time.

Figure 8 presents the IFT performance of polymeric methyl ester sulfonates (PMES) as a function of surfactant (SMES) to polymer (acrylamide) ratio and measuring time. The PMES showed a significant reduction of the IFT of the system crude oil-aqueous solution; IFT decreases as the surfactant to acrylamide ratio increases. As shown in Figure 8, the interfacial tensions between crude oil and surfactant solution reduces from 13.6 mN/m to 0.461 mN/m at a surfactant:acrylamide ratio of 1:0.8 (1.25 surfactant/acrylamide) , and the IFT of the system reaches 0.296 mN/m at a surfactant:acrylamide ratio of 1:0.4 (2.5 surfactant/acrylamide ratio). This demonstrates the aggregative properties of the attached sulfonated group to the polymer chains. As the surfactant to acrylamide ratio increases, the more surfactant is being attached to the polymer backbone and thereby lower IFT values are reached.

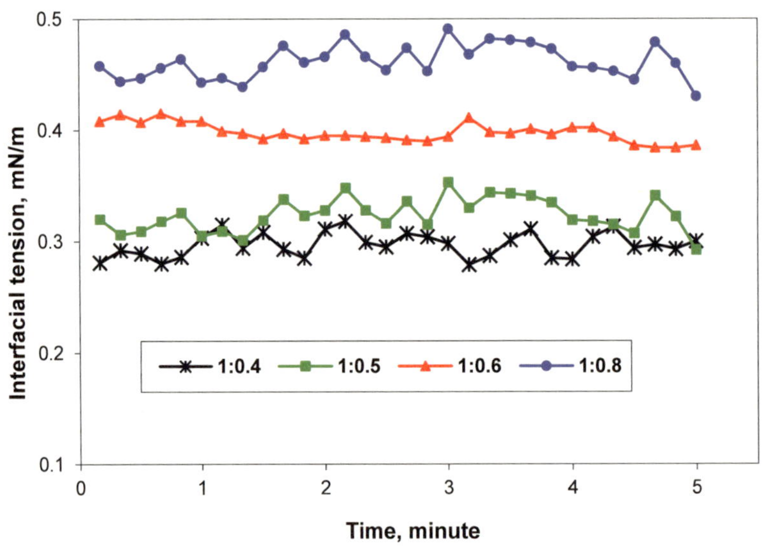

Fig. 8. IFT of the system crude oil - aqueous solution as a function of surfactant to acrylamide ratios and measurement time.

7.3.4 Viscosity measurements

The kinematic viscosity of the polymeric surfactants was measured using a Tamson viscometer. All the measurements were performed at a reservoir temperature of 90°C. In the polymeric surfactant mixtures, surfactant concentration was fixed at 0.2wt%, while the concentration of acrylamide was changed. The purpose of this test was to screen the polymeric surfactant based on performance for the subsequent core flood tests.

The effect of each surfactant to acrylamide ratio on the viscosity performance is illustrated in Figure 9. The viscosity of the polymeric surfactant significantly increases as the surfactant to acrylamide ratio decreases. This is due to the increasing amount of polymer chains attached to the surfactant. The viscosity of the polymeric surfactant having a surfactant:acrylamide ratio of 1:0.4 (2.5 surfactant/acrylamide) is lower than the viscosity of the crude oil (1.654 mm/sec) and therefore this polymeric surfactant (SURF 1) was not selected.

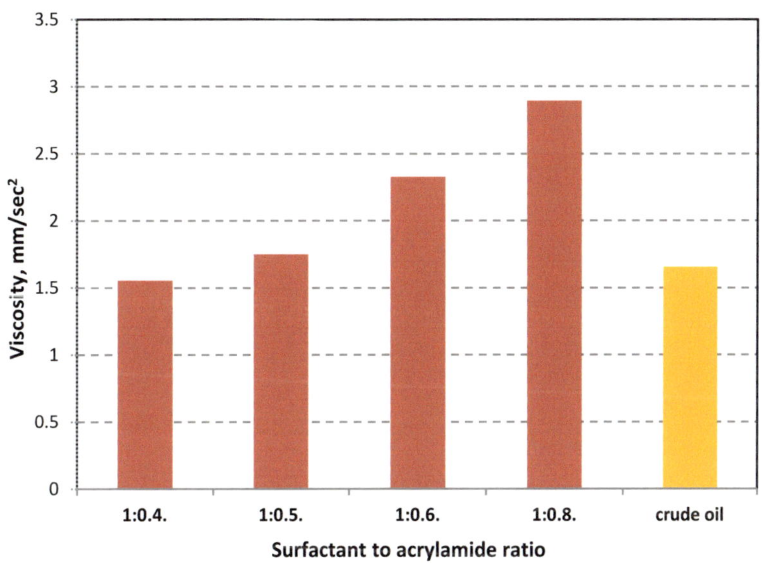

Fig. 9. Viscosity as a function of different surfactant to acrylamide ratios.

The selection of the optimum surfactant:acrylamide ratio was based on several factors including production cost, IFT, and viscosity. Therefore, the polymeric surfactant having a surfactant/acrylamide ratio of 1:0.5 was chosen as the optimum ratio. Then, the viscosity of the chemical slug can be adjusted by increasing the polymeric surfactant concentration to yield a suitable viscosity and an ultra low IFT. Unlike the surfactant:acrylamide ratios of 1:0.6 and 1:0.8, the viscosity of these solutions were very high compared to the viscosity of the crude oil. If the concentration of these polymeric surfactants is increased in aqueous solutions to render low values of IFT, then the high viscosities of the polymeric surfactant solutions will cause injectivity problems during the injection of the chemical slug into the porous media. Therefore, the polymeric surfactant with a surfactant:acrylamide ratio of 1:05 was selected as the optimum PMES for IFT reduction and viscosity control. Similarly, this PMES allows achieving an effective chemical slug that is able to propagate into the rock formation upon injection without losing its integrity.

7.3.5 Viscosity and IFT performance for the optimum polymeric surfactant

The viscosity and IFT performances of the optimum polymeric surfactant were investigated using different concentrations of PMES. Figure 10 shows that the viscosity of the solution significantly increases as PMES concentration increases. The viscosity of the solution was approximately 1.75 mm/sec for a 0.2wt% PMES solution concentration, 2.533 mm/sec for 0.4wt% concentration, and 5.124 mm/sec with the highest PMES solution concentration (0.7wt%). The latter solution viscosity is very high as compared to the viscosity of the crude oil (1.654 mm/sec). So in order to design a cost-effective polymeric surfactant slug that offers a favourable mobility ratio, a polymeric surfactant concentration of 0.4wt% was chosen as the optimum concentration for the chemical flooding displacement of the crude oil used in this work.

Figure 10 also shows the interfacial tension as a function of different PMES concentrations. PMES shows excellent results in terms of IFT reduction. IFT between the crude oil and surfactant solution is reduced from 13.6 mN/m to 0.323 mN/m using 0.2wt% of PMES concentration. And the IFT reduces drastically as the concentration of polymeric surfactant increases. At the optimum PMES concentration of 0.4%, the IFT decreases to 0.192 mN/m.

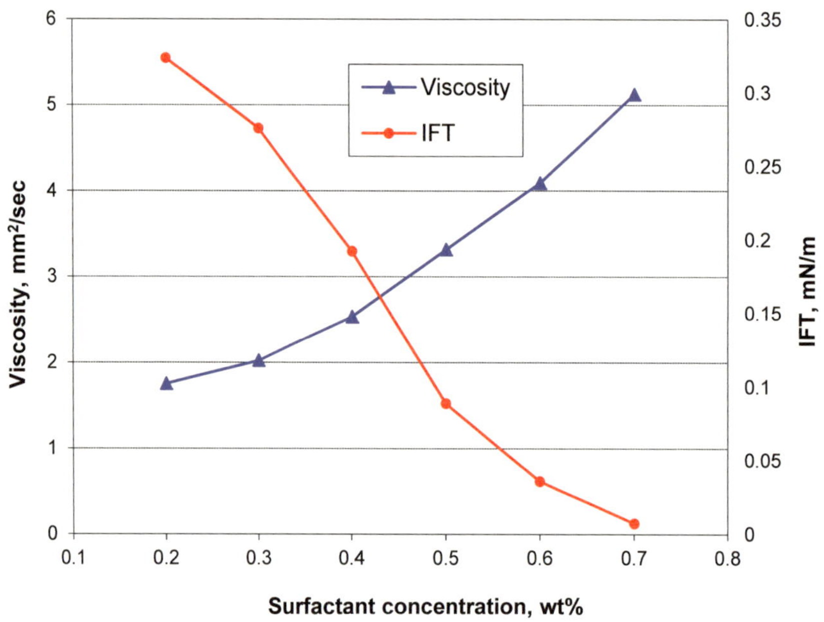

Fig. 10. IFT and viscosity as a function of different polymeric surfactant concentration

7.3.6 Effect of alkali on the PMES viscosity and IFT performance

Since alkali has a significant impact on ASP flooding performance, the effect of alkali on the performance of the PMES was investigated using different sodium carbonate concentrations at a fixed concentration of polymeric surfactant (0.4wt%) in the aqueous solution. The purpose of these measurements was not only to study the effect of the alkali on the IFT reduction, but also to determine if the presence of sodium carbonate in the system would affect the viscosity of the polymeric surfactant.

Figure 11 shows the viscosity performance in the absence and presence of sodium carbonate at 90°C. The presence of alkali at concentrations ranging from 0.2wt% to 1wt% does not affect the viscosity of the system; the viscosity of the polymeric surfactant remains constant at 2.533 mm²/sec. This shows the stability of the viscosity of the new polymeric surfactant in the presence of sodium carbonate if compared to the conventional ASP formula where its viscosity is greatly affected by the added alkali (Nasr-El-Din et al., 1992).

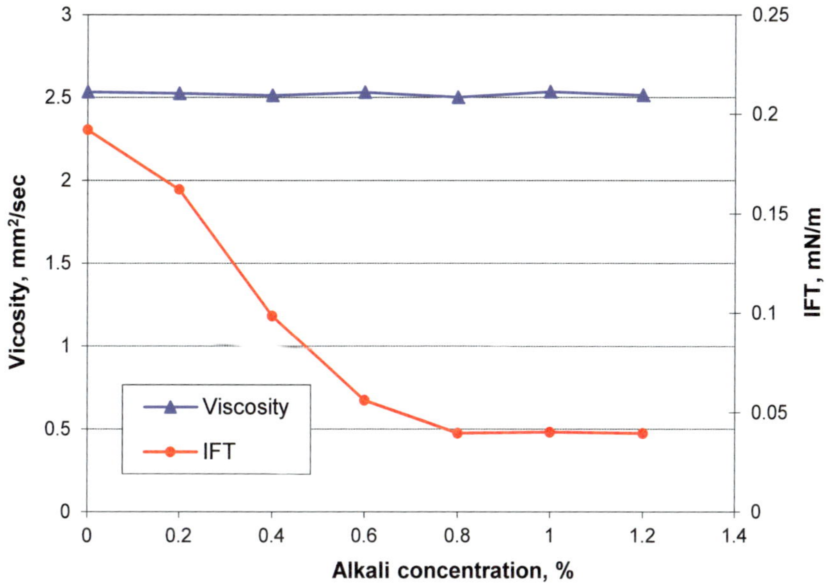

Fig. 11. Viscosity and IFT performance of PMES solution as a function of different Na_2CO_3 concentrations.

Figure 11 shows the effect of alkali concentrations on the IFT performance of the polymeric surfactant. The IFT decreases significantly with the increase of alkali concentration until it levels off (0.024 mN/m) when the concentration of Na_2CO_3 reaches 0.8wt%. At an alkali concentration of 0.2wt%, the IFT of the system slightly decreases. However, significant IFT redaction is observed when the alkali concentration increases from 0.2wt%to 0.8wt%. This rapid decrease in the IFT value can be explained by the production of in-situ surfactants due to saponification reactions between the alkali and the acidic groups in the crude oil. These natural surfactants are associated with the polymeric surfactant to produce synergistic mixtures adsorbed at the oil/brine interface. As a result, a concentration of 0.8wt% of sodium carbonate was selected as the optimum alkali concentration in the presence of 0.4wt% of polymeric surfactant concentration.

7.3.7 Static surfactant adsorption

Surfactant adsorption is detrimental as it results in surfactant loss and reduces surfactant activity. The adsorption of surfactant from aqueous solution to sandstone surface was investigated in the absence and presence of different alkali concentrations. The sandstone was collected from Lumut Beach, Malaysia. The adsorption of surfactant for each case was determined by comparing the obtained refractive index before and after equilibrium (Elraies et al., 2011).

Figure 12 presents the adsorption isotherm as a function of polymeric surfactant concentration. Surfactant adsorption increased as the surfactant concentration increased. At low surfactant concentration, surfactant adsorption occurs mainly due to ion exchange. As surfactant concentration exceeds 0.4wt%, the adsorption increment progressed slowly with the increase of surfactant concentration. This indicates that the adsorption has to overcome the electrostatic repulsive forces between the PMES and the similar charged sandstone until saturation adsorption is reached. The adsorption isotherm in figure 12 also shows that when the polymeric surfactant concentration reaches 0.6%, the maximum adsorption of the PMES on the sand surfaced is reached. The maximum adsorption of the polymeric surfactant on the sand is estimated to be 1.31 mg/g-sand. These results indicate that the adsorption of PMES is a function of polymeric surfactant concentration. Thus, if a dilute surfactant concentration is used, the corresponding loss of polymeric surfactant due to adsorption on the sand will be minimized.

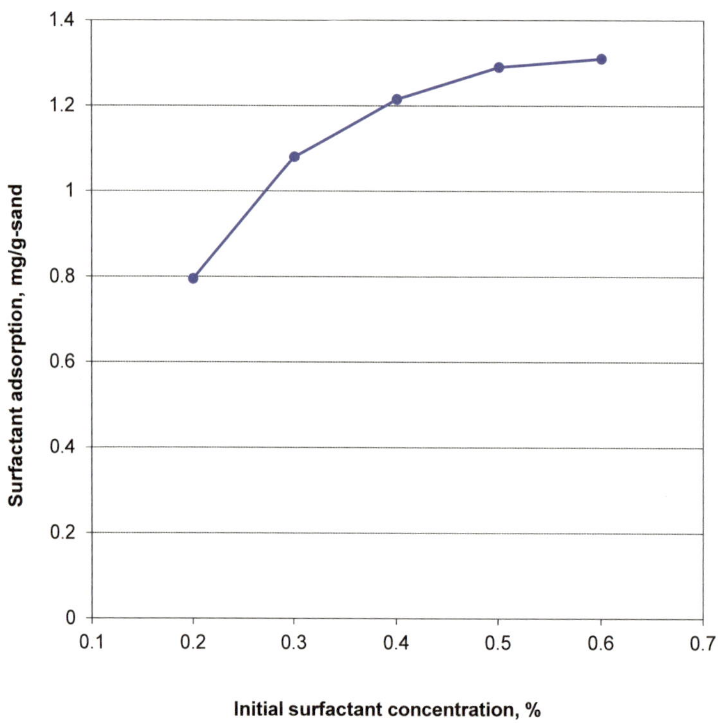

Fig. 12. Surfactant adsorption isotherm as a function of different polymeric surfactant concentrations.

Figure 13 shows the effect of alkali on surfactant adsorption isotherm. Different sodium carbonate concentrations ranging from 0.2wt% to 0.8wt% were used in this test. The polymeric surfactant concentration was kept fixed at 0.4wt%. The polymeric surfactant adsorption decreased considerably with the addition of alkali to the polymeric surfactant solution. This is due to the fact that high pH makes the sand surface more negative, and the electrostatic repulsive forces drive more surfactant to solution. Figure 13 shows that when 0.2wt% alkali is introduced to the system, the polymeric surfactant adsorption was immediately reduced from 1.21 mg/g-sand to 0.79 mg/g-sand. And when the added alkali concentration is over 0.6%, the saturation adsorption of the surfactant on sand levels off and the saturation adsorption is estimated to be about 0.4 mg/g-sand.

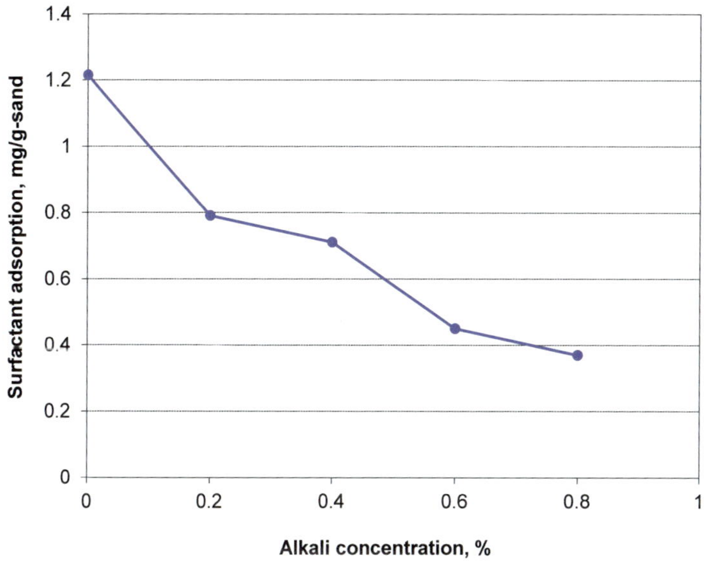

Fig. 13. The effect of alkali concentration on the polymeric surfactant adsorption isotherm

7.3.8 Coreflood test

Although the proposed Alkali/Polymeric/Surfactant (APS) formulation has shown promising potential in these preliminary screening tests, this is a relatively new technology for chemical EOR that requires more research. In order to design a cost-effective injection strategy to recover residual oil, core flood testing at reservoir conditions is essential. In this study, five core flood tests were performed to determine the effects of chemical concentration and slug size on oil recovery performance. For all coreflood experiments, the fluid injection sequence was as follows first waterflooding: followed by the injection of 0.5 PV of chemical slug, and finally waterflooding was resumed. Consolidated Brea Sandstone cores were used for evaluating the proposed procedure. Table 4 summarizes the physical core properties and coreflood results.

Property	Run 1	Run 2	Run 3	Run 4	Run 5
Length, cm	7.6	7.5	7.5	7.5	7.5
Diameter, cm	3.8	3.8	3.8	3.8	3.8
Permeability, md	88.4	113	84.9	82	94
Porosity, %	15.7	16.4	16.9	16.5	16.4
Pore volume, cc	13.3	13.2	14.5	13.9	13.2
Surfactant concentration, %	0.4	0.6	1	0.6	0.6
Alkali concentration, %	0.8	0.8	0.8	0.2	1
Waterflood recovery (% OOIP)	48.1	53.7	56.2	50.0	54.2
APS recovery (% OOIP)	12.6	16.2	20.7	12.8	9.0
Total recovery (% OOIP)	60.7	70.0	77.0	62.8	63.2

Table 4 Summary of core samples properties and coreflood tests

Effect of Surfactant Concentration

Preliminary testing indicated that the optimum formulation for the alkali-polymeric surfactant system was obtained for a composition of 0.8wt% alkali (Na_2CO_3) and 0.4wt% of polymeric surfactant. However, in order to examine the effectiveness of the new polymeric surfactant for enhanced oil recovery application, three additional concentrations of the polymeric surfactant (0.4%, 0.6%, 1%) were evaluated to confirm the optimum formulation of the new APS system. For all coreflood runs # 1, #2, and # 3 the alkali (sodium carbonate) concentration was kept constant at 0.8wt%, while for coreflood runs # 4 and # 5 Na_2CO_3 concentration was 0.2 wt% and 1wt% respectively.

Figure 14 shows the recovery performance as a function of pore volume injected for coreflood tests # 1, # 2, and # 3. Figure 14 shows that Run # 3 with the highest polymeric surfactant concentration (1wt%) had accomplished a better performance in recovering oil than Run # 2 with a polymeric surfactant concentration of 0.6wt% and Run # 1 with a polymeric surfactant concentration of 0.4wt%. After the injection of only 0.5 PV of APS, the percentage of oil recovery for Run # 2 and Run# 3 was 16.2% and 20.7% of the OOIP respectively. While, in Run # 1, with the lowest polymeric surfactant concentration, oil recovery was only 12.6 % of the OOIP after the injection of 0.5 PV of APS slug followed by extended waterflooding. Based on the IFT measurements, the high oil recovery from Run # 2 and Run # 3 was due to the synergistic action of the polymeric surfactant and the alkali causing the emulsification and mobilization of the crude oil. However, in Run # 1 with 0.4% of polymeric surfactant concentration, the recovery mechanism is only due to the formed microemulsion as a result of the low IFT observed during IFT test. In addition to the low surfactant concentration, the viscosity of the polymeric surfactant slug might not be high enough to efficiently displace emulsified crude oil. Based on these results, a chemical slug having a concentration of 0.6wt% of the polymeric surfactant was selected as the optimum APS formulation.

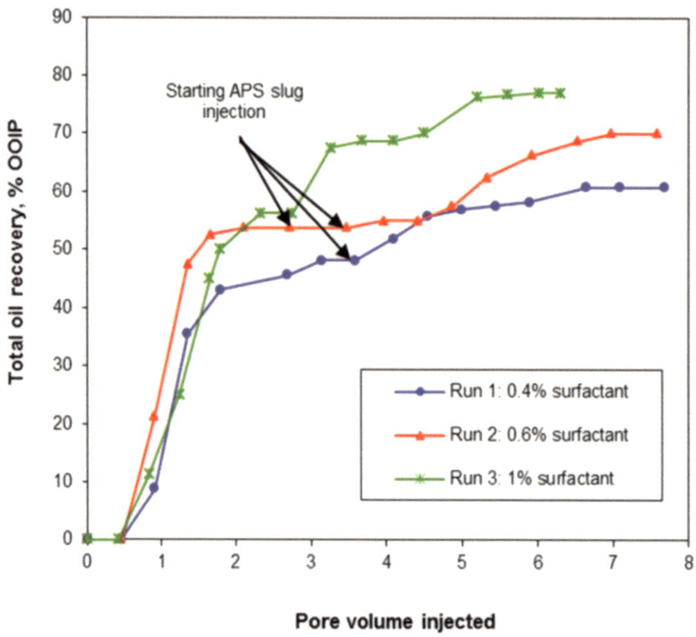

Fig. 14. Effect of surfactant concentration on oil recovery performance

Effect of Alkali Concentration

Three core flood tests were performed to assess the effect of alkali concentration on residual oil recovery. The polymeric surfactant concentration was kept constant at 0.6wt% (optimum concentration). Figure 15 shows oil recovery as a function of alkali concentration. The oil recovery profile in Figure 15 shows that Run # 2 with 0.8wt% alkali rendered the highest oil recovery. The APS slug recovered 16.2% of the OOIP in Run # 2, which is higher than the 12.8% of the OOIP in Run # 4 and 9% of the OOIP in Run # 5. Although, Run # 5 had the highest alkali concentration, the oil recovery achieved in this run was lower than in Run # 2 and Run # 4. This is because of the large amount of oil-in-water emulsion caused by the high alkali concentration used during this run. Figure 16 shows the amount of the oil-in-water emulsion formed during Run # 5 and Run # 4. When a high alkali concentration is used in Run # 5, more oil-in-water emulsion is observed due to the low salinity. Most of the surfactant remained in the aqueous phase, resulting in a very low water-microemulsion IFT and a high oil-microemulsion IFT (Flaaten et al., 2008). This type of emulsion makes the aqueous phase more viscous. The extended waterflood would bypass this viscous phase, resulting in a poor sweep efficiency. On the other hand, Run # 2 with 0.8wt% of alkali concentration shows a better synergistic effect in forming emulsion with a suitable viscosity. As a result, a concentration of 0.8wt% of alkali and 0.6wt% of polymeric surfactant were selected as the optimum composition for the APS system that was used in subsequent core flood tests to investigate the effect of slug size on recovery performance.

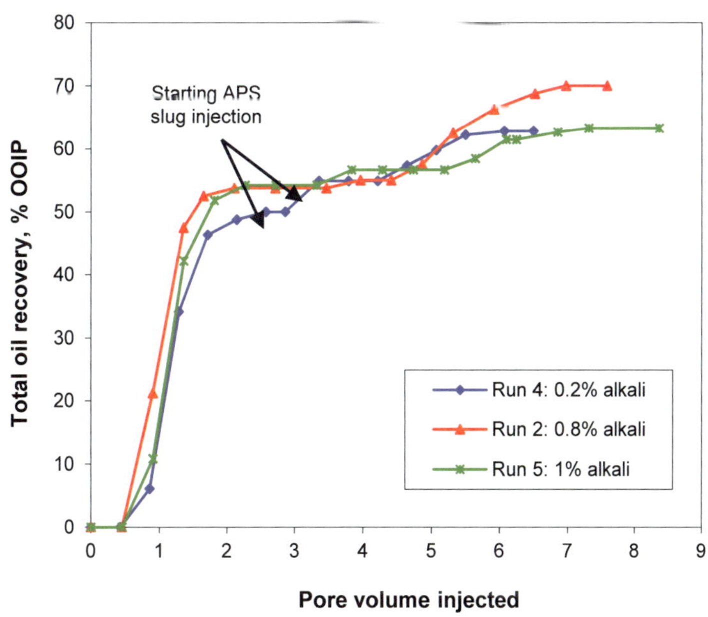

Fig. 15. Effect of alkali concentration on oil recovery performance

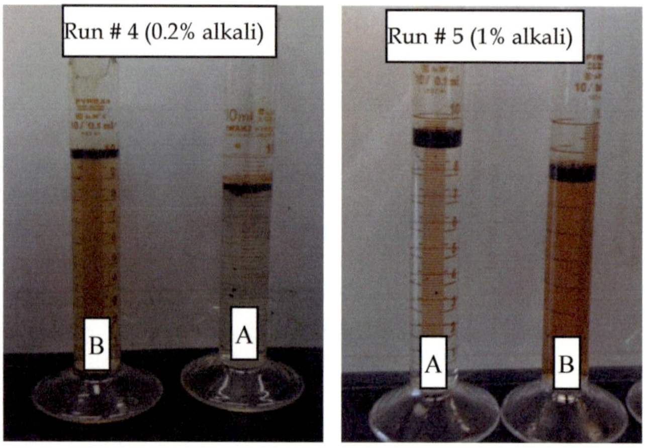

Fig. 16. Oil-in-water emulsion formed during Run # 4 and Run # 5 (A & B show the produced oil and water during waterflooding and after APS slug injection respectively)

Effect of slug size

Determining the most effective chemical slug size which renders the minimum chemical consumption and maximizes oil recovery is one of the most important criterions in the optimization process. To investigate the effect of slug size, the optimum APS formulation was used. In this experimental phase the effect of three different chemical slug sizes on oil recovery were evaluated.

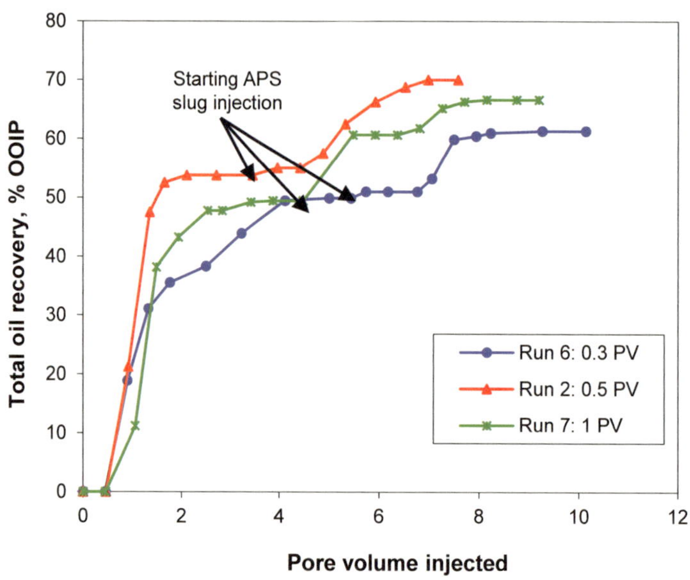

Fig. 17. Effect of slug size on oil recovery performance

Figure 17 shows the oil recovery performance as a function of APS slug size. The APS slug size was varied from 0.3 PV in Run # 6, 0.5 PV in Run # 2, and 1 PV in Run # 7. Figure 17 shows that the recovery performance is much improved as the APS slug size was increased from 0.3 PV to 0.5 PV. However, only small increment recovery (0.9%) was observed when the APS slug size was increased from 0.5 PV in Run # 2 to 1 PV in Run # 7. This indicates that the injection of 0.5 PV of APS slug is effective and therefore more economical than the other relatively larger slug size. .

8. Conclusion

Based on the findings and results illustrated in this study, it can be concluded that the non-edible Jatropha oil has great potential as raw material for the production of surfactants. Production of sodium methyl ester sulfonate (SMES) based on Jatropha oil can satisfy EOR requirements, because it is an inexpensive, natural, and renewable raw material. SMES provides appropriate surfactant properties at low cost, and therefore it offers a strong economic incentive to substitute SDS and other commercial surfactants for EOR applications.

On the basis of the results obtained from IFT and viscosity measurements, the polymeric methyl ester sulfonate (PMES) shows excellent properties for IFT reduction and viscosity control. The grafting of SMES onto acrylamide polymers to produce PMESs offers many benefits as compared to the existing chemical EOR methods. The presence of both the surfactant and polymer in one component system makes the PMES easier to handle especially in offshore applications.

The major contribution of this new APS combination is its ability to maintain the desired viscosity in the presence of sodium carbonate. The optimum alkali-polymeric formulation in terms of oil recovery performance obtained from the coreflood tests corresponds to a concentration of sodium carbonate of 0.8wt% and 0.6wt% of polymeric surfactant. The injection of 0.5 PV of APS slug rendered an oil recovery of 16.2% of the OOIP. These experimental results show the potential of the new alkali-polymeric surfactant system as a promising chemical flooding formulation if compared to the conventional ASP flooding formulation.

9. References

Abdala, A. A. (2002). Solution Rheology & Microstructure of Associative Polymers, *Ph.D. thesis,* North Carolina State University, Raleigh.

American Oil Chemists' Society (2003). Sampling and analysis of commercial fats and oils. Cd 3d-63,

Awang, M. & Goh, M. S. (2008). Sulfonation of phenols extracted from the pyrolysis oil of oil palm shells for enhanced oil recovery, *ChemSusChem,* Vol. 1, pp. 210-214.

Austad, T. & Milter, J. (2000). Surfactant Flooding in Enhanced Oil Recovery. Surfactants, In: *Fundamentals and Applications in the Petroleum Industry.* UK: Cambridge University Press, pp. 203-250.

Berger, P.D. & Lee, C.H. (2006). Improved ASP process using organic alkali, *the SPE/DOE symposium on improved oil recovery,* SPE 99581, Tulsa, April, 2006.

Bosswell, M.J. (2003). Plant Oils: Wealth, health, energy and environment. *Proc. International Conference of Renewable Energy Technology for rural Development,* Kathmandu, Nepal, June, 2003.

Clara, H., Larry, J.C., Lorenzo, A., Abel, B., Jie, Q., Phillip, C.D., &, Malcolm, J.P. (2001). ASP system design for an offshore application in the La Salina field, Lake Maracaibo, *SPE Latin American and Caribbean petroleum engineering conference*, SPE 69544 Buenos Aires, March, 2001.

Chonlin, L., Orear, E. A., Harwell, J.H., & Sheffield, J.A. (1990). Synthesis and characterization of a simple chiral surfactant sodium S-(-)-β-citronellyl sulfate, *Journal of colloid and interface science*, Vol. 137, pp. 296-299.

Craft, B.C., Hawkins, M., & Terry, R.E. (1991). Applied Petroleum Reservoir Engineering. Second Edition, Englewood, Cliffs NJ: Prentice Hall PTR. 4-6, pp. 376-384.

De Groot, M. (1929). Flooding process for recovering oil from subterranean oil-bearing strata, *U. S. Patent* 1 823 439.

Elraies, K. A., Tan, I., Awang, M., & Saaid, I. (2010a). Synthesis and Performance of Sodium Methyl Ester Sulfonate for Enhanced Oil Recovery. *Petroleum Science and Technology*, Vol. 28, No. 17, pp. 1799 - 1806.

Elraies, K. A., & Tan, I. (2010a). Design and Application of a New Acid-Alkali-Surfactant Flooding Formulation for Malaysian reservoirs, *the SPE Asia Pacific Oil & Gas Conference and Exhibition*, SPE 133005, Brisbane, October, 2010.

Elraies, K. A., Tan, I., Fathaddin, M., and Abo-Jabal, A. (2011). Development of a New Polymeric Surfactant for Chemical Enhanced Oil Recovery, *Petroleum Science and Technology*, Vol. 29, No. 14, pp. 1521 - 1528.

Energy Information Administration (2003). U.S. Crude Oil, Natural Gas, and Natural Gas Liquids Reserves, (2002) Annual Report. *Office of Oil and Gas, U.S. Department of Energy*. Washington. 20.

Falls, A.H., Thigpen, D.R., Nelson, R.C., Ciaston, J.W., Lawson, J.B., Good, P.A., & Shahin, G.T. (1992). A Filed Test of Cosurfactant-Enhanced Alkaline Flooding, *the 8th Symposium on Enhanced Oil Recovery*, SPE/DOE 24117, Tulsa, April, 1992.

Flaaten, A.K., Nguyen, Q.P., Pope, G.A., & Zhang, J. (2008). A systematic laboratory approach to low-cost, high-performance chemical flooding, *the improved oil recovery symposium*, SPE/DOE 113469, Tulsa, April, 2008.

Green, D.W., & Willhite, G.P. (1998). Enhanced Oil Recovery. Richardson Taxis: Society of Petroleum Engineers, SPE Textbook Series, 6, pp. 1-7,

Gregorio, C.G. (2005). Fatty Acids and Derivatives from Coconut Oil, *Bailey's Industrial Oil and Fat Products*, Sixth Edition, Six Volume Set,.

Hou., et al. (2001). Study of the Effect of ASP Solution Viscosity on Displacement Efficiency, *the 2001 SPE annual technical conference and exhibition*, SPE 71492, New Orleans, LA, September, 2001.

Laishun, S. (2000). An approach to the flam retardation and smoke suppression of ethylene-vinyl acetate copolymer by plasma grafting of Acrylamide, *Reactive and Functional Polymers*, Vol. 45, pp. 85-93.

Li, G., Mu, J., Li, Y., & Yuan, S. (2000). An experimental study on alkaline/surfactant/polymer flooding systems using nature mixed carboxylate, *Colloids and Surfaces A: Physicochemical and Engineering Aspects*, Vol. 173, pp. 219-229.

Liu, S. 2008.Alkaline Surfactant Polymer Enhanced Oil Recovery Process, *Ph.D. thesis*, Rice University, Houston, Texas.

Mayer, E.H., Berg, R.L., Carrnichale, J.D., & Weinbrandt, R.M. (1983). Alkaline injection for enhanced oil recovery - a status report, *the joint SPE/DOE enhanced oil recovery symposium*, SPE 8848, Tulsa, April, 1983.

McCafferty, J.F., & McClaflin, G.G. (1992). The Field Application of a Surfactant for the Production of Heavy, Viscous Crude Oils, *the 67th Annual Technical Conference and Exhibition*, SPE 24850, Washington, DC, October, 1992.

Mohan, K. (2009). Alkaline Surfactant Flooding for Tight Carbonate Reservoirs, *the Annual Technical Conference and Exhibition*, SPE 129516, New Orleans, October, 2009.

Nasr-El-Din, H.A., Hawkins, B.F., & Green, K.A. (1992). Recovery of Residual Oil Using the Alkaline/Surfactant/Polymer Process: Effect of Alkali Concentration, *Journal of Petroleum Science and Engineering*, Vol. 6, pp. 381-388.

Nelson, R.C., Lawson, J.B., Thigpen, D.R., & Stegemeier, G.L. (1984). Cosurfactant-enhanced alkaline flooding, SPE 12672, *enhanced oil recovery symposium*, Tulsa, April, 1984.

Pant, K.S., Kumar, D. & Gairola, S. (2006). Seed oil content variation in Jatropha curcas L. in different altitudinal ranges and site conditions in H.P, *India. Lyonia*, Vol. 11, pp. 31-34.

Pingping, S., Jialu, W., Shiyi, Y., Taixian, Z., & Xu, J. (2009). Study of Enhanced-Oil-Recovery Mechanism of Alkali/Surfactant/Polymer Flooding in Porous Media from Experiments, *SPE Journal*, Vol. 14, No. 2, pp. 237-244.

Pitts, M.J., Dowling, P., Wyatt, K., Surkalo, H. & Adams, C. (2006). Alkaline-Surfactant-Polymer Flood of the Tanner Field, SPE 100004, *the Symposium on Improved Oil Recovery*, Tulsa, April, 2006.

Pramanik, K. (2003). Properties and use of Jatropha curcas oil and diesel fuel blends in compression ignition engine, *Renewable Energy*, Vol. 28, pp. 239–248.

Pratap, M., & Gauma M.S. (2004). Field Implementation of Alkaline-Surfactant-Polymer (ASP) Flooding: A maiden effort in India, SPE 88455, *the Asia Pacific Oil and Gas Conference and Exhibition*, Perth, October, 2004.

Reisberg, J., & Doscher, T.M. (1956). Interfacial phenomena in crude-oil-water systems, *Production Monthly*, pp. 43-50.

Rudin, J., & Wasan, D.T. (1992). Mechanisms for Lowering of Interfacial Tension in Alkali/Acidic Oil Systems: Effect of Added Surfactant, *Industrial & Engineering Chemistry Research*, Vol. 31, pp. 1899-1906.

Sarin, R., Sharma, M., Sinharay, S., & Malhotra, R.K. (2007). Jatropha-Palm biodiesel blends: An optimum mix for Asia, *Fuel*, Vol. 86, pp. 1365-1371.

Silverstein, R. M., Webster F. X., & Kiemle, D. J. (2005). *Spectrometric identification of organic compounds*, USA: John Wiley & Sons, Inc.

Wang, D., Zhang, Z., Cheng, J., Yang, J., Gao, S., & Li, L. (1997). Pilot test of alkaline/surfactant/polymer flooding in Daqing oil field, *the Annual Technical Conference and Exhibition*, SPE 36748, Denver, October, 1997.

Wang, D., Cheng, J., Wu, J., Yang, Z., Yao, Y., & Li, H. (1999). Summary of ASP Pilots in Daqing Oil Field, SPE 57288, *the Asia Pacific Improved Oil Recovery Conference*, Kuala Lumpur, October, 1999.

Wang, D., Han, P., & Shao, Z. (2006). Sweep Improvement Options for Daqing Oil Field. *SPE/DOE symposium on improved oil recovery*, SPE 99441, Tulsa, OK. April, 2006.

Wangqi, H.D., & Dave, F. (2004). Surfactant Blends for Aqueous Solutions Useful for Improving Oil Recovery," U. S. Patent 6 828 281 B1.

Wuest, W., Eskuchen, R., & Richter, B. (1994). Process for the Production of Surfactant Mixtures Based on Ether Sulfonates and Their Use," U. S. Patent 5 318 709.

Tiwari, A. K., Kumar, A., & Raheman, H. (2007). Biodiesel production from Jatropha oil (Jatropha curcas) with high free fatty acids: an optimized process, *Biomass and Bioenergy*, Vol. 31, pp. 569-575.

Touhami, Y., Rana, D., Hornof, V., & Neale, G.H. (2001). Effects of Added Surfactant on the Dynamic Interfacial Tension Behavior of Acidic Oil/Alkaline Systems, *Journal of Colloid and Interface Science*, Vol. 239, pp. 226-229.

Vicente, G., Martinez, M., & Aracil, J. (2004). Integrated biodiesel production: a comparison of different homogeneous catalysts systems, *Bioresource Technology*, Vol. 92, No. 3, pp. 297-305.

Xuejun, L., Huayang, H., Yujun, W., Shenlin, Z., & Xianglan, P. (2008). Transesterification of soybean oil to biodiesel using CaO as a solid base catalyst, *Fuel*, Vol. 87, pp. 216-221.

Ye, L. Huang, R., Wu, J., & Hoffmann, H. (2004). Synthesis and Rheological Behaviour of Poly[Acrylamide–Acrylic Acid–N-(4-Butyl) Phenylacrylamide] Hydrophobically Modified Polyelectrolytes, *Colloid Polym Sci*, Vol. 282, pp. 305–313.

3

Microbial Enhanced Oil Recovery

Hamid Rashedi[1,2], Fatemeh Yazdian[3,2] and Simin Naghizadeh[1,2]
[1]Department of Chemical Engineering, Faculty of Engineering, University of Tehran, Tehran,
[2]Research Center for New Technologies in Life Science Engineering, University of Tehran, Tehran,
[3]Department of Life Science Engineering, Faculty of Interdisciplinary New Sciences and Technologies, University of Tehran, Tehran,
Iran

1. Introduction

Nowadays the majority of the world's energy comes from crude oil. A large proportion of this valuable and non-renewable resource is left behind in the ground after the application of conventional oil extraction methods. Moreover, there is a dire need to produce more crude oil to meet the worldwide rising energy demand which illustrates the necessity of progressing Enhanced Oil Recovery (EOR) processes. These methods try to overcome the main obstacles in the way of efficient oil recovery such as the low permeability of some reservoirs, the high viscosity of the crude oil, and high oil-water interfacial tensions that may result in high capillary forces retaining the oil in the reservoir rock (Bubela, 1987).

Microbial enhanced oil recovery (MEOR) is one of the EOR techniques where bacteria and their by-products are utilized for oil mobilization in a reservoir. In principle, MEOR is a process that increases oil recovery through inoculation of microorganisms in a reservoir, aiming that bacteria and their by-products cause some beneficial effects such as the formation of stable oil-water emulsions, mobilization of residual oil as a result of reduced interfacial tension, and diverting of injection fluids through upswept areas of the reservoir by clogging high permeable zones. Microbial technologies are becoming accepted worldwide as cost- effective and environmentally friendly approaches to improve oil production (Sarker et al., 1989). This chapter provides an inclusive review on MEOR mechanisms, its advantages over conventional EOR methods, its operational problems and engineering challenges. Furthermore the mathematical modeling of MEOR process is also presented.

2. Primary production

Oil exists in the small pores and in the narrow fissures and interstices within the body of the reservoir rocks underneath the surface of the earth. The natural pressure of the reservoir causes the oil to flow up to the surface and provide the so-called primary production, which depends upon the internal energy and the characteristics of the reservoir rock and the properties of the hydrocarbon fluids. In some reservoirs, which are the part of a much larger aquifer system, a natural flow of underground waters may be the drive force (aquifer drive) to push and displace oil. The initial reservoir pressure is usually high enough to lift the oil up to surface; however as oil production progresses, the reservoir pressure is continually depleted to a point in which artificial lift or pumping is required to maintain an economical oil production rate. In other reservoirs, there may be other recovery mechanisms, such as the expansion of dissolved gas during the pressure decline. As the reservoir pressure falls below the bubble point during production, some of the more volatile components are released and come out of solution to form small gas bubbles. Initially the bubbles are trapped in the pores and then their expansion causes oil displacement (dissolved gas drive). Furthermore in some reservoirs, as the pressure fall, gas bubbles increase in size and eventually coalesce forming a continuous gas phase that flows towards the upper part of the reservoir forming a gas cap. The gas cap constantly expands as the reservoir pressure continually decreases displacing more oil (gas cap drive) to the production wells.

3. Secondary production

As the reservoir pressure declines during primary production, a critical point is reached when it is necessary to provide external energy for the reservoir to achieve additional oil recovery, which is termed secondary recovery. The extra energy can be introduced by injecting gas (gas injection) and/or water (water flooding).

Gas injection is usually only applied to reservoirs which have a gas cap where gas drive would be an efficient displacement mechanism. In Water flooding, which nowadays is one of the most common methods of oil recovery, keeps the reservoir pressure around the bubble point, thus preventing the pores to be blocked by dissolved gases. Also, according to the hydrocarbon thermodynamics, at the bubble point, the oil will have its lowest viscosity. So that, for a specific pressure gradient, the maximum amount of the oil will be displaced under this condition. After some years of operation in a field, due to the reservoir heterogeneity, the injected fluids (water or gas) flow preferentially along high permeable layers that cause these fluids to by-pass oil saturated areas in the reservoir. Therefore, an increasingly large quantity of water (or gas) rises with the oil, and by decreasing the ratio of oil to water, eventually it becomes uneconomic to continue the process and the field must be abandoned. In this situation, due to the low proportion of the oil production in both primary and secondary stages (about 30%), attention will be focused on the third stage of the oil recovery, so-called tertiary production or Enhanced Oil Recovery (EOR) for recovering more oil from the existing and abandoned oil fields (Singer & Finnerty, 1984).

4. Tertiary production or Enhanced Oil Recovery (EOR)

Generally, tertiary or enhanced oil recovery involves the extraction of residual oil after the primary and secondary phases of production. At this stage, modern and technically advanced methods are employed to either modify the properties of reservoir fluids or the reservoir rock characteristics, with the aim of gaining recovery efficiencies more than those obtained by conventional recovery methods (primary and secondary recovery stages). This can be achieved based on different mechanisms such as reducing the interfacial tension between oil and water, reducing oil viscosity (thermal methods), creating miscible displacement and increasing viscosity of the displacing fluid to be more viscous than the oil. The applied EOR method for each reservoir depends on its specifications, and requires a great deal of rocks and fluids sampling and also laboratory investigations. In general, EOR processes can be classified into four main categories as thermal methods, chemical methods, miscible or solvent injection, and microbial methods.

4.1 Thermal processes

The general principle of thermal processes which are mostly used for recovery of heavy or viscous oils is to supply the reservoir with heat energy in order to increase the oil temperature and reduce its viscosity increasing the mobility of the oil towards production wells. Thermal processes can be conducted by two different methods: steam flooding and in-situ combustion. In steam flooding, steam at about 80% quality is injected into an oil reservoir, in which by condensing the steam, its heat energy transfers to reservoir rocks and fluids. This leads to the thermal expansion of the oil and the consequently reduction in its viscosity, and the release of dissolved gases. Steam flooding is the most widely used EOR method and probably the most profitable from an economic standpoint. In the in-situ combustion method (fire flood), which is theoretically more efficient than steam flood, burning some of the reservoir oil results in heating the reservoir and displacement of the remaining oil to the producing wells. But generally, due to the complex operational problems of this method, it is not widely applied.

4.2 Chemical methods

Chemical methods (chemical flooding) are claimed to have significant potential based on successful laboratory testing, but the results in field trials have not been encouraging. Furthermore, these methods are not yet profitable. In these processes, chemicals such as surfactants, alkaline solutions, and polymers are added to the displacing water in order to change the physicochemical properties of the water and the contacted oil making the displacement process more effective. In surfactant flooding, by reducing the interfacial tension between the oil and the displacing water and also the interfacial tension between the oil and the rock interfaces, residual oil can be displaced and recovered. Moreover, in caustic flooding, the reaction of the alkaline compounds with the organic acids in the oil forms in-situ natural surfactants that lower the oil-water interfacial tension. In addition to surfactant and alkaline flooding, polymers are used to increase the viscosity of the displacing water to improve the oil swept efficiency.

4.3 Miscible displacement processes

The underlying principle behind miscible displacement processes is to reduce the interfacial tension between the displacing and displaced fluids to near zero that leads to the total miscibility of the solvent (gas) and the oil, forming a single homogeneous moving phase. The displacing fluid (injected solvent or gas) could be carbon dioxide, nitrogen, exhaust gases, hydrocarbon solvents, or even certain alcohols.

4.4 Microbial processes (MEOR)

Another tertiary method of oil recovery is microbial enhanced oil recovery, commonly known as MEOR, which nowadays is becoming an important and a rapidly developed tertiary production technology, which uses microorganisms or their metabolites to enhance the recovery of residual oil (Banat, 1995; Xu et al., 2009).

In this method, nutrients and suitable bacteria, which can grow under the anaerobic reservoir conditions, are injected into the reservoir. The microbial metabolic products that include biosurfactants, biopolymers, acids, solvents, gases, and also enzymes modify the properties of the oil and the interactions between oil, water, and the porous media, which increase the mobility of the oil and consequently the recovery of oil especially from depleted and marginal reservoirs; thus extending the producing life of the wells (Lazar et al., 2007; Belyaev et al. 2004; Van et al. 2003). In MEOR process, different kinds of nutrients are injected to the reservoirs. In some processes, a fermentable carbohydrate including molasses is utilized as nutrient (Bass & Lappin-Scott, 1997). Some other reservoirs require inorganic nutrients as substrates for cellular growth or as alternative electron acceptors instead of oxygen. In another method, water containing a source of vitamins, phosphates, and electron acceptors such as nitrate, is injected into the reservoir, so that anaerobic bacteria can grow by using oil as the main carbon source (Sen, 2008). The microorganisms used in MEOR methods are mostly anaerobic extremophiles, including halophiles, barophiles, and thermophiles for their better adaptation to the oil reservoir conditions (Brown, 1992; Khire & Khan, 1994; Bryant & Lindsey, 1996; Tango & Islam, 2002). These bacteria are usually hydrocarbon-utilizing, non-pathogenic, and are naturally occurring in petroleum reservoirs (Almeida et al. 2004). In the past, the microbes selected for use, had to have a maximum growth rate at temperatures below 80°C, however it is known that some microorganisms can actually grow at temperatures up to 121°C (Kashefi & Lovley, 2003). Bacillus strains grown on glucose mineral salts medium are one of the most utilized bacteria in MEOR technologies, specifically when oil viscosity reduction is not the primary aim of the operation (Sen, 2008).

5. History of MEOR

MEOR was first described by Beckman in 1926. Few studies were conducted on this topic, between 1926 and 1940 (Lazar et al., 2007). In 1944, ZoBell patented a MEOR method and continued researching on this subject. In 1947, ZoBell initiated a new era of investigation in petroleum microbiology with applications for oil recovery. ZoBell explained that the major MEOR mechanisms which are responsible for oil release from porous media, involve processes such as dissolution of inorganic carbonates by bacterial metabolites; production of bacterial gases, which reduces the oil viscosity supporting its flow; production of surface-active substances or wetting agents, and the high affinity of bacteria for solids (Lazar et al., 2007). The first MEOR field test was conducted in the Lisbon field, Union County, AR, in 1954 (Yarbrough and Coty, 1983). The improvement of MEOR in field trials was based on the injection of mixed anaerobic or facultative anaerobic bacteria such as *Clostridium, Bacillus, Pseudomonas, Arthrobacterium, Micrococcus, Peptococcus,* and *Mycobacterium* among others; selected on their ability to generate high quantities of gases, acids, solvents, polymers, surfactants, and cell-biomass. More details on bacteria's specific abilities were reviewed by Lazar (Lazar, 1991, 1996 to 1998).

The application of MEOR as a tertiary recovery technique and a natural step to decrease residual oil saturation has been reported (Behesht et al. 2008). A complete review (692 references) of the microbiology of petroleum was published by Van Hamme et al. (2003), which covered a literature review up to 2002. This publication is mainly focused on the description of the molecular-biological characteristics of the aerobic and anaerobic hydrocarbon exploitation, with some citations on the application of the microbial action on petroleum waste, microbial oil recovery, and biosensors. The aspect of petroleum microbiology that is perhaps the most important for MEOR is the ability of microbes to use hydrocarbons as the carbon and energy source. Biotechnology research has improved, which has influenced the oil industry to be more open to the evaluation of microorganisms to enhance oil production. Both indigenous and injected microorganisms are used depending on their adaptability to the specific reservoirs. In microbial enhanced oil recovery (MEOR), bacteria are regularly used because they show several practical features (Nielsen et al., 2010). Several publications state that oil recovery through microbial action takes place due to several mechanisms as follows (Jenneman et al. 1984; Bryant et al. 1989; Chisholm et al. 1990; Sarkar et al. 1994; Desouky et al. 1996; Delshad et al. 2002; Feng et al. 2002; Gray et al. 2008; Nielsen et al., 2010):

- Reduction of oil/water interfacial tension and modification of porous media wettability by surfactant production and bacterial action.
- Selective plugging of porous media by microorganisms and their metabolites.
- Oil viscosity reduction caused by gas solution in the oil due to bacterial gas production or degradation of long-chain saturated hydrocarbons.
- Production of acids that dissolve rock improving porous media permeability.

Particularly, the two first mechanisms are believed to have the greatest effect on improving oil recovery (Jenneman et al., 1984; Bryant et al., 1989; Chisholm et al., 1990; Sarkar et al., 1994; Desouky et al., 1996; Delshad et al., 2002; Feng et al., 2002; Gray et al., 2008; Nielsen et al., 2010).

6. MEOR mechanisms

Improvement of oil recovery through microbial actions can be performed through several mechanisms such as reduction of oil-water interfacial tension and alteration of wettability by surfactant production and bacterial presence, selective plugging by microorganisms and their metabolites, oil viscosity reduction by gas production or degradation of long-chain saturated hydrocarbons, and production of acids which improves absolute permeability by dissolving minerals in the rock, however, the two first mechanisms are believed to have the greatest impact on oil recovery (Nielsen et al., 2010). So that, microorganisms can produce many of the same types of compounds that are used in conventional EOR processes to mobilize oil trapped in reservoirs and the only difference between EOR and some of the MEOR methods probably is the means by which the substances are introduced into the reservoir (Bryant & Lockhart, 2000). Table 1 summarizes different microbial consortia, their related metabolites and applications in MEOR (Sen, 2008).

6.1 Biosurfactant application

Chemical surfactants are hazardous and costly compounds which are not biodegradable and can be toxic to the environment (Bordoloi et al., 2008; Suthar et al., 2008). In recent years, the increase concern regarding environment protection has caused the development of cost-effective bioprocesses for biosurfactant production (Morita et al., 2007; Fax & Bala, 2000; Abalos et al., 2001). Biosurfactants are high value products that due to their superior characteristics, such as low toxicity, ease of application, high biodegradability and tolerance even under extreme conditions of pH, temperature, and salinity, are efficient alternatives to

Microbial product	Example microbes	Application in MEOR
Biomass	*Biomass Bacillus, Leuconostoc, Xanthomonas*	Selective plugging and wettability alteration
Surfactants	*Acinetobacter, Arthrobacter, Bacillus, Pseudomonas*	Emulsification and de-emulsification through reduction of IFT
Polymers	*Bacillus, Brevibacterium, Leuconostoc, Xanthomonas*	Injectivity profile and viscosity modification, selective plugging
Solvents	*Clostridium, Zymomonas, Klebsiella*	Rock dissolution for better permeability, oil viscosity reduction
Acids	*Clostridium, Enterobacter, Mixed acidogens*	Permeability increase, emulsification
Gases	*Clostridium, Enterobacter Methanobacterium*	Increased pressure, oil swelling, IFT and viscosity reduction

Table 1. Microorganism, their metabolites and applications in MEOR.

chemically synthesized surface-active agents with potential applications in the petroleum industry (Banat et al., 2000; Cameotra & Makkar, 2004; Desai & Banat, 1997). Generally, there are three major strategies for the application of biosurfactants in oil recovery (Banat, 1995): (i) injection of biosurfactant producing microorganisms into the reservoir through the well, with subsequent multiplication of microorganisms in-situ through the reservoir rocks, (ii) injection of selected nutrients into a reservoir, thus stimulating the growth of indigenous biosurfactant producing microorganisms, (iii) production of biosurfactants ex-situ and their subsequent injection into the reservoir.

Biourfactants can contribute positively to improve oil recovery by dramatically reducing interfacial tension and also by altering the wettability of reservoir rock to displace more oil from the capillary network. When the interfacial tension at the oil–rock interface is lowered, the capillary forces that prevent oil from moving through rock pores are reduced. Therefore, the detachment of oil films from rocks can occur (Dyke, 1991; Li, 2002). Although a reduction in interfacial tension will help to mobilize the oil, a change in wettability of the pore surfaces to a more water-wet state, will release more oil from the surfaces and consequently can improve oil recovery. It has been estimated (Lake, 1989) that the interfacial tension must be lowered in the range from 0.01 to 0.001 mN/m to achieve significant oil recovery. Biosurfactants, produced anaerobically which are capable of reducing IFT to such low values, have been reported (Brown et al. 1986).

6.2 Application of gases and solvents

Some strains of anaerobic bacteria such as clostridia can produce hydrogen, carbon dioxide, methane, acetate, and butyrate by carbohydrates fermentation during the initial growth phase of the fermentation process. Additional products of this microbial process are some kinds of solvents including acetone, butanol, ethanol, isopropanol, and other solvents in lesser amounts, produced during the stationary growth phase. In an oil reservoir, these gaseous and liquid metabolites, which are produced in-situ, are dissolved in the crude oil resulting in lower oil viscosity and reduction of the capillary forces contributing to oil retention. Moreover microbes increase the pressure in the reservoir by producing gases in the pore spaces that would have been normally bypassed with conventional gas flooding operations (Bryant & Douglas, 1988). Both gases and solvents can dissolve the carbonate rock, thereby increasing its permeability and porosity (Bordoloi & Konwar 2008).

6.3 Clogging mechanism

One method of microbial improving oil recovery is by modifying the fluid flow through the reservoir by shifting fluid flow from the high permeability zones in a reservoir to the moderate or low permeability zones thus increasing the sweep efficiency by forcing the injected water to pass through previously by-passed oil zones of the reservoir (Bryant et al., 1998). The changes in flow pattern can be achieved by an increase in microbial cell mass within the reservoir. Stimulating either indigenous microbial populations or injecting microorganisms together with nutrients produce biomass and hence microbial plugging. The injected nutrient and microbes preferentially flow into the high permeability zones of the reservoir and as a result of cell growth, the biomass selectively plugs these zones to a greater extent than the moderate or low permeability zones (Crawford, 1961 & 1962). Experiments using brine-saturated sandstone cores showed that injecting nutrients and viable bacterial cells resulted in clogging of 60-80% of the pore space (Jenneman et al., 1984).

6.4 Biopolymers application

Some water insoluble biopolymers produced by certain bacteria can increase the oil recovery by the same mechanism of plugging by cell growth (Jack & Diblasio, 1985). In water-flooding operations, in which water is pumped into injection wells in the reservoir in order to force the oil up to the surface, biopolymers plug high-permeability zones to redirect the water-flood to oil-rich zones in the reservoir and the sweep efficiency increases by equalizing the permeability across the reservoir (Casellas et al., 1997; Yakinov et al., 1995; Abu- Ruwaida et al., 1991)

7. MEOR advantages

The most outstanding advantages of MEOR over other EOR technologies are listed below (Lazar, 2007):

1. The injected bacteria and nutrient are inexpensive and easy to obtain and handle in the field.
2. MEOR processes are economically attractive for marginally producing oil fields and are suitable alternatives before the abandonment of marginal wells.
3. Microbial cell factories need little input of energy to produce the MEOR agents.
4. Compared to other EOR technologies, less modification of the existing field characteristics are required to implement the recovery process by MEOR technologies, which are more cost-effective to install and more easily applied.
5. Since the injected fluids are not petrochemicals, their costs are not dependent on the global crude oil price.
6. MEOR processes are particularly suited for carbonate oil reservoirs where some EOR technologies cannot be applied efficiently.
7. The effects of bacterial activity within the reservoir are improved by their growth with time, while in EOR technologies the effects of the additives tend to decrease with time and distance from the injection well.

8. MEOR products are all biodegradable and will not be accumulated in the environment, therefore are environmentally compatible.
9. As the substances used in chemical EOR methods are petrochemicals obtained from petroleum feedstock after downstream processing, MEOR methods in comparison with conventional chemical EOR methods, in which finished commercial products are utilized for the recovery of raw materials, are more economically attractive.

8. Field trials

Microbial enhanced oil recovery methods were developed from laboratory-based studies in the early 1980s to field applications in the 1990s (Ramkrishna, 2008). In 2010, various countries allocated one-third of their oil recovery plans toward MEOR techniques. Although it has been constantly observed that the effects of MEOR projects applied to one well had positively affected oil recovery in neighboring wells, it has been recognized that several MEOR process variables must be optimized before it develops into a practical method for common field applications. These variables include a better description of the candidate reservoirs, better knowledge of the biochemical and physiological characteristics of the microbial consortia, a better handle of the controlling mechanisms, and an unambiguous estimation of the process economics. Most of the MEOR processes leading to field trials have been completed in the last two decades and now the knowledge has advanced from a laboratory-based assessment of microbial processes, to field applications globally (Ramkrishna, 2008). Portwood (1995) reported an analysis of the data based on the information gathered from 322 MEOR projects, led the evaluation of the technical efficiency and economics of MEOR, which is useful for forecasting treatments outcome in any given reservoir. A collection of significant information from field trials in the USA and Romania was considered as well in the analysis reported. Likewise, several reports discussing in-situ uses of MEOR in field trials with analysis of the results are published elsewhere (Portwood, 1995; Clark et al., 1981; Jenneman et al., 1984; Dennis, 1998; Kleppe, 2001; Youssef et al., 2007). For example, in an MEOR field trial in the Southeast Vassar Vertz Sand Unit salt-containing reservoir in Oklahoma, nutrient injection motivated the growth of the indigenous microbial populations, which reduced the effective permeability by 33% (Jenneman et al., 1996). A biosurfactant flooding process using a very low concentration of biosurfactant, which was produced by the *Bacillus mojavensis* strain JF-2, was reported to be very effective in recovering residual oil from Berea sandstone cores (Bailey et al., 2001). Also, a new model for enhanced oil production was developed for using ultra microbacteria generated from indigenous reservoir microbiota through nutrient treatment (Lazar et al., 2007). The external cell layers of such ultra microbacteria had surface-active properties. Such a microbial scheme was successfully verified in increasing oil production in the Alton oil field in Queensland, Australia (Sheehy, 1991 and 1992).

The activity of MEOR field experiments after 1990 is based on the foundation that successful MEOR applications must be conducted on water floods, where a continuous water phase facilitates the application of well stimulation procedures and the low cost of MEOR makes it a preferable option. At the same time, specific microbial applications such as microbial paraffin removal, microbial skin damage removal, microbial control souring and clogging, and those based on using ultra microbacteria are potential technologies for the additional growth of MEOR (Lazar et al., 2007). Worldwide experience in MEOR field trials during the last 40 years has been discussed by Lazar et al., (2007).

9. MEOR problems

MEOR techniques face some common problems that are outlined as follows (Lazar, 2007):

1. Injectivity lost due to microbial plugging of the wellbore—to avoid wellbore plugging, some actions must be taken such as filtration before injection, avoid biopolymers production, and minimize microbial adsorption to rock surface by using dormant cell forms, spores, or ultra-micro-bacteria.
2. Dispersion or transportation of all necessary components to the target zone.
3. Optimization of the desired in-situ metabolic activity due to the effect of variables such as pH, temperature, salinity, and pressure for any in-situ MEOR operation.
4. Isolation of microbial strains, adaptable to the extreme reservoir conditions of pH, temperatures, pressure and salinity (Sen, 2008).
5. Low in-situ concentration of bacterial metabolites; the solution to this problem might be the application of genetic engineering techniques (Xu & Lu, 2011).

10. Mathematical modeling

The current need of maximizing oil recovery from reservoirs has prompted the evaluation of various Improved Oil Recovery (IOR) methods and EOR techniques, including the use of microbial processes. MEOR is a driving force behind the efforts to come up with different and cost-efficient recovery processes (Kianipey and Donaldson, 1986). Bryant and Lockhart, (2002) examined the quantitative correlations between microbial activity, reservoir features, and operating conditions such as injection rates, well spacing, and residual oil saturation (Bryant and Lockhart, 2002). Marshall (2008) stated that a mathematical model could be used to recognize the most important parameters and their practical relationships for the application of MEOR.

Improvement of detailed mathematical models for MEOR is an exceptionally demanding task, not only as a consequence of the natural difficulty of the microbes, but also because of the diversity of physical and chemical variables that control their activities in subsurface porous media. Specific or general aims can be foreseen for modeling by researchers. In specific cases, it is desired to employ the models to maximize the yield and minimize the costs of the MEOR procedure. Main physical insights of the process can be obtained from quite simple analytical models; whereas the exact models regularly require thorough numerical computation. The important point claimed by researchers is that modeling of microbial reactions still faces strict limitations. Models are based on the relation between the residence time (τ_{res}) of the bacteria in a cylindrical reaction zone of radius r_m, depth h, and porosity φ, which is:

$$\tau_{res} = \frac{\pi r_m^2 h \phi (1 - S_{or})}{Q} \tag{1}$$

Where Q is the volumetric flow rate and S_{or} is the residual oil saturation, and the time τ_{rsd} required for the microbial reaction to produce a desired concentration c_{req} of some metabolite.

To estimate the reaction time, Marshall (2008) posed the following assumptions: isothermal plug flow through the reactor, nutrient consumption is first order and irreversible, and that nutrients initial concentration is n_0.

The physical model on which the above argument is based is very basic, but the analysis draws interest to the important issue of reaction kinetics that has to be addressed by more complex treatments. It is possible to write a balanced chemical equation for the production of a given metabolite, but the rate of production can only be determined experimentally, and must be given by actual bacterial growth velocities (Marshall, 2008).

Several mathematical models were developed to simulate MEOR processes. The models usually included multidimensional flow of the multiphase fluid consisting of water and oil in porous media along with specific equations for adsorption and diffusion of metabolites, microorganisms, and nutrients (Islam, 1990; Behesht et al., 2008). The main multidimensional transport equations were combined with equations of different microbial features such as growth, death, and nutrient consumption.

The majority of the published mathematical models for performance of bacteria and viruses in porous media were initially stimulated by problems arising in water filtration and wastewater treatment (Corapcioglu and Haridas, 1984; Stevik et al., 2004). Such models have three major parts: Transport Properties, Conservation Law (Local Equilibrium, Breakdown of Filtration Theory, and Physical Straining), Biofilm Clogging and Related Phenomena such as the theoretical description of the biological clogging of pores. The clogging agent is coupled nonlinearly not only to the growth of the bacteria, but moreover to the flux of nutrients transported by the fluid. The origin of the earliest approaches to the development of models of this phenomenon is the idea that medium can be characterized as a bundle of independent capillary tubes (Marshall, 2008).

The first is an approximate of the transport properties of the bacteria in the fluid. In the treatment given by Corapcioglu and Haridas (1984), bacteria diffusivity was achieved by function of the Stokes-Einstein equation, which effectively treats the microbe as if it were a particle that is undergoing Brownian movement.

The second is conservation law. If chemotaxis is neglected, the concentration of bacteria in the fluid phase of a small constituent of the porous medium is defined by a partial differential equation expressing the rate of change of the concentration as the sum of terms resulting from diffusion (or dispersion), advection, and transfer between the fluid phase and the surface of the solid grains. Numerical solution of systems of equations of this general type is at the heart of computational hydrology and simulation of oil wells. The other parts of the model that must be considered are biofilm clogging and related phenomena.

For analysis of MEOR, it is interesting to present the characteristics of the water phase saturation profiles and the corresponding oil recovery curves. A mathematical model for MEOR was considered by Islam (1990), where bacterial growth resulted in plugging, decrease of oil viscosity, reduction of interfacial tension between oil and water, and gas production. In the model, interfacial tension was correlated with bacteria concentration to avoid adding another variable to account for surfactant production. In this model, it was clear that the reduction of surface tension between water and oil was the most important factor on the improvement of oil recovery (Islam, 1990).

Chang et al., (1991) improved a mathematical model depicting adsorption, growth and decomposition of microorganisms, consumption of nutrients, and other physical procedures. Due to microorganisms' organic build up, porosity and permeability were altered. Moreover, the model showed that the oil recovery increased by microbial plugging (Chang et al., 1991). Additional development of MEOR mathematical models is essential because none of the present models account for all of the variables involved in microbial growth. For instance, wettability modification and changes in interfacial tension (IFT) are two vital variables of microbial growth which are ignored in previous models. Moreover, some important physico-chemical features such as surfactant and polymer adsorption, and the effects of salinity and polymer viscosity on the mobility of the aqueous phase are ignored in these models. Finally, all of them are limited to transport in porous media. Simulation efforts to consider the effect of various parameters on the efficiency of MEOR using the current deficient models may not lead to successful results in the field (Behesht et al, 2008).

Surfactant production and adsorption, salinity effects, adsorption of microorganisms, reduction of interfacial tension, and wettability changes were taken into account in a MEOR model presented by Behesht et al. (2008). In this work, polymer was also injected in order to reduce permeability of the porous media and to increase the viscosity of the displacing water. The use of these two techniques resulted in an increase of oil recovery. Behesht et al., (2008) developed a three-dimensional multi-component transport model in a two-phase oil-water system. The model accounted for the effects of dispersion, convection, injection, growth and death of microbes, and accumulation of microbial debris. For the first time, effects of both porous media wettability modification from oil wet to water wet and the reduction of interfacial tension (IFT) on the relative permeability and capillary pressure curves were included in a MEOR simulation model. Transport equations were considered for the bacteria, nutrients, and metabolite in the matrix, reduced interfacial tension on phase trapping, surfactant and polymer adsorption, and the effect of polymer viscosity on mobility of the aqueous phase. The model was used to simulate the effects of parameters such as: flooding time schedules, washing water flow rate, substrate concentration, permeability, polymer and salinity concentration on the recovery of original oil in place (OOIP) in a hypothetical reservoir (Behesht et al., 2008).

Several methods were used to model relative permeability changes as a function of interfacial tension. Nielsen et al. (2010) used a correlation between surfactant concentration and interfacial tension (Nielsen et al., 2010). Usually, a reduction of interfacial tension decreases residual oil saturation affecting the relative permeability curve endpoints, but it also straightens the relative permeability curves approaching full miscibility (Coats, 1980; Al-Wahaibi et al., 2006). Nielsen et al. (2010) investigated three methods: (1) capillary number and normalized residual oil saturation correlations; (2) Coats interpolation between relative permeability curves; and (3) interpolation of factors of Corey type relative permeability curves (Coats, 1980; Green and Willhite, 1998). They recommend the third method, in which more parameters can be estimated in order to obtain a better fit with experimental data. Moreover, different distributions of surfactant between phases, the effect of bacterial growth rate, and the effect of injection concentrations of substrate and bacteria were considered as well. The saturation curves with specific MEOR characteristics were stated mutually with the oil recovery curves. Nielsen et al. (2010) developed a mathematical model describing the process of microbial enhanced oil recovery. The one-dimensional isothermal model comprised dislocation of oil by water containing bacteria and substrate as energy source. The bioproducts were both bacteria and metabolites. In the situation of MEOR modeling, a novel approach was partitioning of metabolites between the oil and the water phases. The partitioning was considered by a distribution coefficient. The portion of metabolite transferred to the oil phase was termed as vanishing so that the total amount of metabolite in the water phase was reduced. The metabolite produced was biosurfactant that reduced the oil–water interfacial tension, which resulted in oil mobilization. Different methods of incorporating surfactant-induced reduction of interfacial tension into models were also investigated. Reactive transport models were used to describe convection, bacterial growth, substrate consumption, and metabolite production, where the metabolite was a surfactant. The model was based on two-phase flow comprising five components; oil, water, bacteria, substrate, and metabolite/surfactant. The water phase comprised water, bacteria, substrate, and metabolite. The following assumptions were used in this model (Nielsen et al. 2010):

- Fluid flow was one-dimensional.
- The microorganisms were anaerobic bacteria, and they were injected into the reservoir. It was assumed that there was no local microorganism in the reservoir.
- Bacterial growth rate could be explained by Monod-kinetics being independent of temperature, pressure, pH, and salinity (Nielsen et al. 2003).
- The major metabolite was surfactant and other possible metabolites were considered insignificant.
- Surfactant could be distributed between both phases (water and oil). Surfactant sharing was instantaneous and the distribution kinetics was neglected.
- Adsorption of any component was neglected
- No substrate and metabolite adsorption on pore walls.
- Partial flow function was exploited, because capillary pressure was considered negligible.
- Negligible diffusion and chemotaxis.
- Isothermal method with incompressible flow.
- No volume change on mixing.

Therefore, the transport equation for each component was given as (Nielsen et al. 2010):

$$\frac{\delta}{\delta t}\left(\phi\sum_{j=1}^{n_p} w_{ij}.\rho_j.s_j\right)+\frac{\delta}{\delta x}\left(v\sum_{j=1}^{n_p} w_{ij}.\rho_j.f_j\right)=\phi q_i \tag{2}$$

where j is the phase, i is the component, n_p is the number of phases, ω_{ij} are component mass fractions in phase j, v is the linear velocity, ρ_j is the phase density, f_j is the fractional flow function of phase j, x is the length variable, t is the time, φ is the porosity, and q_i is the source expression for component i also comprising the reaction terms.

Growth rate expressions for microorganisms are regularly the Monod- expression based on the Michaelis–Menton enzyme kinetics and Langmuir expressions for heterogeneous catalysis (Islam, 1990; Chang et al., 1991; Nielsen et al., 2003).

The relative permeability curves for oil k_{ro} and water k_{rw}, and the Corey correlations were used (Lake, 1989). Moreover, the capillary number N_{ca} (ratio of viscous to capillary forces) are applied, which depend on changes in interfacial tension σ.

11. Challenges in MEOR

In spite of the various advantages of MEOR over other EOR methods, MEOR has not gained credibility in the oil industry because the value of MEOR can only be determined by the results of field trials. MEOR literature is mainly based on laboratory data and a shortage of field trials can be seen in this field. Also, because of reservoir heterogeneity, it is so difficult to extrapolate laboratory results into what is to be expected in the field or predict what will happen in a new field based on the results obtained from another field. Furthermore, few of the tests explain the mechanisms of oil recovery or offer a reasonable analysis of the application outcome. In addition, as Moses (1991) pointed out, the follow-up time of most field trials was not long enough to determine the long-term effects of the process. Finally, the precise mechanisms of in-situ MEOR operations are still unclear. Thus more research is required in this field (Xu & Lu, 2011).

12. Conclusion

MEOR is a cost effective and eco-friendly process that shows several advantages over other EOR processes. MEOR has great potential to become a viable alternative to the traditional EOR chemical methods. Although MEOR is a highly attractive method in the field of oil recovery, there are still uncertainties in meeting the engineering design criteria required by the application of microbial processes in the field, which has led to its current low acceptance by the oil industry. Therefore, a better understanding of the MEOR processes and its mechanisms from an engineering standpoint are required; as well as the systematic evaluation of the major factors affecting this process such as reservoir characteristics and microbial consortia, to improve the process efficiency.

13. References

Abalos, A., Pinazo, A., Infante, M.R., Casals, M., García, F., & Manresa, A. (2001). Physicochemical and antimicrobial properties of new rhamnolipids produced by *Pseudomonas aeruginosa* AT10 from soybean oil refinery wastes. *Langmuir*, Vol. 17, pp. (1367–1371).

Abu-Ruwaida, A.S., Banat, I.M., Haditirto, S., Salem, A., & Kadri, M. (1991). Isolation of biosurfactant producing bacteria. Product characterization and evaluation. *Acta Biotechnol.*, Vol. 11, pp. (315–324)

Almeida, P.F., Moreira, R.S., Almeida, R.C.C., Guimaraes, A.K., Carvalho, A.S., & Quintella, C. (2004). Selection and application of microorganisms to improve oil recovery. *Eng. Life Sci.*, Vol. 4, pp. (319–325).

Al-Wahaibi, Y.M., Grattoni, C.A., & Muggeridge, A.H. (2006). Drainage and imbibition relative permeabilities at near miscible conditions. *J. Petroleum Sci. Eng.*, Vol. 53, pp. (239–253).

Bailey, SA., Kenney, TM., & Schneider D. (2001). Microbial enhanced oil recovery: diverse successful applications of biotechnology in the oil field. SPE J., Paper no. 72129.

Banat, I.M. (1995). Biosurfactants production and possible uses in microbial enhanced oil recovery and oil pollution remediation: a review. *Biores. Technol.*, Vol. 51, pp. (1-12)

Banat, I.M., Makkar, R.S., & Cameotra, S.S., (2000). Potential commercial applications of microbial surfactants. Appl. Environ. *Microb.*, Vol. 53, pp. (495–508).

Bass, C., & Lappin-Scott, H. (1997). The bad guys and the good guys in petroleum microbiology. *Oilfield Rev.*, pp. (17–25).

Behesht, M., Roostaazad, R., Farhadpour, F., & Pishvaei M.R. (2008). Model development for meor process in conventional non-fractured reservoirs and investigation of physico-chemical parameter effects. *Chem. Eng. Technol.*, Vol. 7, pp. (953–963).

Belyaev, S.S., Borzenkov, I.A., Nazina, T.N., Rozanova, E.P., Glumov, I.F., & Ibatullin, R.R. (2004). Use of microorganisms in the biotechnology for the enhancement of oil recovery. *Microbiol (Maik Nauka Interperiodica)*, Vol. 73, pp. (590–598).

Bordoloi, N.K., & Konwar, B.K. (2008). Microbial surfactant-enhanced mineral oil recovery under laboratory conditions. *Colloids and Surfaces B: Biointerfaces*, Vol. 63, pp. (73–82).

Brown, F.G., (1992). Microbes: the practical and environmental safe solution to production problems enhanced production and EOR. SPE J, Paper No. 23955.

Brown, M.J., Moses, V., Robinson, J.P., & Springham, D.G. (1986). Microbial enhanced oil recovery: Progress and Prospects. *Critical Rev. Biotechnol.*, Vol. 3, pp. (159-197).

Bryant, R., Burchfield, T., Chase, K., Bertus, K., & Stepp A. (1989). Optimization of oil mobilization, transport of microbes and metabolites, and effects of additives. In: SPE 19686 presented in 64th Annual Technical Conference and Exhibition of the Society of Petroleum Engineers held in San Antonio, TX, pp. (567–578).

Bryant, S.L., & Lockhart, T.P. (2002). Reservoir engineering analysis of microbial enhanced recovery [Paper SPE 79179]. SPE Reservoir Engineering and Evaluation, pp. (365-374).

Bryant, R.S., & Douglas, J. (1988). Evaluation of Microbial Systems in Porous Media for EOR. *Res. Eng.*, pp. (489–495).

Bryant, R.S., Bailey, S.A., Stepp, A.K., Evans, D.B., Parli, J.A., & Kolhatkar, A.R. (1998). Biotechnology for Heavy Oil Recovery. BDM Petroleum Technology, Bartlesville, Oklahoma, USA, No. 110.

Bryant, S., & Lockhart, T.P., (2000). Reservoir engineering analysis of microbial enhanced oil recovery. In Proceedings of the SPE Annual Technical Conference and Exhibition, Dallas, TX.

Bryant, S.R., & Lindsey, R.P. (1996). World-wide applications of microbial technology for improving oil recovery. SPE J, Paper No. 35356.

Bubela, B. (1987). A comparison of strategies for enhanced oil recovery using in situ and ex situ produced biosurfactants, Surfact. *Sci. Ser.*, Vol. 25, pp. (143–161).

Cameotra, S.S., & Makkar R.S., (2004). Recent applications of biosurfactants as biological and immunological molecules. *Curr. Opin. Microbiol.*, Vol. 7, pp. (262–266).

Casellas, M., Grifoll, M., Bayona, J.M., & Solanas, A.M. (1997). New metabolites in the degradation of fluorene by *Arthrobacter* sp. strain F101. *Appl. Environ. Microbiol.*, Vol. 63, No. 3, pp. (819–826).

Chang, M.M., Chung, F., Bryant, R., Gao, H., & Burchfield T. (1991). Modelling and laboratory investigation of microbial transport phenomena in porous media. In: SPE 22845 presented at 66thAnnual Technical Conference and Exhibition of SPE in Dallas Texas.

Chisholm, J., Kashikar, S., Knapp, R., McInerney, M., & Menzie D. (1990). Microbial enhanced oil recovery: Interfacial tension and gas-induced relative permeability effects. In: 65th Annual Technical Conference and Exhibition of the Society of Petroleum Engineers on September 23–26, New Orleans, LA.

Clark, JB., Munnecke, DM., & Jenneman, GE. (1981). In situ microbial enhancement of oil production. *Dev. Ind. Microbiol.*, Vol. 22, pp. (695–701).

Coats, K.H. (1980). An equation of state compositional model. *Soc. Pet. Eng. J.*, Vol. 20, pp. (363–376).

Corapcioglu, M. Y., & Haridas A. (1984). Transport and fate of microorganisms in porous media: a theoretical investigation. *J. Hydrology*, 72, pp. (149-169).

Crawford, P.B. (1961). Possible Bacterial Collection of Stratification Problems. Producer's Monthly, Vol. 25, pp. (10–11).

Crawford, P.B. (1962). Water Technology: Continual Changes in Bacterial Stratification Rectification. Producer's Monthly, Vol. 26, pp. (12–14).

Delshad, M., Asakawa, K., Pope, G.A., & Sepehrnoori, K. (2002). Simulations of chemical and microbial enhanced oil recovery methods. In: SPE 75237 at SPE/DOE Improved Oil Recovery Symposium held in Tulsa, Oklahoma.

Dennis, DM. (1998). Microbial production stimulation. Rocky mountain oilfield testing center project test results. DOE: Website: /http://www.rmotc.com/pdfs/97pt25.pdfS.

Desai, J.D., & Banat, I.M., (1997). Microbial production of surfactants and their commercial potential. Microbiol. *Mol. Biol. Rev.*, Vol. 61, pp. (47–64).

Desouky, S.M., Abdel-Daim, M.M., Sayyouh, M.H., & Dahab A.S. (1996). Modeling and laboratory investigation of microbial enhanced oil recovery. *J. Petroleum Sci. Eng.*, Vol. 15, pp. (309–320).

Dyke, M.I., Lee, H., & Trevors, J.T. (1991), Applications of microbial surfactants. *Biotech. Adv.*, Vol. 9, pp. (241–252).

Feng, Q., Zhou, J., Chen, Z., Wang, X., Ni, F., & Yang, H. (2002). Study on EOR mechanisms by microbial flooding. In: 26th Annual SPE International Technical Conference and Exhibition on Abuja, Nigeria.

Fox, S.L., & Bala, G.A. (2000). Production of surfactant from Bacillus subtilis ATCC 21332 using potato substrates. *Bioresour. Technol.*, Vol. 75, pp. (235–240).

Gray, M.R., Yeung, A., Foght, J.M., & Yarranton, H.W. (2008). Potential microbial enhanced oil recovery processes: a critical analysis. In: SPE 114676 at the 2008 Annual Technical Conference and Exhibition held in Denver, Colorado, USA.

Green, D.W., & Willhite, G.P. (1998). Enhanced Oil Recovery. 6. SPE Textbook Series.

Islam, M. (1990). Mathematical modeling of microbial enhanced oil recovery. In: 65th Annual Technical Conference and Exhibition of the Society of Petroleum Engineers on September 23-26, New Orleans, LA.

Jack, T.R., & Diblasio, E. (1985). Selective Plugging for Heavy Oil Recovery. *Int. Biores. J.*, Vol. 1, pp. (205-212).

Jenneman, G., Knapp, R., McInerney, M., Menzie, D., & Revus D. (1984). Experimental studies of in-situ microbial enhanced recovery. Soc. *Pet. Eng. J.*, Vol. 24, pp. (33-38).

Jenneman, GE., Moffitt, PD., & Young, GR. (1996). Application of a microbial selective plugging process at the North Burbank Unit: Prepilot tests. SPE Prod Facil., pp. (11-17).

Jenneman, G.E., Knapp, R.M., Mclnerney, M.J., Menzie, D.E., & Revus, D.E. (1984). Experimental studies of in situ microbial enhanced oil recovery, Soc. *Petr. Engin. J.*, pp. (33-37).

Kashefi, K., & Lovley, D.R. (2003). Extending the upper temperature limit for life. Science, pp. (301-934).

Khire, J.M., & Khan, M.I. (1994). Microbially enhanced oil recovery (MEOR) Part 2: Microbes and subsurface environment for MEOR. *Enz. Microb. Technol.*, Vol. 16, pp. (258-259).

Kianipey, S.A., & Donaldson E C. (1986). 61st Annual Technical Conference and Exhibition, New Orleans, LA.

Kleppe, TS. (2001). Enhanced oil recovery—an analysis of the potential for EOR from known fields in the United States 1976-2000. USA: National Petroleum Council Report.

Lake, L.W. (1989). Enhanced Oil Recovery. Prentice-Hall Inc, Englewood Cliffs, NJ, USA.

Lake, L.W. (1989). Enhanced oil recovery. Peactice Hall, Englewood cliffs, New jerky,

Lazar, I. (1991). MEOR field trials carried out over the world during the past 35 years. In: Microbial Enhancement of Oil Recovery. Recent Adv., Donaldson, E. C. (Ed.). Amsterdam: Elsevier Science, pp. (485-530).

Lazar, I. (1996). Microbial systems for enhancement of oil recovery used in Romanian oil fields. In: Mineral Proc. *Extractive Metal. Rev.*, Vol. 19, pp. (379-393).

Lazar, I. (1997). International and Romanian experience in using the suitable microbial systems for residual oil release from porous media. *Annual Sci. Session Institute Bio., Bucharest*, pp. (225-234).

Lazar, I. (1998). International MEOR applications for marginal wells. *Pakistan J. Hydrocarbon Res.*, Vol. 10, pp. (11-30).

Lazar, I., Petrisor, I.G., & Yen, T.F. (2007). Microbial Enhanced Oil Recovery (MEOR). *Petrol. SciTechnol.*, Vol. 25, No. 11, pp. (1353-1366).

Li, Q., Kanga, C., Wang, H., Liu, Ch., & Zhang, Ch. (2002). Application of microbial enhanced oil recovery technique to Daqing Oilfield. Biochem. Eng. J., Vol. 11, pp. (197-199).

Marshall, S.L. (2008). Fundamental aspects of microbial enhanced oil recovery: A Literature Survey, CSIRO Land and Water Floreat, Western Australia, pp. (1-42).

Morita, T., Konishi, M., Fukuoka, T., Imura, T., & Kitamoto T. (2007). Microbial conversion of glycerol into glycolipid biosurfactants, mannosylerythritol lipids, by a basidiomycete yeast *Pseudozyma antarctica* JCM 10317. *J. Biosci. Bioeng.*, Vol. 104, pp. (78–81).

Moses, V. (1991). MEOR in the field: why so little? Microbial Enhancement of Oil Recovery. *Recent Adv.*, pp. (21–28).

Nielsen, S.M., Shapiro, A.A., Michelsen, M.L., & Stenby, E.H. (2010). 1D Simulations for Microbial Enhanced Oil Recovery with Metabolite Partitioning. *Transp. Porous Med.*, Vol. 85, pp. (785–802).

Portwood, JT. (1995). A commercial microbial enhanced oil recovery technology: evaluation of 322 projects. SPE J., Paper no. 29518.

Ramkrishna S. (2008). Biotechnology in petroleum recovery: The microbial EOR. Prog. Energy Comb. Sci., Vol. 34, pp. (714– 724).

Sarkar, A., Georgiou, G., & Sharma M. (1994). Transport of bacteria in porous media: II. A model for convective transport and growth. *Biotechnol. Bioeng.*, Vol. 44, pp. (499–508).

Sarker, A.K., Goursaud, J.C., Sharma, M.M., & Georgiou, G. (1989). A critical evaluation of MEOR processes. In Situ 13, pp. (207–238)

Sheehy, A. J. (1991). Microbial physiology and enhancement of oil recovery recent advances. In: Develop. Petrol. Sci., 31, Donaldson, E. C. (Ed.). Amsterdam: Elsevier, pp. (37–44).

Sheehy, J. A. (1992). Recovery of Oil from Oil Reservoirs. U.S. Patent No. 5.083.610.

Singer, M.E., & Finnerty, W.R. (1984). Microbial metabolism of straight and branched alkanes. Petrol. Microbiol., ecl. R. Atlas., Collier MacMillan, New York, pp. (1-59).

Stevik, Tor K., Aa, K., Ausland, G. & Fredrik, H.J. (2004). Retention and removal of pathogenic bacteria in wastewater percolating through porous media: a review. *Water Res.*, Vol. 38, pp. (1355-1367).

Suthar, H., Hingurao, K., Desai, A., & Nerurkar, A. (2008). Evaluation of bioemulsifier mediated Microbial Enhanced Oil Recovery using sand pack column. *J. Microbiol. Methods*, Vol. 75, pp. (225–230).

Tango, M.S.A., & Islam, M.R. (2002). Potential of extremophiles for biotechnological and petroleum applications. *Energy Sources*, Vol. 24, pp. (543–59).

Van, H.J.D., Singh, A., & Ward, O.P. (2003). Recent advances in petroleum microbiology. *Microbiol Mol Biol Rev.*, Vol. 67, No. 4, pp. (503–549).

Xu, T., Chen, Ch., Liu, Ch., Zhang, Sh., Wu, Y., & Zhang, P. (2009). A novel way to enhance the oil recovery ratio by Streptococcus sp. BT-003. *J. Basic Microbiol.*, Vol. 49, pp. (477-481).

Xu, Y., & Lu, M. (2011). Microbially enhanced oil recovery at simulated reservoir conditions by use of engineered bacteria. *J. Petrol. Sci. Eng.*, Vol. 78, pp. (233–238).

Yakinov, M.M., Timmis, K.N. Wray, V., & Fredrickson, H.L. (1995). Characterization of a new lipopeptide surfactant produced by thermo tolerant and halotolerant subsurface Bacillus licheniformis BA50. *Appl. Environ. Microbiol.*,Vol. 61, pp. (1706–1713).

Yarbrough, H. F., & Coty, F. V. (1983). Microbially enhancement oil recovery from the Upper Cretaceous Nacafoch formation Union County, Arkansas. Proceedings of

1982 InternationalConference on MEOR, Donaldson, E. C. and Benett Clark, J. B. (Eds.), Afton, Oklahoma, pp. (149-153).

Youssef, N., Simpson, DR., Duncan, KE., McInerney, MJ., Folmsbee, M., & Fincher T. (2007). In situ biosurfactant production by *Bacillus* strains injected into a limestone petroleum reservoir. *Appl. Environ. Microbiol.*, Vol. 73, pp. (1239-1247).

Zobell, C.E. (1947). Bacterial release of oil from sedimentary materials. *Oil Gas J.*, Vol. 2: pp. (62-65).

4

Enhanced Oil Recovery in Fractured Reservoirs

Martin A. Fernø
Department of Physics and Technology, University of Bergen
Norway

1. Introduction

In this chapter oil recovery mechanisms in fractured reservoirs will be reviewed and discussed. Most attention will be devoted to experimental studies on fluid flow in fractured reservoirs and imaging techniques to visualize fluid flow in-situ. Special focus will be on complementary imaging in the laboratory, where important processes in fractured reservoirs are studied at different length scales over a range of 5 orders of magnitude. A solid understanding of the flow functions governing fluid flow in fractured reservoirs provides the necessary foundation for upscaling laboratory results to the field scale using numerical simulators. The fact that numerical models and reservoir simulators are based on observations from hydrocarbon producing field and laboratory tests demonstrates the need to study the same process at different length scales. It also illustrates the close link between experimental and numerical efforts and the need for interdisciplinary knowledge to constantly improve the representation of fractured reservoirs and the predications made.

2. Naturally fractured carbonate reservoirs

Naturally fractured carbonate reservoirs are geological formations characterized by a heterogeneous distribution of porosity and permeability. A common scenario is low porosity and low permeability matrix blocks surrounded by a tortuous, highly permeable fracture network. In this case, the overall fluid flow in the reservoir strongly depends on the flow properties of the fracture network, with the isolated matrix blocks acting as the hydrocarbon storage. Most reservoir rocks are to some extent fractured, but the fractures have in many cases insignificant effect on fluid flow performance and may be ignored. In naturally fractured reservoirs, defined as reservoirs where the fractures have a significant impact on performance and oil recovery, fracture properties should be evaluated because they control the efficiency of oil production. Fractures are usually caused by brittle failure induced by geological features such as folding, faulting, weathering and release of lithostatic (overburden) pressure (Miller, 2007). Fractured reservoirs may be divided into categories characterized by the relationship between matrix and fracture properties such as permeability and porosity. Allen and Sun, 2003 defined four categories of fractured reservoirs based on the ratio between permeability and porosity in their comprehensive study of fractured reservoirs in the US as follows.

- Type I - little to no porosity and permeability in the matrix. The interconnected fracture network constitutes the hydrocarbon storage and controls the fluid flow to producing well.

- Type II - low matrix porosity and permeability. Some of the hydrocarbons are stored in matrix. Fractures control the fluid flow, and fracture intensity and distribution dictates production.
- Type III - high matrix porosity and low matrix permeability. Majority of the hydrocarbons are stored in matrix. Matrix provides storage capacity, the fracture network transport hydrocarbons to producing wells.
- Type IV - high matrix porosity and permeability. The effects of the fracture network are less significant on fluid flow. In this type category reservoir fractures enhance permeability instead of dictating fluid flow.

The four types of fractured reservoir defined above honors the geological features related to hydrocarbon storage and the relationship between permeability and porosity. Furthermore, the production characteristics of fractured reservoirs differ from conventional reservoirs in many fundamental ways. Some of the most pronoun differences are listed below (Allen and Sun, 2003).

- Due to high transmissibility of fluids in the fracture network, the pressure drop around a producing well is lower than in conventional reservoirs, and pressure drop does not play as important role in production from fractured reservoirs. Production is governed by the fracture/matrix interaction.
- The GOR (gas-oil ratio) in fractured reservoirs generally remains lower than conventional reservoirs, if the field is produced optimally. The high permeability in the vertical fractures will lead the liberated gas towards the top of the reservoir in contrast to towards producing well in conventional reservoirs. This is to some degree sensitive to fracture spacing and orientation and the position of producers. Liberated gas will form a secondary gas gap at the top of reservoir or will expand the existing cap.
- Fractured reservoirs generally lack transition zones. The oil-water and oil-gas contacts are sharp contrasts prior to and during production. The high fracture permeability allows the rapid re-equilibration of the fluid contacts.

In the following section the main focus will be on the production of hydrocarbons from fractured reservoirs, more specifically production of oil from fractured carbonate reservoirs.

3. Recovery mechanism in fractured reservoirs

The presence of fractures dramatically influences the flow of fluids in a reservoir because of the large contrast in transmissibility between the fracture and the matrix. High permeable fractures carry most of the flow, and therefore limit the buildup of large differential pressures across the reservoir. The limited viscous forces are negative for production during e.g. a waterflood, where most of the water flows in the fracture network only and does not displace oil from the matrix blocks, leading to poor sweep efficiency and low recoveries. In this scenario, the recovery mechanism is capillary imbibition rather than viscous displacement. Counter-current spontaneous imbibition, where water in the fracture spontaneously enters a water-wet rock and oil is displaced in the opposite direction, is a key recovery mechanism in fractured reservoirs during waterflooding. The amount of water imbibed from the fracture network depends on the capillary pressure curve, which is closely correlated to the pore structure and wettability preference of the rock surface. The amount of spontaneously imbibed water into an oil saturated rock is ultimately controlled by the capillary pressure curve, or more accurately, the positive part of the imbibition capillary pressure curve. The shape and range of the positive capillary pressure curve is dictated by the wettability.

3.1 Scaling laws and shape factor

The performance of a waterflood in a given field may be tested in the laboratory in a spontaneous imbibition test. A standard imbibition test where an oil saturated rock sample is immerged in brine and the production of oil is measured as a function of time has been used to estimate production by spontaneous imbibition in fractured reservoirs. A method to upscale the laboratory imbibition curves to the production of oil from isolated reservoir matrix blocks, with various sizes and shapes, has been studied extensively. Several scaling groups to readily use laboratory results to estimate reservoir behavior are proposed. Aronofsky *et al.*, 1958 proposed an exponential form of the matrix-fracture transfer function, and formulated an important usage of scaling groups to increase computational efficiency of simulators by several orders of magnitudes. The time to complete a simulation of oil recovery from a fractured reservoir decreased dramatically when the transfer functions for fluid flow between fracture and matrix contained the rate of imbibition with scaled dimensionless time. The use of shape factors (Barrenblatt *et al.*, 1960) related to the geometric shape of the matrix block (Kazemi *et al.*, 1976; Zimmerman and Bodvarsson, 1990) was introduced to increase efficiency of numerical simulations. Experimental evidence for the validity of the shape factor was presented by Mattax and Kyte, 1962 with a dimensionless group to scale the imbibition behavior from matrix block of the same rock type with different geometries. According to Morrow and Mason, 2001, the application of the scaling group by Mattax and Kyte are subject to the following six conditions:

- gravity effects may be neglected
- sample shapes and boundary conditions must be identical
- oil/water viscosity is duplicated
- initial fluid distributions are duplicated
- the relative permeability functions must be the same
- capillary pressure functions must be directly proportional

The effect of viscosity was included by Ma *et al.*, 1997, who experimentally showed that the rate of spontaneous imbibition was proportional to the square root of the viscosity ratio for systems with similar geometry. They introduced a new definition of characteristic length, and used the geometric mean of viscosities to modify the dimensionless scaling group proposed by Mattax and Kyte, 1962, defined as:

$$t_D = t\left(\sqrt{\frac{k}{\Phi}}\frac{\sigma}{\mu_{gm}}\frac{1}{L_C^2}\right) \quad (1)$$

where t_D is the dimensionless time, k is the permeability, Φ is the porosity, σ is the interfacial tension between the wetting and non-wetting phases, t is imbibition time, μ_{gm} is the geometric mean of the viscosities and L_C is the characteristic length defined as:

$$L_C = \sqrt{\frac{V}{\sum_{i=1}^{n}\frac{A_i}{X_{A_i}}}} \quad (2)$$

where V is the bulk volume of the matrix, A_i is the area open to imbibition at the ith direction and X_{A_i} is the distance traveled by the imbibition front from the open surface to the no-flow boundary. A limitation to this scaling group is that its validity was only tested for strongly water-wet systems.

3.2 The influence of wettability

Understanding the wettability effect on the spontaneous imbibition process during waterflooding in fractured reservoirs is crucial because most of the worlds known oil reservoirs are not strongly water-wet. Clean carbonate rocks (and sandstones) are naturally water-wet, even though most reservoir rocks show some oil-wet characteristics. Enhanced oil recovery techniques successfully implemented in water-wet reservoirs may not necessarily perform as well in oil-wet reservoirs as the waterflood performance is strongly dependent on the wettability of the reservoir. In an oil-wet, fractured reservoir, water will not spontaneously displace oil from the matrix, and only the oil in the fractures will be displaced, resulting in poor recoveries and early water breakthrough. In water-wet fractured reservoirs, imbibition can lead to significant recoveries. The recovery of oil from fractured reservoirs is controlled by the interaction between brine/oil/rock interaction, which again depends on the wetting and two-phase flow, the chemical and physical properties of all of the three components, fracture geometry and pore structure of the matrix (Morrow and Mason, 2001).

The impact of matrix wettability on imbibition potential is well-known and the rate of spontaneous imbibition is highly sensitive to wettability (Zhou *et al.*, 2000). A general scaling law that included the effect of wettability was proposed by Li and Horne, 2002 to predict the oil recovery by water injection in fractured reservoirs. Morrow and Mason, 2001 cautioned against implementing the wettability effect in scaling laws directly, and pointed out the difficulty to scale the rate of imbibition even in simple systems like a cylindrical tube where the issue of static vs. dynamic angles arises. Although it is appealing to represent the wettability in terms of the cosine to the contact angle and apply this directly in the scaling law, the assignment of a single effective average contact angle is not physically correct for systems where there is a distribution of contact angles (Jackson *et al.*, 2003, Behbahani and Blunt, 2005).

The impact of wettability on oil recovery by waterflooding was demonstrated by Jadhunandan and Morrow, 1995, who found a maximum recovery at moderately water-wet conditions, with an Amott-Harvey index (Amott, 1959) of $I_{w-o} = 0.2$, in Berea sandstone core plugs. Zhou *et al.*, 1995 also reported the highest recovery by long-term spontaneous imbibition (~50 days) in moderately water-wet Berea core plugs. Additional evidence for the importance of wettability on spontaneous imbibition was provided by Johannesen *et al.*, 2006 that reported a similar trend in chalk. In this crude oil/brine/rock type system the maximum recovery was shifted to a wettability index of $I_{w-o} = 0.4$ measured by the Amott test.

To understand the physics behind the wettability impact, Behbahani and Blunt, 2005 used pore-scale modeling to explain the decrease in imbibition recovery and production rates with increasing aging times reported by Zhou *et al., 2000*. They used a topologically equivalent Berea network model and adjusted the distribution of contact angles at the pore scale. They found that the increase in imbibition time for mixed-wet samples was a result of very low water relative permeability caused by low connectivity of water at intermediate saturation demonstrating that pore-scale modeling is a useful tool to understand the physics involved with e.g. spontaneous imbibition in mixed-wet cores. The aim of pore-scale modeling is to predict properties that are difficult to measure, such as relative permeability, from more readily available data, such as drainage capillary pressure (Valvatne *et al.*, 2005).

Additional areas where the use of pore-scale modeling is beneficial was demonstrated by Jackson *et al.*, 2003 investigated the effect of wettability variations on flow at the reservoir scale using a pore-scale network model in conjunction with conventional field-scale reservoir simulators. They successfully predicted experimental relative permeability and waterflood recovery data for water-wet and mixed-wet Berea sandstone, and found that the traditionally used empirical models for predicting hysteresis in the transient zone above the oil-water-contact (OWC) were insufficient if wettability varied with height. A significant increase in oil recovery using scanning curves generated by pore-scale network rather than the empirical models was also demonstrated.

4. Complementary imaging techniques

Designing an enhanced oil recovery project is only possible when the governing processes that control fluid flow and transport within a multi compositional oil reservoir is fully understood. Performing advanced experiments that shed light on these processes is vital for validation of numerical simulators and theoretical calculations needed to evaluate the project. An important objective for experimental reservoir physics is to contribute to numerical simulation of fluid flow within a petroleum reservoir. The governing physics that describe fluid interface interaction and flow on the pore-scale (10^{-3}m) is used as input in numerical simulators to predict flow on the reservoir scale (10^{3}m). However, the parameters observed to be important on the pore-scale may or may not be dominant on the field scale. By experimentally investigating the same problem at increasing length scales one is well equipped to decide how to upscale from the pore-scale to the length scale of a grid block in a numerical simulation of e.g. a waterflood in an oil reservoir. Figure 1 illustrates this approach and demonstrates the importance of a strong link and interaction between numerical simulations and experiments at different length scales to fully understand the process.

The key issue for simulating flow in fractured rocks is to model the fracture-matrix interaction correctly under conditions such as multiphase flow. Combining visualization tools with different spatial resolutions and size capabilities may dramatically increase the overall understanding of the studied phenomena. The use of three visualization techniques is described below, and demonstrates the increased knowledge when applying complementary imaging.

4.1 The micro scale

Direct visualization of fluid flow and displacement mechanisms on the pore-scale is possible by using mircromodels. Micromodels are porous structures in two dimensions that are based on a real rock and are normally made of glass. The micromodel used here was etched silicone micromodels designed and manufactured by Dept. of Energy Resources

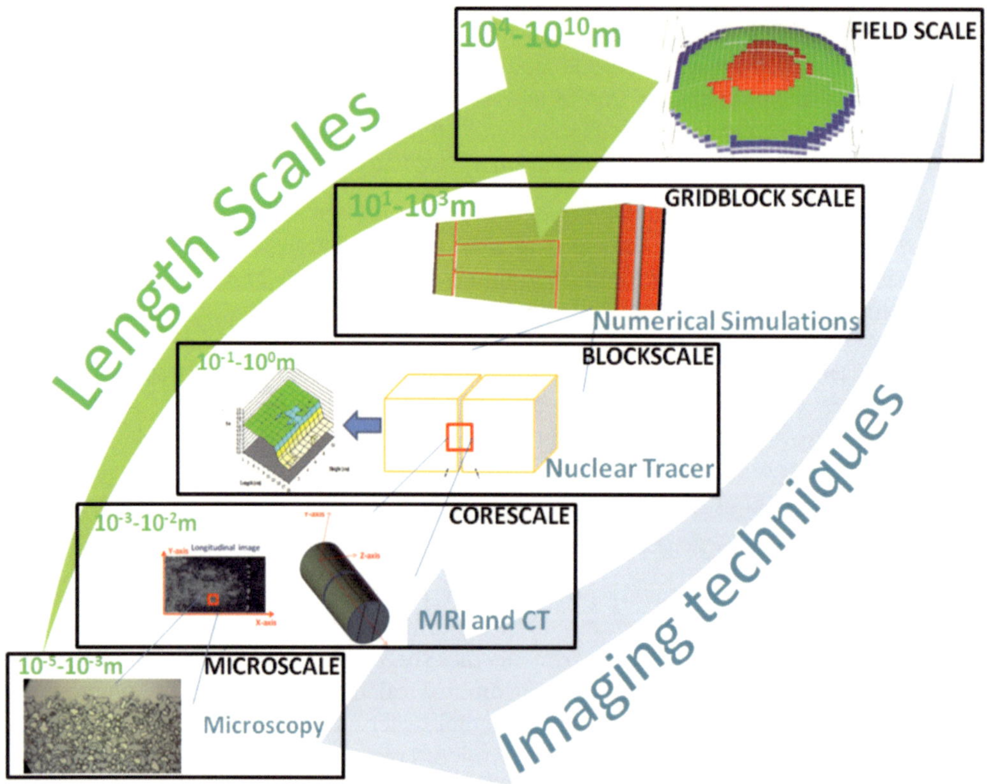

Fig. 1. Complementary imaging at different length scales.

Engineering, Stanford University. The structure of the micromodel was a 1:1 realization of the pore size and pore shapes based on a Berea sandstone thin section. The structure was etched in silicon for increased control of etching depth and accurate reproduction of fine-scale details that result in sharp, unrounded corners (Kovscek *et al.*, 2007). Average channel depth was 25μm and sand grain features ranged between 50-300μm. The total area of the micromodel was 50 x 50 mm^2, representing approximately 600 x 600 pores (Buchgraber *et al.*, 2011). The absolute permeability was 950mD and the porosity was 0.47, resulting in a pore volume (PV) of 0.013ml. For further detail on fabrication process the reader is referred to Hornbrook *et al.*, 1991.

4.2 The core scale

Magnetic Resonance Imaging is a versatile visualization tool frequently used in hospitals and medical applications. The applications of MRI for characterizing core samples and flow properties of porous materials are also known and a considerable amount of literature has been published on the topic. MRI provides a visualization tool to study the movements of fluids inside a fracture network, and provides high spatial resolution and fast data acquisition. Previous work have discussed MRI imaging for core characterization purposes (Baldwin and Spinler, 1998; Baldwin and King, 1999), the monitoring of imbibition and displacement processes (Baldwin, 1999) and the application of MRI to the study of formation damage. Recent studies have reported flow behavior and production mechanisms in fractured chalk and their dependency of wettability, rate and fracture aperture and fracture configurations (Graue *et al.*, 2001;Aspenes *et al.*, 2002;Aspenes *et al.*, 2008).

4.3 The block scale

The nuclear tracer imaging (NTI) technique was developed by Bailey *et al.*, 1981 and improved by Graue *et al.*, 1990, and utilizes the emitted radiation from radioactive isotopes, individually labeling the fluids to measure the *in situ* fluid saturation profiles during core flood tests. The large dimensions of the rock samples imaged is a strong advantage of this method, for instance enabling the simultaneous study of the impacts from viscous, capillary and gravity forces in a controlled fractured system. One of the advantages of using nuclear tracers is its inertness nature with respect to the delicate network of pores, assuming that adsorption is minimized by preflushing with non-radiation brine. The possibility to perform multiple experiments on the same rock sample allows for experimental reproduction and investigation of impacts on flow- and recovery mechanisms from a single parameter (e.g. injection rate, wettability or fracture). Also, the NTI technique has the capability of imaging the 1D *in-situ* oil production in cores up to 2 m in length, thus minimizing the disturbance from capillary end effects, and enabling large scale gravity drainage experiments with local saturation measurements. Details on fluid saturation calculations and experimental procedures are found in Ersland *et al.*, 2010.

5. Complementary imaging in fractured reservoirs

The production of oil is challenging in fractured reservoirs due to the large transmissibility contrast between matrix and fracture, and primary recovery is often low. The recovery efficiency depends on the relationship between the fracture and matrix permeabilities, and is strongly dependent on the wettability of the matrix, which reflects the imbibition potential of the reservoir. High demands and rising oil prices has increased focus on improved oil recovery from large, low recovery oil fields. Some of the world's largest remaining oil reserves are found in oil-wet, fractured, carbonate reservoirs. The understanding of multiphase fluid flow in oil-wet fractured reservoirs has been studied, especially the influence of capillary pressure. The presence of capillary pressure is important in recovery mechanisms like spontaneous imbibition, waterflooding, and gravity drainage. Complementary imaging techniques used to study Enhanced Oil Recovery (EOR) processes in fractured oil reservoirs have provided new and improved fundamental understanding waterflood oil recovery. MRI provides high spatial resolution and fast data acquisition necessary to capture fluid displacements that occur inside fractures less than 1 mm wide, whereas NTI provides information on macro-scale saturation distribution in larger fractured systems.

The oil recovery mechanisms involved with waterflooding fractured chalk blocks were found to be dependent on the wettability of the chalk, as the wettability had great impact on the fracture/matrix hydrocarbon exchange. The wettability was altered by dynamic aging (Fernø *et al.*, 2010) in a North Sea crude oil. After aging, the crude oil was displaced from the core at elevated temperature by injecting 5PV of decahydronaphthalene followed by 5PV decane to avoid asphaltene precipitation, to stop the aging and to establish more reproducible experimental conditions by using decane as the oleic phase throughout the experiments (Graue *et al.*, 1996;Graue *et al.*, 1998,1999a;Graue *et al.*, 1999b). The wettability after aging was measured with the Amott method (Amott, 1959).

Increased oil recovery during waterflooding was reported by Viksund *et al.* (1999) in fractured chalk. They used the NTI method to observe that fractures significantly affected water movement during waterfloods at strongly water-wet conditions, whereas the fractures had less impact on the waterfront movement at moderately water-wet conditions, where the injected water crossed fractures more uniformly, apparently through capillary contacts. The MRI images of oil saturation development inside the fractures (Graue *et al.*, 2001) demonstrated transport mechanisms for the wetting phase across the fracture at several wettability conditions, and provided new and detailed information on fluid fracture crossing previously observed in block scale experiments investigated by NTI. The high spatial resolution MRI images revealed water droplets forming on the fracture surface at moderately water-wet conditions, transporting water across the fracture.

A plausible explanation to the increased recovery above the spontaneous imbibition potential in fractured reservoirs during waterflooding was an added viscous pressure drop exerted by the wetting phase bridges across isolated matrix blocks surrounded by fractures.

5.1 Wettability effects in core plugs

Mechanistically similar to water droplets forming during waterfloods in water-wet chalk, oil was injected in oil-wet limestone cores to observe the forming of oil droplets on the surface (Fernø *et al.*, 2011). Stacked core plugs separated by 1 mm space were waterflooded, and the *in-situ* development in oil and water saturations was monitored with MRI. Figure 2 shows the wettability effect on the forming of oil droplets on the fracture surface during a waterflood in a stacked core system consisting of an inlet core, a 1mm space between the core plugs, and an outlet core plug. The space between the core plugs constitutes the fracture. Oil appears bright on the images, whereas a reduction in signal indicates increased water saturation. Each image represents a snapshot in time of the fracture, and time increases from left to right.

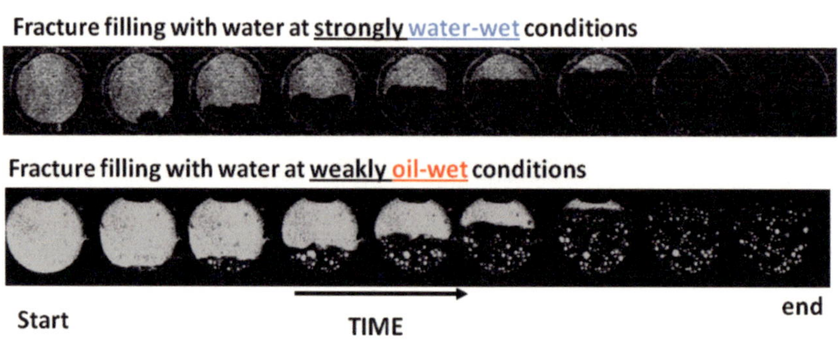

Fig. 2. The effect of wettability on the fracture-matrix fluid transfer during waterfloods in stacked core systems.

Water was injected into two stacked systems with different wettabilities. The fracture was initially oil filled, and water was injected in the core plug upstream of the fracture. Water breakthrough from the inlet plug to the fracture was observed at the bottom of the fracture.

Water displaced oil upwards and into the outlet core plug to the producer because there was no alternative escape path for the oil phase. At strongly water-wet conditions, the water filled the fracture from the bottom to the top, displacing oil in a horizontal oil-water interface. No residual oil was observed in the fracture after the water filled the fracture and invaded the outlet core plug. At weakly oil-wet conditions, water displaced the oil from the fracture similarly to strongly water-wet conditions, with a horizontal oil-water interface moving upwards. However, unlike strongly water-wet conditions, droplets of oil remained on the fracture surface after the water advanced into the outlet plug.

The locations where the droplets emerged at the fracture surface were not arbitrary, and were likely to be controlled by clusters of larger pores throats. The larger pore throats will reach zero capillary pressure before pores with narrower throats and, hence, the droplets will emerge at these locations. In the conceptual model proposed by Gautam and Mohanty (2004), the oil droplets emerged at the matrix-fracture interface in clusters of large pore throats, and, in accordance with the experimental observations made by Rangel-German and Kovscek (2006), the water transport from fracture to the matrix occurred via narrow pore throats in the vicinity of the oil producing locations. The oil droplet growth process was intermittent, i.e. a blob starts to grow, is displaced by a new blob that starts to grow at the same location and detaches from the fracture surface. Oil droplets forming on the fracture surface during displacements may be an important recovery mechanism if they bridge the fracture to create liquid bridges for oil transport and reduce the capillary retained oil in the hold-up zone during gravity drainage in oil-wet, fractured reservoirs (Fernø, 2008).

Oil was also injected through stacked strongly water-wet limestone core plugs to investigate the impact from wettability and the significance of wetting affinity between injected fluid and fracture surface on the forming of oil droplets. No oil droplets were observed in the fracture during oil injections at water-wet conditions. The hydraulic pressure in the fracture was measured during both waterfloods and oilfloods at oil-wet conditions, and the pressure development demonstrates the mechanistic difference when there is a wetting affinity between the fracture surface and the injecting fluid phase. During waterfloods, the fracture hydraulic pressure demonstrated the need to overcome a threshold value before the water invaded the outlet core, whereas no pressure increase was observed during oil injection, demonstrating the spontaneous nature of the transport of oil.

5.2 Wettability effects in block samples

The impact of wettability on the larger scale was also experimentally investigated using the NTI method and the MRI during waterfloods in fractured systems with different wettabilities. The presence of fractures dramatically changed the oil recovery, flow dynamics, and displacement processes. Three wettability conditions were tested: strongly water-wet, weakly water-wet, and weakly oil-wet. In combination with volumetric production data, the *in-situ* fluid saturation data provided information about the flow pattern with the presence of fractures to better understand the recovery mechanisms (Haugen *et al.*, 2010b). Figure 3 demonstrate the difference in waterflood behavior for three wettabilities and list the main recovery mechanisms and implications in the fractured systems. Large fractured block samples (LxWxH= 16x3x9 cm^3) were waterflooded to study the effect of wettability on the waterflood behavior. Warmer colors indicate higher oil saturations.

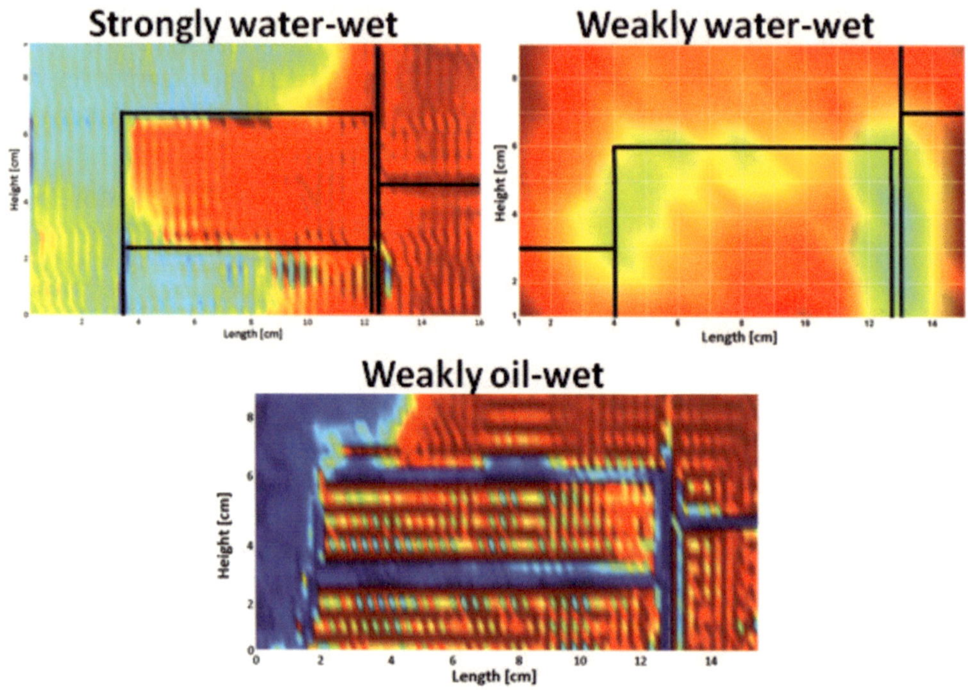

Fig. 3. The effect of wettability on the fracture-matrix fluid transfer during waterfloods in fractured block systems. Water is injected at the boundary on the left, and the direction of flow is from left to right. Fractures are indicated in black where not directly observed.

5.2.1 Strongly water-wet

In strongly water-wet systems containing several disconnected matrix blocks the fractures were barriers to flow. The displacement process was governed by capillary forces, corroborated by the visualized displacement pattern with a block-by-block displacement attributed to discontinuity in the capillary pressure curve at the matrix-fracture interface. The positive capillary pressure in the matrix block during oil displacement trapped the water phase in the matrix block until the capillary pressure at both sides of the matrix-fracture interface were equal. This occurred at the end of spontaneous imbibition, when Pc=0 in the matrix block. Water in the fracture network may imbibe into the adjacent downstream matrix block. Consequently, the fluid flow dynamics were strongly influenced by the presence of fractures at strongly water-wet conditions, but residual oil saturation and recovery were similar to the waterflood without fractures present. Waterflood recovery without fractures was R_F=45%OOIP compared with R_F=42%OOIP with fractures. The displacement pattern was in most cases determined by the location of the water in the fractures, and due to gravity segregation the lower blocks closer to the inlet imbibed water first. The combination of high capillary imbibition and the applied water injection rate led to the *filling regime* (Rangel-German and Kovscek, 2002) and co-current imbibition.

5.2.2 Weakly water-wet

At weakly water-wet conditions oil was recovered by capillary imbibition of water from the fractures to the matrix blocks. The interconnected fracture network limited the viscous forces in the system as the injector and producer was in direct contact, see Figure 3. The mechanism for oil recovery was purely capillary imbibition of water from the fracture. The strength of the capillary pressure to transport water from the fracture network to the matrix was reduced compared with strongly water-wet condition. The weaker capillary forces led to a reduction in matrix-fracture transfer rate and increased the likelihood of *instantly filled fracture* regime (Rangel-German and Kovscek, 2002), with counter-current imbibition of water. The presence of fractures dramatically reduced the oil recovery. Waterflood oil recovery without fractures was R_F=63%OOIP compared with R_F=22%OOIP with the presence of fractures. This demonstrates the potential to increase oil recovery in fractured reservoirs by increasing the viscous forces.

5.2.3 Weakly oil-wet

The fracture system for the weakly oil-wet block was similar to the strongly water-wet block (see Figure 3). Waterfloods were performed both before and after fractures were present, injecting water with the same rate (2 cc/hr) in both cases. The waterflood oil recovery at fractured state was R_F=15%OOIP, compared with R_F=65%OOIP without the presence of fractures. In addition to a strong reduction in recovery, the fractures changed the time of water breakthrough to the producer. Water breakthrough without fractured was observed after 0.47PV injected, in contrast to only 0.10PV injected with fractures. The oil-wet wettability preference of the matrix suppressed oil recovery by capillary imbibition of water from the fracture network. Recovery was dictated by the ability to generate a differential pressure across the system. Displacement of oil from matrix only took place in the un-fractured inlet block, once water entered the fracture network it rapidly filled the fractures, first the lower horizontal fracture, then the remaining fracture network before it reached the outlet. No transport of water from fracture to matrix was observed.

6. Enhanced oil recovery in fractured reservoirs

Oil production from fractured oil reservoirs poses great challenges to the oil industry, particularly because fractures may exhibit permeabilities that are several orders of magnitude higher than the permeability of the rock matrix. Low viscosity fluids used for enhanced oil recovery, such as gases or supercritical fluids may channel into the high-permeable fractures, potentially leading to early breakthrough into the production well and low sweep efficiency. Carbonate reservoirs usually exhibit low porosity and may be extensively fractured. The oil-wet nature of the matrix reduces capillary imbibition of water. Carbonate reservoirs contributes substantially to US oil reserves (Manrique *et al.*, 2007), and the low primary recovery and the large number of carbonate reservoirs in the US and around the world makes them good targets for EOR efforts.

Foam has the potential to increase oil recovery by improving areal sweep, better vertical sweep (less gravity override), less viscous fingering, and diversion of gas away from higher-permeable or previously swept layers (Bernard and Holm, 1964, Holm, 1968, Hanssen *et al.*, 1994, Schramm, 1994; Rossen, 1995). Diversion of gas into lower-permeable layers using foams have previously been reported (e.g. Casteel and Djabbarah, 1988, Llave *et al.*, 1990, Zerhboub *et al.*, 1994, Nguyen *et al.*, 2003 *et al.*). This may be important for fractured systems, where a very large permeability contrast exists and cross-flow between the zones occur (Bertin *et al.*, 1999).

The application of foam to enhance oil recovery was studied at different length scales in fractured systems with complementary imaging techniques. Experiments were performed on the micro scale, core plug scale, and the block scale to study the use of foam in a fractured system to improve oil recovery. At the *micro scale* the mechanism for gas and liquid transport from the fracture to the matrix was investigated. Liquid snap off at the pore throat was observed as a mechanism for foam generation within the matrix, whereas large pressure fluctuations along the fracture lead to foam invasion from the fracture to the matrix. At the *core scale* the added oil recovery during foam injection compared to gas injection was demonstrated in a fracture core plug with the presence of oil in the matrix. The increased pressure drop across the fracture contributed to fluid transport from the fracture to the matrix and displacement of oil. The wettability of the core plug was weakly oil-wet reflecting the reservoir wetting preference in several large carbonate oil fields. The same process was studied in at the *block scale*, where three forces (gravity, viscous, and capillary) were active simultaneously. The injection of pregenerated foam greatly increased recovery by increasing the differential pressure across the system. Foam injection was compared to waterfloods, surfactant injection, pure gas injection and co-injection of surfactant and gas (Haugen *et al.*, 2010a). Experimental results at each length scale are described in detail in the sections below.

6.1 EOR at the micro scale

Foam injection in fractured reservoirs was studied at the micro scale using etched silicon micromodels manufactured at the Department of Earth Sciences, Stanford University (Hornbrook *et al.*, 1991; Buchgraber *et al.*, 2011). Experiments were designed to study the transport of gaseous phase from the fracture into the matrix during foam flow at a pore level. The micromodel represents an actual Berea sandstone pore space with respect to pore size distributions. The model was shaped as a square, with injection ports in each corner. Injection ports were paired and connected with a conduit that constituted a fracture. All injections were performed horizontally, without the influence of gravity. A NIKON microscope, in combination with a cam recorder, was used to visualize and store images of the displacement processes in the fracture and in the matrix during injections. The injected gas phase was N_2, and the foam was pregenerated outside the micromodel by co-injecting gas and a standard 1wt% active AOS surfactant solution in a Berea sandstone core plug. Foam generation in silicon micromodels was previously reported by Kovscek *et al.*, 2007.

Figure 4 shows the displacement process during a pure gas injection (left column) and foam injection (right column) in a fractured reservoir. The fracture was located at the top of the image and in direct contact with a porous structure. Both fracture and matrix were initially fully saturated with surfactant solution. In each image sand grains are white, the aqueous surfactant solution is blue, gas is red, and the interface between the gas and the surfactant is black. A slight backpressure was applied to reduce gas compressibility effects. Total injection rate was 4 cc/hr during both the pure gas injection and during the foam injection. The gas fraction during foam injection was 0.95. The results show that during pure gas injection, the gas only displaced surfactant in the fracture and was not able to penetrate the matrix to displace water. During pregenerated foam injection into the fracture; the foam was able to penetrate the matrix to displace water. The gas pressure in the fracture fluctuated highly as a result of the forming and breaking of gas bubbles in the fracture. The pressure fluctuations controlled the local gas rates in the fracture, as the gas phase switched between stationary to high flow rates continuously. The fluctuating pressure and flow contributed to the amount of gas invaded into the matrix. Surfactant in the matrix resided in the smallest pores and in corners.

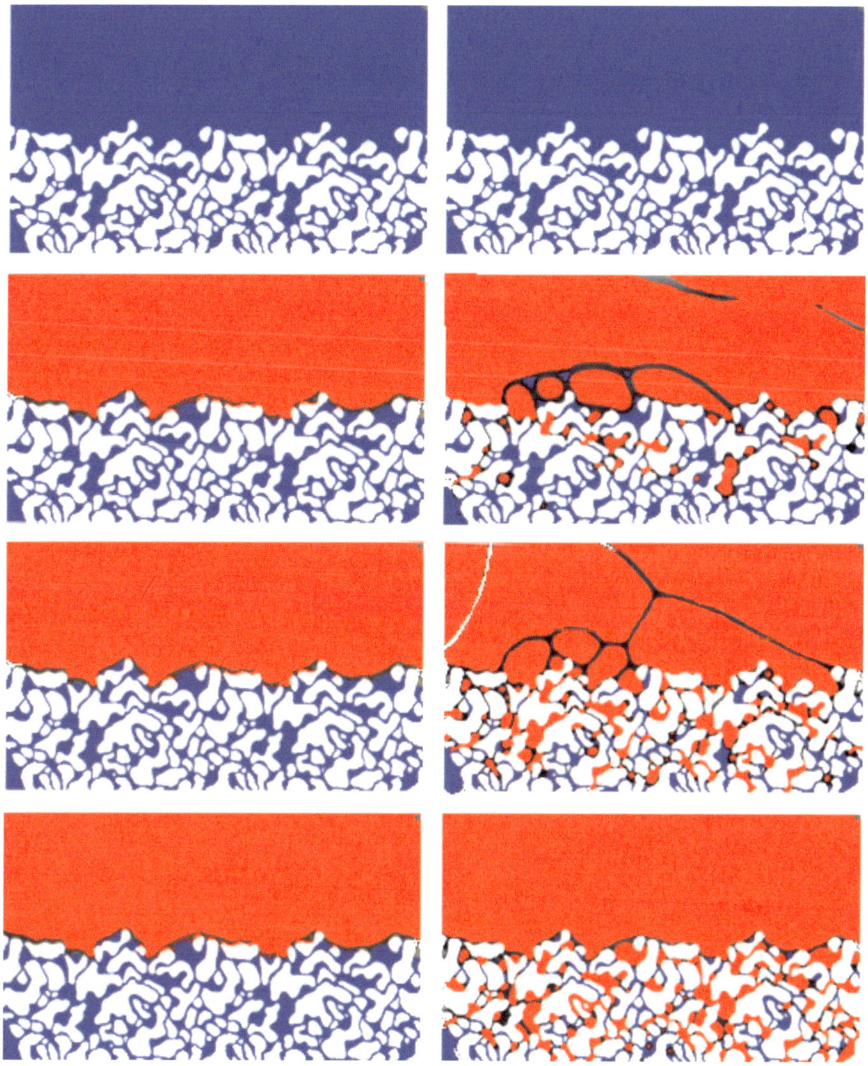

Fig. 4. The effect of foam on the transport of gas from the fracture to the matrix. Left column: pure gas injection with gas flow was only in fracture. Right column: foam injection displacing water in the fracture and the matrix. In each image, sand grains are white, the aqueous surfactant solution is blue, gas is red, and the interface between the gas and the surfactant is black.

6.2 EOR at the core plug scale

Foam experiments in a fractured limestone core plug (length: 7.07 cm, diameter: 4.8cm) was performed to study the use of foam as an EOR technique in fractured reservoirs. The performance of foam was compared to pure gas injection. The wettability of the matrix was weakly oil-wet. The vertical fracture (top to bottom) in the horizontally positioned cylindrical core plugs constituted a rectangular slit. The fracture aperture was not measured, but was calculated from the empirical relationship between aperture and the measured permeability. Because the core plug was cut using a circular saw with a diamond coated blade, the fracture surface was assumed to be smoother than naturally-fractured reservoirs. The two half-cylindrical core plugs were inserted into a core holder and the confinement pressure was increased until a desired fracture aperture or system permeability was established.

Figure 5 shows the oil recovery and differential pressure buildup during the gas and foam injections. The core geometry is seen in the lower right corner, with the fracture plane indicated. Two injections were performed: 1) gasflood with N_2 gas with 30 ml/hr injection rate, and 2) injection of pre- generated foam at 33ml/hr (N_2 gas fraction at 0.91).

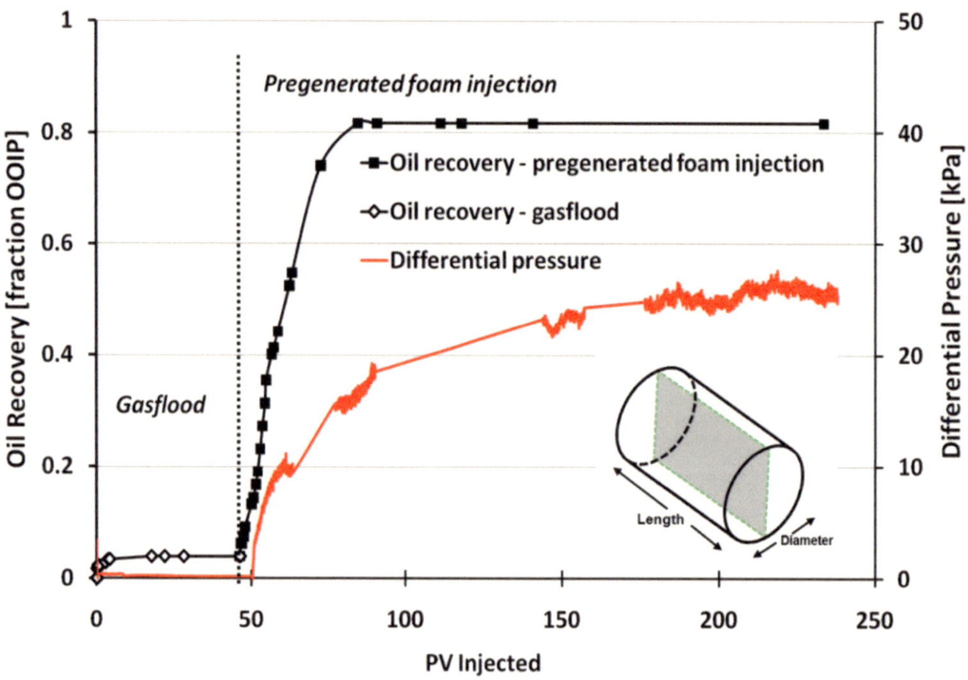

Fig. 5. The effect of foam as an EOR technique in fractured, oil-wet carbonate core plug.

Gasflood oil recovery was R_F=4%OOIP, attributed to the oil in the fracture volume of the core. No pressure build up was recorded as the pressure remained at 0.15kPa during the injection (42hrs, 49PV). Significant increase in oil recovery was observed during the pre-generated foam injection, with R_F=77%OOIP additional recovery. Oil recovery ceased after 30hrs injection (total 48PV, 43.7PV N_2 gas and 4.3PV surfactant solution), but the injection was maintained for 170hrs (total 272PV, 247.5PV N_2 gas and 24.5PV surfactant solution). A sharp initial increase and slowly rising differential pressure, reaching maximum at 25kPa, was observed. Foam was not observed at the effluent, but rater single menisci dividing alternate bubbles of liquid and gas. Final recovery after the foam injection demonstrate the potential of this technique, but the high number of pore volumes injected to reach ultimate recovery must be improved for the technique to be successfully applied in the field.

6.3 EOR at the block scale

Foam injection as an EOR technique in fractured, oil-wet carbonate rocks was further investigated in a block sample. Oil recovery during waterflood, pure gasflood, and surfactant injection were compared with co-injection of gas and surfactant in a pre-generated foam injection. Three injections were performed: 1) a waterflood with 10 cc/hr injection rate, 2) a surfactant flood with 10 cc/hr injection rate, and 3) a pre-generated foam injection with an injection rate of 130 ml/hr (N_2 gas fraction was 0.92). Figure 6 shows the experimental setup used with a Bentheim sandstone core plug as a foam generator. The fracture network in the block sample can also be seen.

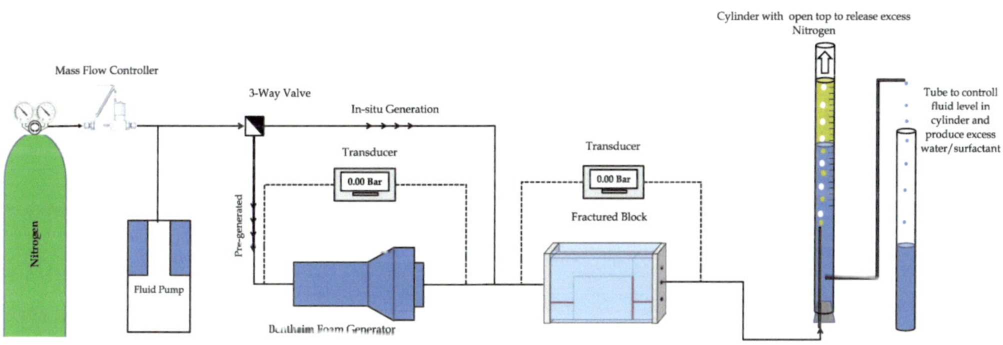

Fig. 6. Experiment setup used during pregenerated foam injection in a fractured block sample (Haugen *et al.*, 2010a).

Figure 7 shows the oil recovery for each injection and the development in differential pressure. Waterflood produced R_F=10%OOIP with a clean water-cut (i.e. no two-phase production). The subsequent surfactant injection produced an additional R_F=2.5%OOIP during 9PV injected. The pre-generated foam injection produced additional oil, amounting to R_F=75.7%OOIP, with a fast initial recovery followed by a slowly reducing oil production. Differential pressure across the block was fluctuating, with an overall increasing trend. Foam was not observed in the effluent, but rather as single menisci dividing alternate slugs of liquid and gas.

During pre-generated foam injection the differential pressure increased in all tests, and the oil production increased significantly. Foam was not observed at the outlet suggesting that foam collapsed or changed its configuration when advancing through the fracture(s). The destabilization of foam inhibited further foam front movement (Bernard and Jacobs, 1965). The mechanism(s) behind foam collapse cannot be distinguished from this work alone, but plausible mechanisms include:

- Oil intolerance
- Gravity segregation in the fracture
- Fracture wettability
- Divalent ions in the brine

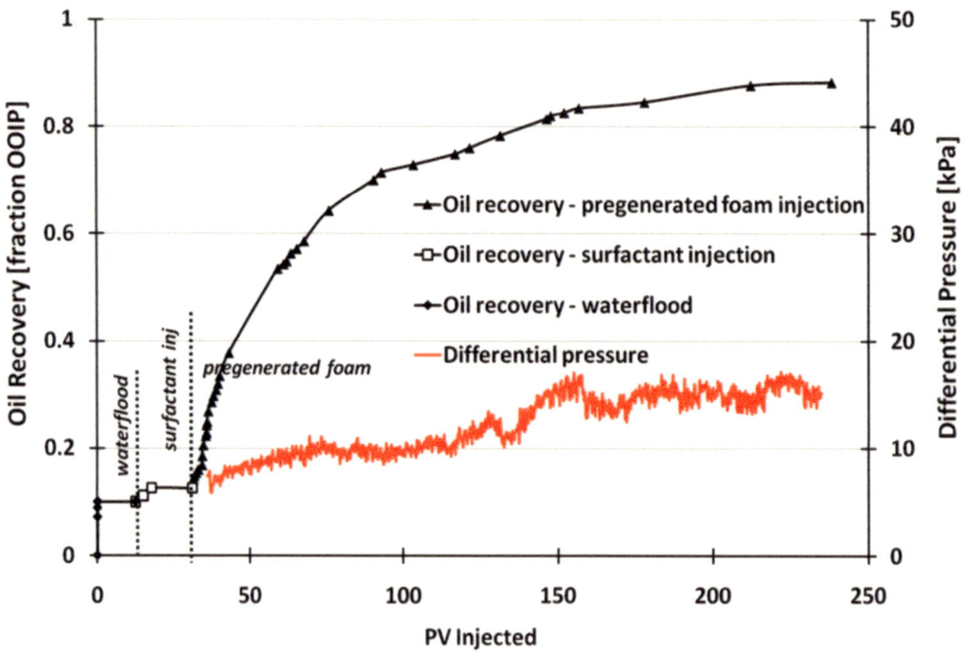

Fig. 7. The effect of foam as an EOR technique in fractured, oil-wet carbonate block sample.

6.4 EOR by foam injection in fractured reservoirs

Enhanced oil recovery by foam injection during gravity drainage was studied numerically by reducing the gas mobility (Farajzadeh *et al.*, 2010). Water and oil were produced at the bottom of the fracture at the instance they were displaced from the matrix, resulting in the fracture always being gas-filled. With decreased fracture permeability, the mobility of all fluids was reduced, increasing the viscous forces in the system. Similar results were found by Haugen *et al.*, 2008 and Fernø *et al.*, 2008 where decreasing fracture transmissibility increased oil recovery during waterfloods. In the case of foam flow in fractured media, the apparent foam viscosity is more important for oil recovery than the reduction in fracture transmissibility because increased foam viscosity leads to increased differential pressure and the increased oil recovery.

The application of foam as an EOR technique in fractured reservoirs will also lead to improved sweep efficiency. This is essential in highly heterogeneous reservoirs, where the majority of fluid flow is concentrated in the high permeable zones. Continuous foam injection at fixed injection rate may improve sweep efficiency by raising the injection-well pressure (Shan and Rossen, 2004), but the pressure increase may be undesirable in conventional, un-fractured reservoirs, where a rise in injection-well pressure could damage the well (Shi and Rossen, 1998). For fractured reservoirs, the pressure drop generated by the low-mobility foam, extending from the displacement front back to the injection well, is an important mechanism to improve the sweep efficiency. The injection of pre-generated foam reported at the core and block scale was successful with respect to oil production, which significantly increased recovery in all samples. The process was inefficient in terms of the number of pore volumes required to recover the oil. Economically viable production rates and reduction of surfactant cost could be achieved by taking the necessary steps to reduce foam collapse, which could be caused by any of the mechanisms previously outlined. The increase of the foam tolerance to oil (Hanssen and Dalland, 1990;Dalland and Hanssen, 1993;Dalland and Vassenden, 1998;Hanssen and Dalland, 2000) is a proven method to make foam injections more efficient.

Foam injection as an EOR technique at field scale should be considered to reduce gravity override and injectivity issues. Foam injection was successfully implemented on the Norwegian Continental Shelf in the Snorre field as a Foam-Assisted-Water-Alternating-Gas (FAWAG). This project revealed that important parameters for the success of foam injectivity were surfactant adsorption, critical surfactant concentration, foam drying effect, foam oil tolerance, and foam strength (Blaker *et al.*, 2002). Foam efficiency could also be improved with smaller well-spacing (Awan *et al.*, 2008). A surfactant preflush that alters the wettability of the fracture surface (removes oil) may benefit foam stability in the fracture. In vertical fractures, care should be taken to limit gravity segregated fluid flow during co-injection of aqueous and gaseous phases. During gas injection projects, foam should be injected as early as possible after gas breakthrough (Surguchev *et al.*, 1996). Foam injection in combination with other EOR efforts may also improve economics, for instance foam in combination with ASP flooding or polymer gels (Wang *et al.*, 2006).

7. Conclusion

Complementary imaging techniques used to study EOR techniques such as waterflooding, gas injection, and foam injections in fractured oil reservoirs provide new and improved fundamental understanding of the oil recovery process. The use of complementary imaging techniques allows studying EOR processes in porous media from the micro scale (10^{-5}-10^{-3}m) to block scale (10^{-1}-10^{0}) covering 5 orders of magnitude. The combination of two imaging techniques such as Magnetic Resonance Imaging (MRI) and Nuclear Tracer Imaging (NTI) enables a complementary investigation on materials and processes, where large scale (~meters) phenomena are controlled by small scale (micrometer) heterogeneities.

MRI provides high spatial resolution and fast data acquisition necessary to capture the processes that occur inside fractures less than 1 mm wide, whereas NTI provides information on macro-scale saturation distribution in larger fractured systems. The oil recovery mechanisms involved with waterflooding fractured carbonate blocks were found to be dependent on the wettability of the matrix, as the wettability had great impact on the fracture/matrix hydrocarbon exchange. The MRI images of oil saturation development inside the fractures revealed transport mechanisms for the wetting phase across fractures at several wettability conditions, and provided new and detailed information on fluid fracture crossing previously observed in block scale experiments investigated by NTI. The forming of oil droplets of the fracture surface at oil-wet conditions mechanistically corroborated the results previously observed when waterflooding chalk at water-wet conditions. Pore scale modeling should be considered as a tool to calculate more accurately the capillary pressure and relative permeability curves in the fracture in the presence of growing liquid droplets. This would increase the confidence in the applied multiphase functions and strengthen the overall physical understanding of flow in fractured rocks. The derived multiphase functions should be implemented in the numerical model, and the errors by not accounting for this should be understood and quantified.

The effect of capillary pressures at various wettabilities was experimentally studied at different length scales, which was used as the basis for numerical simulations of multiphase flow in fractured reservoirs. This review demonstrates the influence of capillary pressure as a vital input parameter in simulations to describe fluid flow in the reservoirs and as the dominant multiphase function for important recovery mechanisms in fractured reservoirs; such as spontaneous imbibition and gravity drainage.

Injection of pre-generated foam greatly enhanced oil recovery, with recoveries up to 80%OOIP. Increased apparent foam viscosity increased oil recovery in similar fractured systems. Enhanced oil recovery by pre-generated foam injection in fractured, oil-wet limestone was observed only after high pore volumes throughput. Oil recovery efficiency may increase by reducing the rate of foam collapse in the fractures by:

- Improving the foam oil tolerance.
- Changing the wettability of the fracture surface to water-wet.
- Optimizing the injection of foam (e.g. FAWAG).

8. Acknowledgement

The Royal Norwegian Research Council, BP, Statoil and ConocoPhillips for financial support.

9. References

Allen, J. and Sun, S.Q. "Controls on Recovery Factor in Fractured Reservoirs: Lessons Learned from 100 Fractured Fields", SPE ATCE, Denver, CO, USA, October 4-7, 2003

Amott, E. "Observations Relating to the Wettability of Porous Rock", *Trans. AIME* 216:pp.156-162, 1959

Aronofsky, J.S., Masse, L. and Natanson, S.G. "A model for the mechanism of oil recovery from the porous matrix due to water invasion in fractured reservoirs", *Trans. AIME* 213:pp.17-19, 1958

Aspenes, E., Graue, A., Baldwin, B.A., Moradi, A., Stevens, J. and Tobola, D.P. "Fluid Flow in Fractures Visualized by MRI During Waterfloods at Various Wettability Conditions - Emphasis on Fracture Width and Flow Rate", SPE ATCE, San Antonio, TX, USA, Sept. 29-Oct.3, 2002

Aspenes, E., Ersland, G., Graue, A., Stevens, J. and Baldwin, B.A. "Wetting Phase Bridges Establish Capillary Continuity Across Open Fractures and Increase Oil Recovery in Mixed-Wet Fractured Chalk", *Transport in Porous Media* 74(1):pp.35-47, 2008

Awan, A.R., Teigland, R. and Kleppe, J. "A Survey of North Sea Enhanced-Oil-Recovery Projects Initiated During the Years 1975 to 2005", *SPE Reservoir Evaluation & Engineering* 11(3):pp.497-512, 2008

Bailey, N.A., Rowland, P.R. and Robinson, D.P. "Nuclear Measurements of Fluid Saturation in EOR Flood Experiments", European Symposium on Enhanced Oil Recovery, Bournemouth, UK, September 21-23, 1981

Baldwin, B.A. and Spinler, E.A. "A direct method for simultaneously determining positive and negative capillary pressure curves in reservoir rock", *Journal of Petroleum Science and Engineering* 20(3-4):pp.161-165, 1998

Baldwin, B.A. and King, R. "Why Would an Oil Company use MRI", Spatially Resolved Magnetic Resonance (Methods, materials, Medicine, Biology, Rheology, Geology, Ecology, Hard Wave), P. Blumler, B. Blumich, R. Botto and E. Fukushima, Wiley-VCH, New York, 1999

Barrenblatt, G.I., Zheltov, I.P. and Kochina, I.N. "Basic concepts in the theory of the seepage of homogeneous liquids in fissured rocks (strada)", *J. Appl. Math. Mech.* 24:pp.1286-1303, 1960

Behbahani, H. and Blunt, M.J. "Analysis of Imbibition in Mixed-Wet Rocks Using Pore-Scale Modeling", *SPE J.* 10(4):pp.466-473, 2005

Bernard, G.B. and Jacobs, W.L. "Effect of Foam on Trapped Gas Saturation and on Permeability of Porous Media to Water", *SPEJ* 41965

Bernard, G.G. and Holm, L.W. "Effect of Foam on Permeability of Porous Media to Gas", 4(3):pp.267 - 274, 1964

Bertin, H.J., Apaydin, O.G., Castanier, L.M. and Kovscek, A.R. "Foam Flow in Heterogeneous Porous Media: Effect of Crossflow", *SPEJ* 4(2):pp.75-82, 1999

Blaker, T., Aarra, M.G., Skauge, A., Rasmussen, L., Celius, H.K., Martinsen, H.A. and Vassenden, F. "Foam for Gas Mobility Control in the Snorre Field: The FAWAG Project", *SPE Reservoir Evaluation & Engineering* 5(4):pp.317-323, 2002

Buchgraber, M., Clemens, T., Castanier, L.M. and Kovscek, A. "A Microvisual Study of the Displacement of Viscous Oil by Polymer Solutions", *SPE Reservoir Evaluation & Engineering* 14(3)2011

Casteel, J.F. and Djabbarah, N.F. "Sweep Improvement in CO2 Flooding by Use of Foaming Agents", *SPE Reservoir Engineering* 3(4):pp.1186-1192, 1988

Dalland, M. and Hanssen, J.E. "Foam barriers against gas coning: Gas blocking with hydrocarbon foams", 7th European Symposium on Improved Oil Recovery Moscow, Russia, 1993

Dalland, M. and Vassenden, F. "Cost-effective foams for gas influx control: Role of polymer additive", 19th IEA Workshop & Symposium on IOR, Carmel, CA, 1998

Ersland, G., Fernø, M.A., Graue, A., Baldwin, B.A. and Stevens, J. "Complementary imaging of oil recovery mechanisms in fractured reservoirs", *Chemical Engineering Journal* 158(1):pp.32-38, 2010

Farajzadeh, R., Wassing, L.B.M. and Boerrigter, P.M. "Foam Assisted Gas Oil Gravity Drainage in Naturally-Fractured Reservoirs", SPE Annual Technical Conference and Exhibition, Florence, Italy, 09/19/2010, 2010

Fernø, M.A. "A Study of Capillary Pressure and Capillary Continuity in Fractured Rocks" PhD, Dept. of Physics and Technology, Univeristy of Bergen, Bergen, 2008

Fernø, M.A., Haugen, Å., Howard, J.J. and Graue, A. "The significance of wettability and fracture properties on oil recovery efficiency in fractured carbonates", International Symposium of the Society of Core Analysts, Abu Dhabi, October 29-November 2, 2008

Fernø, M.A., Torsvik, M., Haugland, S. and Graue, A. "Dynamic Laboratory Wettability Alteration", *Energy & Fuels* 24(7):pp.3950-3958, 2010

Fernø, M.A., Haugen, Å. and Graue, A. "Wettability effects on the matrix-fracture fluid transfer in fractured carbonate rocks", *Journal of Petroleum Science and Engineering* 77(1):pp.146-153, 2011

Graue, A., Kolltveit, K., Lien, J.R. and Skauge, A. "Imaging Fluid Saturation Development in Long-Core Flood Displacements", *SPE Form. Eval.* 5(4):pp.406-412, 1990

Graue, A., Tonheim, E. and Baldwin, B.A. "Control and alteration of wettability in low-permeability chalk", 3rd International Symposium on Evaluation of Reservoir Wettability and Its Effect on Oil Recovery University of Wyoming, Laramie, WY, 1996

Graue, A., Viksund, B.G. and Baldwin, B.A. "Reproduce Wettability Alteration of Low-Permeable Outcrop Chalk", SPE/DOE Improved Oil Recovery Symposium, Tulsa, Oklahoma, 04/19/1998, 1998

Graue, A., Viksund, B.G. and Baldwin, B.A. "Reproducible Wettability Alteration of Low-Permeable Outcrop Chalk", *SPE Reservoir Evaluation & Engineering* 2(2):pp.134-140, 1999a

Graue, A., Viksund, B.G., Eilertsen, T. and Moe, R. "Systematic wettability alteration by aging sandstone and carbonate rock in crude oil", *Journal of Petroleum Science and Engineering* 24(2-4):pp.85-97, 1999b

Graue, A., Aspenes, E., Moe, R.W., Baldwin, B.A., Moradi, A., Stevens, J. and Tobola, D. "MRI Tomography of Saturation Development in Fractures During Waterfloods at Various Wettability Conditions", SPE ATCE, New Orleans, LA, USA, September 30-October 3, 2001

Hanssen, J.E. and Dalland, M. "Foams for Effective Gas Blockage in the Presence of Crude Oil", SPE/DOE Enhanced Oil Recovery Symposium, Tulsa, Oklahoma, 04/22/1990, 1990

Hanssen, J.E., Holt, T. and Surguchev, L.M. "Foam Processes: An Assessment of Their Potential in North Sea Reservoirs Based on a Critical Evaluation of Current Field Experience", SPE/DOE Improved Oil Recovery Symposium, Tulsa, Oklahoma, 04/17/1994, 1994

Hanssen, J.E. and Dalland, M. "Increased Oil Tolerance of Polymer-Enhanced Foams: Deep Chemistry or Just "Simple" Displacement Effects?", SPE/DOE Improved Oil Recovery Symposium, Tulsa, Oklahoma, 04/03/2000, 2000

Haugen, A., Ferno, M.A., Graue, A. and Bertin, H.J. "Experimental Study of Foam Flow in Fractured Oil-Wet Limestone for Enhanced Oil Recovery", SPE Improved Oil Recovery Symposium, Tulsa, Oklahoma, USA, 01/01/2010, 2010a

Haugen, Å., Fernø, M.A. and Graue, A. "Numerical simulation and sensitivity analysis of in-situ fluid flow in MRI laboratory waterfloods of fractured carbonate rocks at different wettabilities", SPE ATCE, Denver, CO, USA, September 21-24, 2008

Haugen, Å., Fernø, M.A., Bull, Ø. and Graue, A. "Wettability Impacts on Oil Displacement in Large Fractured Carbonate Blocks", *Energy & Fuels* 24(5):pp.3020-3027, 2010b

Holm, L.W. "The Mechanism of Gas and Liquid Flow Through Porous Media in the Presence of Foam", 8(4):pp.359 - 369, 1968

Hornbrook, J.W., Castanier, L.M. and Pettit, P.A. "Observation of Foam/Oil Interactions in a New, High-Resolution Micromodel", SPE Annual Technical Conference and Exhibition, Dallas, Texas, 01/01/1991, 1991

Jackson, M.D., Valvatne, P.H. and Blunt, M.J. "Prediction of wettability variation and its impact on flow using pore- to reservoir-scale simulations", *Journal of Petroleum Science and Engineering* 39(3-4):pp.231-246, 2003

Jadhunandan, P.P. and Morrow, N.R. "Effect of Wettability on Waterflood Recovery for Crude-Oil/Brine/Rock Systems", *SPE Reservoir Eng.* 10(1):pp.40-46, 1995

Johannesen, E., Steinsbø, M., Howard, J.J. and Graue, A. "Wettability characterization by NMR T2 measurements in chalk", International Symposium of the Society of Core Analysts, Trondheim, Norway, September 12-16 2006

Kazemi, H., Merril Jr., J.S., Poterfield, K.L. and Zeman, P.R. "Numerical simulation of water-oil flow in naturally fractured reservoirs", *SPE J.* 16:pp.317-326, 1976

Kovscek, A.R., Tang, G.Q. and Radke, C.J. "Verification of Roof snap off as a foam-generation mechanism in porous media at steady state", *Colloids and Surfaces A: Physicochemical and Engineering Aspects* 302(1-3):pp.251-260, 2007

Li, K. and Horne, R.N. "A General Scaling Method for Spontaneous Imbibition", SPE ATCE, San Antonio, TX, USA, September 29-October 2, 2002

Llave, F.M., Chung, F.T.-H., Louvier, R.W. and Hudgins, D.A. "Foams as Mobility Control Agents for Oil Recovery by Gas Displacement", SPE/DOE Enhanced Oil Recovery Symposium, Tulsa, Oklahoma, 04/22/1990, 1990

Ma, S., Morrow, N.R. and Zhang, X. "Generalized scaling of spontaneous imbibition data for strongly water-wet systems", *Journal of Petroleum Science and Engineering* 18(3-4):pp.165-178, 1997

Manrique, E.J., Muci, V.E. and Gurfinkel, M.E. "EOR Field Experiences in Carbonate Reservoirs in the United States", *SPE Reservoir Evaluation & Engineering* 10(6):pp.pp. 667-686, 2007

Mattax, C.C. and Kyte, J.R. "Imbibition Oil Recovery from Fractured, Water-Drive Reservoir", *SPE J.* 2(2):pp.177-184, 1962

Miller, M. "Naturally Fractured Reservoir Engineering", Reservoir Engineering Course Handouts, HOT Engineering, Vienna, Austria, 2007

Morrow, N.R. and Mason, G. "Recovery of oil by spontaneous imbibition", *Current Opinion in Colloid & Interface Science* 6(4):pp.321-337, 2001

Nguyen, Q.P., Currie, P.K. and Zitha, P.L.J. "Effect of Capillary Cross-Flow on Foam-Induced Diversion in Layered Formations", SPE European Formation Damage Conference, The Hague, Netherlands, 05/13/2003, 2003

Rangel-German, E.R. and Kovscek, A.R. "Experimental and analytical study of multidimensional imbibition in fractured porous media", *Journal of Petroleum Science and Engineering* 36(1-2):pp.45-60, 2002

Rossen, W.R. and Kumar, A.T.A. "Effect of Fracture Relative Permeabilities on Performance of Naturally Fractured Reservoirs", International Petroleum Conference and Exhibition of Mexico, Veracruz, Mexico, 10/10/1994, 1994

Rossen, W.R. "Foams in Enhanced Oil Recovery", Foams : theory, measurements, and applications, R. K. Prud'homme and S. A. Khan, Marcel Dekker, Inc., New York, 1995

Schramm, L.L. "Foams: Fundamentals and Application in the Petroleum Industry". Washington, Am. Chem. Soc., 1994

Shan, D. and Rossen, W.R. "Optimal Injection Strategies for Foam IOR", *SPE Journal* 9(2):pp.132-150, 2004

Shi, J.-X. and Rossen, W.R. "Improved Surfactant-Alternating-Gas Foam Process to Control Gravity Override", SPE/DOE Improved Oil Recovery Symposium, Tulsa, Oklahoma, 04/19/1998, 1998

Surguchev, L.M., Hanssen, J.E., Johannessen, H.M. and Sisk, C.D. "Modelling Injection Strategies for a Reservoir with an Extreme Permeability Contrast: IOR Qualification", European 3-D Reservoir Modelling Conference, Stavanger, Norway, 04/16/1996, 1996

Valvatne, P.H., Piri, M., Lopez, X. and Blunt, M.J. "Predictive Pore-Scale Modeling of Single and Multiphase Flow", *Transport in Porous Media* 58(1):pp.23-41, 2005

Wang, D., Han, P., Shao, Z. and Seright, R.S. "Sweep Improvement Options for the Daqing Oil Field", SPE/DOE Symposium on Improved Oil Recovery, Tulsa, Oklahoma, USA, 04/22/2006, 2006

Zerhboub, M., Touboul, E., Ben-Naceur, K. and Thomas, R.L. "Matrix Acidizing: A Novel Approach to Foam Diversion ", *SPE Production & Facilities* 9(2):pp.121-126, 1994

Zhou, X., Torsæter, O., Xie, X. and Morrow, N.R. "The Effect of Crude-Oil Aging Time and Temperature on the Rate of Water Imbibition and Long-Term Recovery by Imbibition", *SPE Form. Eval.* 10(4):pp.259-266, 1995

Zhou, X., Morrow, N.R. and Ma, S. "Interrelationship of Wettability, Initial Water Saturation, Aging Time, and Oil Recovery by Spontaneous Imbibition and Waterflooding", *SPE J.* 5(2):pp.199-207, 2000

Zimmerman, R.W. and Bodvarsson, G.S. "Absorption of Water Into Porous Blocks of Various Shapes and Sizes", *Water Resour. Res* 26(11):pp.2797-2806, 1990

Section 2

Environmental Management Through Bioremediation

5

Microorganisms and Crude Oil

Dorota Wolicka and Andrzej Borkowski
University of Warsaw
Poland

1. Introduction

Crude oil is one of the most important energetic resources in the world. It is used as raw material in numerous industries, including the refinery-petrochemical industry, where crude oil is refined through various technological processes into consumer products such as gasoline, oils, paraffin oils, lubricants, asphalt, domestic fuel oil, vaseline, and polymers. Oil-derived products are also commonly used in many other chemical processes.

Although crude oil is a natural resource, in some conditions its presence is unfavorable and causes devastation of the surroundings. Crude oil and formation water in oil reservoirs represent an extreme environment with many groups of autochthonous microorganisms strictly linked with this setting. The relationship between microorganisms and this extreme environment begins when crude oil is formed and it ends when these specialized microorganisms are applied for the bioremediation of the polluted environment by crude oil and oil-derived products.

It is common knowledge that crude oil is formed by biological, chemical, and geochemical transformations of organic matter accumulated in favorable locations. In the first stage crude oil is transformed during sediment diagenesis at moderate temperatures up to 50°C. Due to defunctionalization and condensation, kerogen, which is immature crude oil, is formed. Kerogen accumulations are considered to be the richest coal accumulation on Earth (Widdel & Rabus, 2001). Based on geochemical studies, immature crude oil contains higher volume of hydrocarbons with an odd number of carbon atoms, synthesized in plants. This fact has a practical meaning in determining the Carbon Preference Index (CPI). The organic origin of crude oil is also supplied by biomarkers, i.e. compounds whose carbon skeleton was not changed in geochemical processes, formed by living organisms, e.g. microorganisms. Such compounds include e.g. terpenes, porphyrines, and metalloporphyrines (Surygała, 2001).

The life activity of microorganisms occurring in crude oil has significant influence on its chemical composition and physical-chemical properties, and as a result often changes its economical value or exploitation conditions. This influence can be positive, e.g. decreased viscosity of heavy crude oil favors its exploitation, but also negative, e.g. corrosion of drilling equipment due to bacterial production of hydrogen sulphide. Products from the biological activity of autochthonous microorganisms or microorganisms introduced into the reservoir rock are the basis of biological methods applied to enhance the recovery of oil from already exploited (depleted) reservoirs. At specific conditions, crude oil may flow uncontrolled onto the lithosphere surface and cause significant hazard to the environment. Such cases are often related to spilled oil and oil by-products during oil exploitation, processing, and transportation. This type of pollution is often removed by natural microorganisms occurring in crude oil; these microorganisms have the ability to biodegrade crude oil and oil-derived products (Mokhatab, 2006; Nazina et al., 2007; Wolicka et al., 2009; Wolicka et al., 2011).

2. Crude oil – Environment for microorganisms' growth

2.1 Crude oil composition

Crude oil is a mixture of thousand of various compounds, organic and inorganic, including aliphatic and aromatic hydrocarbons, which in average reaches 75 % of its content, as well as resins and asphalts. Non-hydrocarbon compounds include sulphur compounds (0.01–8 %), mainly as hydrogen sulfide (H_2S), mercaptans (compounds containing the -SH group), sulfides and disulfides, thiophenes, as well as benzothiophenes and naphthothiophenes that prevail in oil fractions (Fig. 1). These compounds are unfavorable due to their chemical recalcitrance, therefore their presence is considered in evaluating crude oil quality (Surygała, 2001).

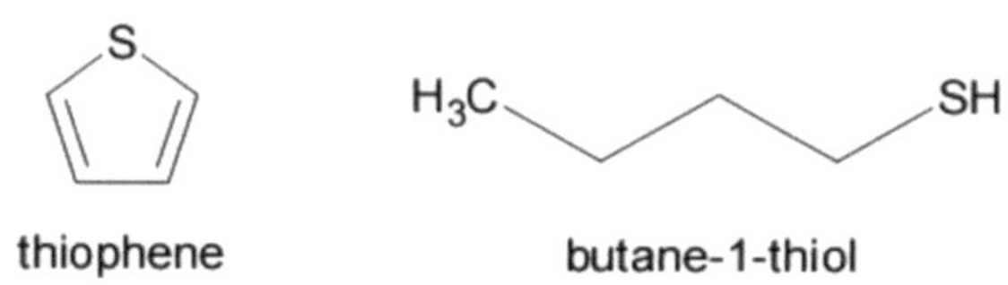

Fig. 1. Chemical structures of thiophene and 1-butanothiol (methyl mercaptan).

Nitrogen compounds represent non-hydrocarbon compounds that occur in crude oil at the level of 0.01–2 % weight, although over 10 % concentrations have been noted. Nitrogen occurs in alkaline and non-alkaline compounds. The first group includes pyridines, acridines, and quinolines, and the second group comprises pyrroles, carbazoles, indoles, and heterocyclic compounds (Fig. 2). Similarly, nitrogen also occurs as sulphur bonds, compounds from this group are concentrated in high-boiling fractions (Surygała, 2001).

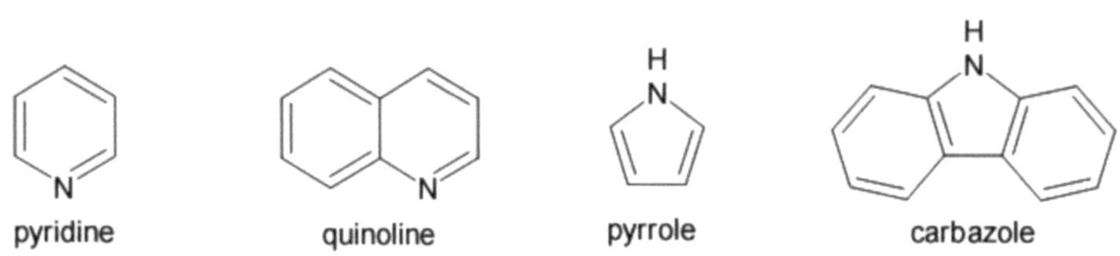

Fig. 2. Chemical structures of pyridine, qinoline, pyrrol, and carbazol

Oxygen compounds such as phenols, carboxylic acids (having the COOH functional groups), furans, and alcohols (Fig.3) occur in the heavy fractions of crude oil (Surygała, 2001).

phenol furan cyclohexanol

Fig. 3. Structures of phenol, furan, and cyclohexanol.

Porphyrins often occur in crude oil. They are products of degradation of dyes produced by living organisms. They are composed of pyrrole rings connected by methine bridges. They often form chelate compounds of nickel, vanadium, and other metals. There is a relationship between the maturity of crude oil and the concentration of porphyrins, which decreases with oil maturity. Oil formed from marine organisms contains more vanadium than nickel porphyrins (Surygała, 2001).

Trace elements are present in crude oil in ppm quantities. Besides porphyrins, trace elements occur as soaps (particularly compounds of Zn, Ti, Ca, and Mg), as well as metal-organic bonds (V, Cu, Ni, Fe). The highest concentrations (up to 1500 ppm) of trace elements that have been determined correspond to vanadium, nickel, and iron (up to 1200 ppm), as well as calcium and cobalt (up to 12 ppm). These compounds are unfavorable during the refining process. Crude oil is naturally enriched with these elements during its migration within the reservoir rock. Particularly high contents of vanadium have been found in crude oils from Venezuela (Surygała, 2001; Swaine, 2000). Also trace quantities of phosphorus, arsenic, and selenium are found in most types of crude oil.

Resin-asphalt substances are present in crude oil, particularly with a low degree of maturity. They have a very complex chemical structure and include most heteroatoms, trace elements, and polycyclic aromatic hydrocarbons. They contain in average up to 80 % C, 10 % H, and 14 % heteroatoms; in which 1 to 2 % correspond to metal-organic compounds.

The majority of these organic compounds may serve as electron donors for various groups of autochthonous microorganisms present in crude oil.

2.2 Microorganisms present in crude oil

The conditions prevailing in a crude oil reservoir significantly differ from environmental settings typical to the occurrence of living organisms on Earth. The red-ox potential is very low, the pressure and temperature are very high, and the salt content may reach up to over 10%. Moreover, this setting lacks electron acceptors, such as oxygen, typical for most microorganisms; while sulfate and carbonate are present (Ortego-Calvo & Saiz-Jimenz, 1998) and the range of electron donors admissible for microorganisms is very wide.

Most hydrocarbons occurring in crude oil have toxic effects resulting mainly from their chemical structure (Gałuszka & Migaszewski, 2007). These toxic hydrocarbons include both aliphatic and aromatic compounds, such as polycyclic aromatic hydrocarbons (PAHs), whose toxicity increases proportionally to the number of carbon atoms in the compound. Particularly in the case of PAHs with more than four-member rings in its structure. Despite the toxicity of the chemical compounds occurring in crude oil, several groups of microorganisms have been found in this setting (Magot et al., 2000; Stetter & Huber, 1999).

The main sources of carbon for microorganisms in crude oil are hydrocarbons, both aliphatic and aromatic, but also organic compounds that are often the products of crude oil biodegradation. These organic compounds include: organic acids such as acetic, benzoic, butyric, formic, propanoic, and naphthenic acids reaching up to 100 mM (Dandie et al., 2004). The electron donors may be H_2, and in the case of immature oil - resins and asphalthenes, whose metabolic availability is confirmed by the fact that anaerobic microorganisms may develop in cultures with crude oil without any modifications of its composition.

Recently, it was discovered that crude oil lenses contained not only bacteria that were supplied from the external environment to the reservoir by infiltration of surface water or introduced through fluid injection during oil recovery operations but also these bacteria which are autochthonous for environment of crude oil and formation waters. Distinguishing these autochthonous microorganisms from other surface microorganisms is very difficult, almost impossible, particularly in settings with low salt content and temperature, because in this environment surface bacterial strains can grow. Generally, only absolutely anaerobic microorganisms, whose physiological characteristics indicate adaptation to the *in-situ* conditions, are considered as really autochthonous for this setting and only these microorganisms demonstrate significant activity in these specific environments. However, so far it is not clear yet if these communities of microorganisms are characteristic of these ecosystems and the factors causing their activation or growth inhibition (Magot et al., 2000).

An important factor influencing microorganism activity is temperature (Stetter & Huber, 1999). Living organisms are not considered to occur theoretically at temperatures above 130–150°C due to the instability of biological compounds (Magot et al., 2000). Such conditions would correspond to deep oil reservoirs at depths ranging from 4030 to 4700 m having a geothermal gradient of 3°C per 100 m and a surface temperature of 10°C (unpublished data). So far, the presence of microorganisms has been detected at a depth of 3500 m (Stetter & Huber, 1999). Previous data indicates the presence of microorganisms at maximum temperatures of 80°C to 90°C, above which autochthonous bacteria do not occur. In some cases, microorganisms that are present at these high temperatures were introduced into the reservoir with sea water through rock fractures and faults. Crude oil is a setting characterized by the presence of many microorganism groups: fermentation bacteria and sulfate reducing bacteria (SRB) that causes complete oxidation of organic compounds to CO_2 or incomplete oxidation of hydrocarbon compounds to acetate groups; as well as iron reducing bacteria and methanogenic archaea.

2.2.1 Fermentation bacteria

Numerous species of fermentation bacteria have been detected in crude oil (Nazina et al., 2007). Strains capable of thiosulfate (S_2O_3) and elemental sulfur (S°) reduction were determined. Electron donors for these microorganisms may be sugars, proteins, H_2, CO_2, and hydrocarbons. The products of metabolic reactions are organic acids and gases, such as H_2 and CO_2, which may cause increase of reservoir pressure. These microorganisms have potential for their application in microbiological methods of oil production enhancement

(Magot et al., 2000; Nazina et al., 2007). Mesophilic fermentation bacteria are more uncommon than the thermophilic ones. The first group comprises such haloanaerobes as *Haloanaerobium acetoethylicum, H. congolense,* and *H. salsugo* that produce acetate or ethanol in the process of carbohydrate fermentation. These microorganisms differ also in the type of substrates used and their tolerance to salt content (up to 10 %). For example, *Spirochaeta smaragdinae* isolated from a Congo oil field prefers salt contents of up to 5 %. The same source, however, yielded also *Dethiosulfovibrio peptidovarans* with specific metabolism. These bacteria have the ability to biodegrade protein extracts and the products of its metabolism are organic acids such as: acetic, isobutyric, isovaleric, and 2-methylbutyric acids. Moreover, it has the ability to reduce thiosulfate and prefers salt contents up to 3 % of NaCl (Magot et al., 2000).

In a hydrocarbon reservoir together with crude oil, there is also water in the formation, in which different groups of microorganisms are known to occur. For example, from formation waters of the Tatarstan and western Siberia reservoirs, microorganisms such as *Acetoanaerobium romaskovii* were isolated (Magot et al., 2000). These microorganisms use acetates, H_2, CO_2, amino acids, and sugars as sources of energy and carbon. It has been reported in the literature that at these high reservoir temperatures a larger number of thermophilic bacteria has been detected than the number of mesophilic bacteria (Magot *et al.*, 2000). Thermophilic microorganisms contain thermostable enzymes that are capable of enduring temperatures exceeding even 100°C. To this group of microorganisms belong species of *Thermotoga: T. subterranean, T. elfii,* and *T. hypogea,* which are capable of reducing thiosulfate to sulfides, as well as bacteria resembling *Thermotoga* that reduce elemental sulfur. Microorganisms as *Thermotoga* occur at low salinities up to 2.4% of NaCl and in the course of glucose degradation, these bacteria produce acetic acid and L-alanine (Magot *et al.*, 2000).

Bacteria representing *Geotoga* and *Petrotoga* from the order *Thermotogales,* which are moderate thermophiles, occur also in a wide range of salt content conditions. They were detected for the first time in crude oil reservoirs in Texas and Oklahoma. Microbiological investigations of numerous high-temperature crude oil reservoirs supplied evidence on the significant biogeochemical role of these bacteria, which in morphologically and physiologically sense resemble representatives of the order *Thermotogales,* such as *Fervidobacterium* and *Thermosipho.* They include *Thermoanaerobacter* and *Thermoanaeobacterium* from the family *Thermoanaerobiaceae,* which are often isolated from hot but poorly salinated reservoirs. The first genus reduces thiosulfate to sulfides, and the second - thiosulfate to elemental sulfur (Davey et al., 1993).

Hyperthermophilic fermenting microorganisms were distinguished in high-temperature reservoirs. They include *Archaea,* such as *Thermococcs celer, T. litoralis,* and *Pyrococcus litotrophicus.* The first two species showed activity during incubation at 85°C, and the latter - above 100°C. These microorganisms used proteins or yeast extract as electron donors, and reduced elemental sulfur to sulfides (Magot et al., 2000; Stetter & Huber, 1999).

2.2.2 Sulfate reducing bacteria (SRB)

Sulfate reducing bacteria (SRB) are some of the oldest microorganisms on Earth. Their initial development and activity goes back to the Proterozoic Era (Rabus et al., 2000). The process of dissimilative sulfate reduction is considered to be one of the few metabolic pathways that did not undergo mutations and horizontal gene transfer (Voordouw, 1992). This fact evidences also that the gene coding the enzyme catalyzing the first stage of dissimilative reduction is strongly conserved evolutionarily and occurs in unchanged form since its formation (Baker et al., 2003). Based on sulfur isotopic studies, bacterial sulfate reduction is believed to have developed earlier than oxygen photosynthesis (Kopp et al., 2005). The first studies on the metabolism and biology of these microorganisms were commenced in 1864. Meyer (1864) and Cohn (1867) first observed the production of hydrogen sulfide of biogenic origin in marine sediments. Bastin (1926) noted the undoubted presence of SRB in areas of crude oil exploitation, and Werkman & Weaver (1927) described the first sporing thermophilic SRB. In addition, these reports indicated the role of microorganisms in the corrosion of drilling equipment. The 1950s and 1960s brought the first attempts to understand metabolic processes conducted by SRB.

SRB are heterotrophic organisms and absolute anaerobes that use sulfates as well as other oxygenated sulfur compounds (sulfites, thiosulfites, trithionate, tetrathionate, and elemental sulfur) as final electron acceptors in respiration processes (Postgate, 1984; Gibson, 1990). All SRB are gram negative with the exception of the species of *Desulfonema.* This group of bacteria is very diverse and depending on the soil and water composition, different kinds of bacteria can be found within this group such as psychro-, meso- and thermophilic, halo- and barophilic. Some species of SRB like *Desulfosporosinus orientis* (Stackebrandt et al., 1997), *Desulfotomaculum halophilum* sp. nov. (Tardy-Jacquenod et al., 1998), and *Desulfosporosinus meridiei* sp. nov. (Robertson et al., 2001) have the ability to develop spores.

The 1980-ties brought new discoveries with regard to the mechanisms of biological sulfate reduction. This allowed a different view on SRB metabolism. Reactions of the entire metabolic trail taking place during sulfate reduction were described in detail and two metabolic trails of SRB were confirmed. The first was linked to the partial oxygenation of organic compounds, i.e. to acetate, and the second metabolic trail corresponds to the complete oxygenation to CO_2 (Laanbroek et al., 1984). The second important scientific activity at that time was research on the SRB genome. Till the 1980s all SRB were classified based on their characteristic phenotype features such as feeding or morphology. However, with wider application of the 16S rRNA gene sequence analysis, a more detailed classification of SRB was possible. It indicated that the genus *Desulfotomaculum* was the only genus belonging to the group of gram positive bacteria, whereas the remaining SRB are gram negative (Madigan et al., 2006).

Stetter (1987) discovered that the ability to reduce sulfates is not only a feature of the SRB but also of some archaea. For instance, the termophilic strain *Archaeoglobus fulgidus,* which can grow in environments at 83°C and is capable of sulfate reduction, was isolated. This strain showed larger similarity to Achaea than to the remaining SRB.

Pure strains of SRB capable of complete oxygenation of some hydrocarbons such as alkanes, xylenes, toluene, and naphthalene to CO_2 were isolated in the 1990s. It was also discovered that SRB may occur and develop in crude oil, whose components are a good source of carbon. This fact would explain the presence of hydrogen sulfide in crude oil reservoirs and in formation waters (Rabus et al., 2000; Wolicka & Borkowski, 2008a; Wolicka, 2008; Wolicka et al., 2010). Samples of isolated sulphidogenic bacterial communities from Carpathian's crude oil from Poland are presented in Fig. 4.

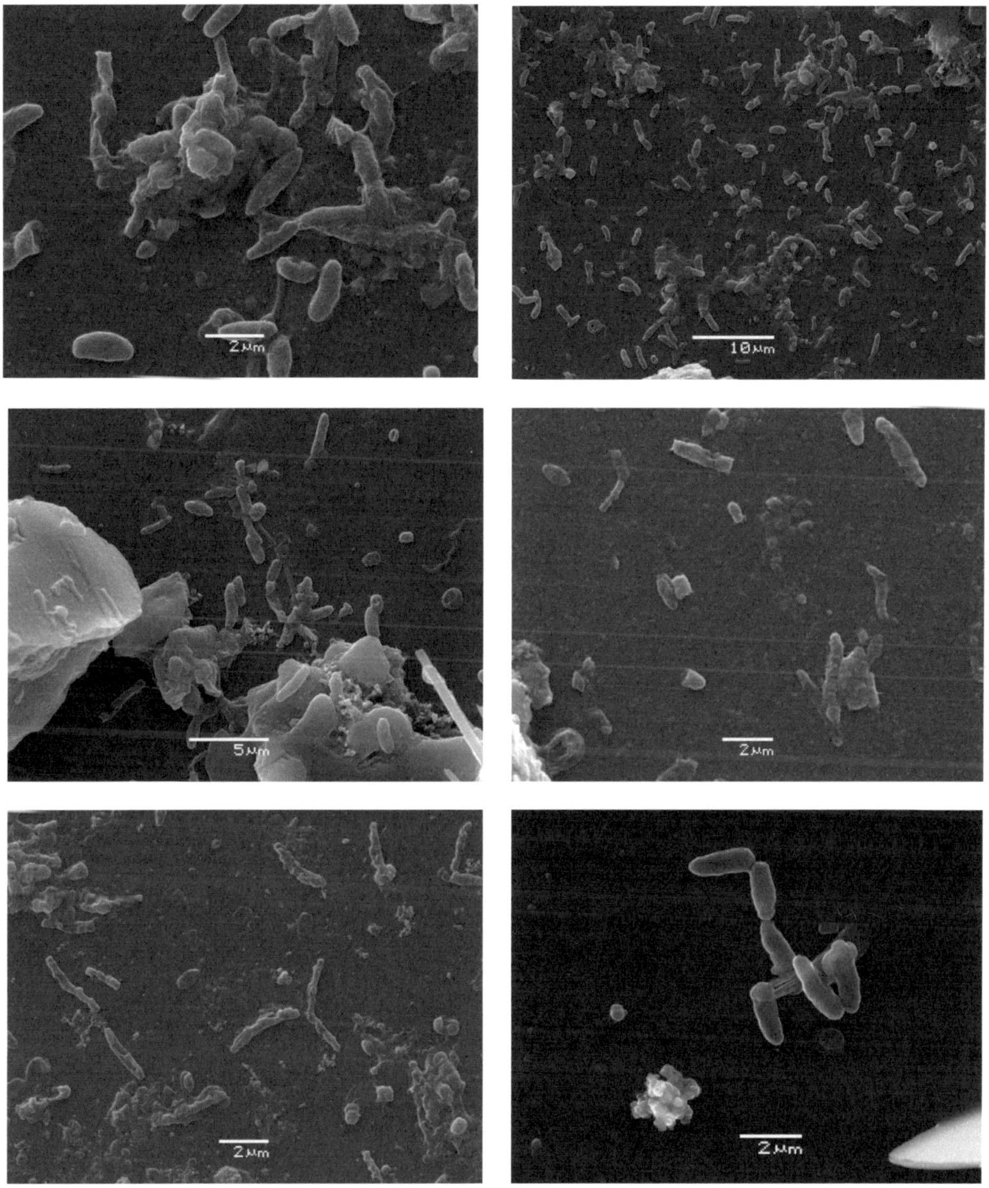

Fig. 4. Sulphidogenic bacterial communities from Carpathian's crude oil from Poland (Wolicka, own studies, not published).

For long time, SRB were thought to occur in environments polluted by crude oil and oil-derived products, and they always were considered to act as the producer of the toxic hydrogen sulfide and the main cause of bio-corrosion (O'Dea, 2007). Thus they were beyond scientific interest and their significant role in the biodegradation of organic compounds in anaerobic conditions was not known. Currently, the ability of SRB to metabolize many different organic compounds including crude oil and oil-derived products has been recognized; even the influence of the biological activity of SRB on oil quality and fluidity (e.g. heavy oils) has been determined. Moreover, the activity of these microorganisms decreases the permeability of reservoir rocks caused by the precipitation of insoluble sulfides, as well as carbonates (Magot et al., 2000; Nemati et al., 2001).

SRB always accompany crude oil and therefore for long time were considered as indicator organisms when searching for new reservoirs (Postgate, 1984). Such cases were only possible when the natural environment was not polluted by oil-derived products as it is nowadays.

SRB are a group of microorganisms that play a significant role in the biodegradation of organic compounds in anaerobic conditions and in the biogeochemical cycle of many elements such as carbon or sulfur (Jørgensen, 1982a; Wolicka, Borkowski, 2007). The content of SRB in the terminal stages of organic matter mineralization in marine sediments exceeds 70% (Jørgensen, 1982b). SRB are introduced in the anaerobic biodegradation of organic compounds at the level of low-molecule compounds such as organic acids, e.g. acetic, propionic acid, formic, or alcohols, e.g. ethanol, propanol, butanol, etc., because most of them do not produce hydrolytic enzymes.

Due to the lack of chemical electron acceptors, such as oxygenated sulfur compounds (e.g. sulfates, sulfites, thiosulfates, or elemental sulfur), the transfer of electrons on a biological acceptor may take place by using hydrogen as is the case for methanogenic archaea. This mechanism enables the persistence of bacteria (e.g. reducing sulfates), in settings with poor availability of electron acceptors. This property is known as syntrophic growth (Nazina et al., 2007). Some evidences that confirm the biological activity of SRB are the concentration of hydrogen sulfide produced, the decreased concentration of SO_4^{2-} ions in relation to their concentration in the injected seawater; as well as the increased concentration of the isotopic sulfur in sulfates and its decreased in hydrogen sulfide occurring in the gas accompanying the reservoir (Rozanova et al., 2001).

The most common mesophilic SRB causing detrimental effects on drilling equipment and oil storage vessels include *Desulfovibrio longus, D. vietnamensis,* and *D. gabonensis*. These species incompletely oxidize organic compounds to acetate and use energy from the oxidation of hydrogen, lactate, and pyruvic acid. However, the *D. longus* bacteria are not considered autochthonous, as is the case of *Desulfotomaculum halophilum* (Magot et al., 2000).

In contrast to the *Desulfovibrio* bacteria, the *Desulfomicrobium apsheronum* bacteria, which belong to the SRB group and are tolerant to high salt content, are autotrophic (Rozanova et al., 1988). In the case of the *Desulfobacter vibrioformis* bacteria, which were isolated from the oil-water separation systems, use acetate as their only source of energy and carbon is used in sulfate reduction; while, the *Desulfobacterium cetonicum* bacteria have the ability to oxidize ketones to carbon dioxide. Research based on 56 samples of oil collected from several oil reservoirs using molecular techniques showed that the SRB may be grouped in communities preferring freshwater or brine (Magot et al., 2000).

Thermophilic SRB are mainly responsible for processes of *in-situ* oil transformation. An important genus from this group is the *Desulfotomaculum* bacteria. The *D. kuznetsovii* bacteria were found in a reservoir in the Paris Basin. In the Norwegian sector of the North Sea, the bacteria D. *thermocisternum* have been detected, which incompletely biodegrade hydrocarbons to compounds such acetate, lactate, ethanol, butanol, and carboxylic acids in the presence of sulfates. New subspecies, the *D. nigrificans – salinus* bacteria that incompletely oxidize lactate and alcohols to acetate were identified in oil samples from western Siberia (Nazina et al., 2005). In oil samples from the North Sea, bacteria such as *Desulfacinum infernum, Termodesulforhabdus norvegicus,* and *Thermodesulfobacterium mobile* were found. These first two bacteria the *Desulfacinum inferno* and *Termodesulforhabdus norvegicus* oxidize completely acetate, butyrate, and palmitate to carbon dioxide. The *T. norvegicus* bacteria also utilize alcohols (Jeanthon et al., 2002). At high temperatures up to 80–85°C (optimum temperature range 60–65°C), the thermophilic bacteria *Thermodesulfobacterium* is capable of growing (Magot *et al.*, 2000; Stetter & Huber, 1999). *T. mobile* bacteria were isolated from a reservoir in the North Sea and the *T. commune* bacteria were isolated from a reservoir located in the eastern part of the Paris Basin. Electron donors for these species include hydrogen, formate, lactate, and pyruvic acid (Magot *et al.*, 2000). In turn, at higher temperatures (at an optimum temperature of 75°C) heterotrophic hyperthermophilic bacteria exist from the genus *Archaeoglobus*, which use lactate, pyruvic acid, and valerate in the presence of hydrogen as the carbon source (Stetter at al., 1987; Stetter & Huber, 1999). Genetically, these microorganisms are close to bacteria occurring near submarine vents. Thermophilic SRB have been discovered in the lower parts of the White Tiger oil reservoir. These SRB are probably autochthonous because they were not detected in the injected water or supplying boreholes. Nonetheless, the same samples yielded mesophilic aerobic bacteria and methanogenic archaea, suggesting the presence of a fractured system running through areas with low temperatures up to the productive horizon. It is thus not clear whether thermophilic SRB are derived from submarine vents or oceanic ridges, where bacterial sulfate reduction takes place at temperatures close to 100°C (Rozanova et al., 2001).

SRB play a significant role in oil reservoirs, mainly due to the ability to metabolize various organic compounds, including aliphatic, aromatic, and polycyclic aromatic hydrocarbons (PAHs). In anaerobic ecosystems, the process of organic matter mineralization is usually much more complex than in aerobic conditions and requires the co-operation of different microorganism groups. Each group has its own specific stage of substrate oxygenation and the final products are metabolized by the next link of the food chain until complete mineralization

Table 1 outlines several species of microorganisms reducing sulfates that have been isolated from crude oil exploitation areas.

Species	**Salt Content (wt%)**	**T (°C)**	**Occurrence**
Desulfotomaculum nigrificans	0–4	40–70	Oil field
Desulfacinum infernum	0–5	40–65	
Thermodesulfobacterium mobile	lack of data	45–85	
Thermodesulforhabdus norvegicus	0–5.6	44–74	
Archaeoglobus fulgidus	0.02–3	60–85	Formation water
Desulfomicrobium apsheronum	0–8	4–40	
Desulfovibrio gabonensis	1–17	15–40	
Desulfovibrio longus	0–8	10–40	
Desulfovibrio vietnamensis	0–10	12–45	
Desulfobacterium cetonicum	0–5	20–37	Formation waters
Desulphomaculum halophilum	1–14	30–40	Drill bit
Desulfobacter vibrioformis	1–5	5–38	Oil-water separator
Desulfotomaculum thermocisternum	0–5	41–75	Marine Sediments

Table 1. Species of sulfate reducing microorganisms isolated from different areas of crude oil exploitation (Magot et. al., 2000)

According to literature data (Magot et al., 2000; Stetter&Huber, 1999) SRB isolated from crude oil and formation waters are characterized by a wide tolerance range in relation to salt content (0–17 %) and temperature (4–85°C).

2.2.3 Methanogenic archaea

Methanogenic archaea bacteria are the next important group of microorganisms occurring in crude oil reservoir settings (Magot et al., 2000; Nazina et al., 2007). The product of their activity is methane; therefore the biological activity of these microorganisms is measured by the methane production rate or by the volume of methane produced.

Methanogenic archaea bacteria occur in diverse settings. Their development and activity is influenced by physical and chemical factors such as temperature, salt content, and pH. Most methanogenes are mesophilic organisms, although extremophiles are also present. The latter include *Methanopyrus kandleri,* which can occur at temperatures of 110°C (Kurr et al., 1991) and *Methanococcus vulcanicus* (Jeanthon et al., 1999). There are very few publications on psychrophilic methanogenic archaea. Important conditions for the development and activity of methanogenic archaea are the salt content of the environment and the lack of oxygen. These microorganisms are very sensitive to very low concentrations of oxygen even in the range of several ppm (Elias et al., 1999).

Methanogenic archaea is physiologically nonuniform. Representatives of this group are known to produce hydrogen sulfide in the process of sulfur reduction (Mikesell & Boyd, 1990). Autotrophy is a common phenomenon among archaea. Many methanogenic archaea may bind carbon dioxide as well as use methanol or acetate as the carbon source to synthesize organic compounds.

Methanogenes include prototrophic species, requiring only CO_2, H_2, and mineral salts for growth. An example of this group is the *Methanobacterium thermoautotrophicum* (Zeikus & Wolfe, 1972). However, most methanogenes utilize hydrogen as the electron donor and carbon dioxide as the electron acceptor. The final product of this process is methane. The process of biogenic methane formation is known as methanogenesis, which is specific from anaerobic respiration with low energy yield. Most species, however, require very specific compounds for methanogenesis to take place such as vitamins (e.g. biotin or riboflavin). Methanogenic archaea take part in the final stage of organic matter degradation at strictly anaerobic conditions and very low reduction potential (–330 mV).

Mesophilic and thermophilic methanogenic archaea may occur in settings with various salt contents, however the simultaneous presence of high temperatures and high salt concentrations may hamper the activity of methanogenic archaea. *Methanococcoides (Methanohalophilus) euhalobius* was detected among methanogenic caryopsis utilizing methylamines in samples taken from oil reservoir drilling operations having salt contents of 140g NaCl/l. The presence of methanogenic archaea using methyl substituents in environments having intermediate to very high salt concentrations has been linked to the amine degradation that took part during osmoregulation; a similar process could occur within oil reservoir environments (Sowers & Ferry, 1983).

The next group of methanogenic archaea is represented by species utilizing acetate as the electron donor. Methane production was observed in samples from oil field reservoirs located in the North Sea and western Siberia that were inoculated with yeast extract or acetate (Magot et. al., 2000). In this work, the production of methane was used to detect the presence of methanogenic archaea, although they were not isolated (Gray et al., 2009; Kotsyurbenko et al. 2004). In these cases, a syntrophic bacterial community involving acetate oxidizing bacteria and methanogens (utilizing hydrogen) could be formed. Similar conclusions were drawn in relation to the Kongdian oil reservoir in China (Nazina et al., 2007). The situation may be explained by the competition between SRB and methanogenic archaea (MA). SRB utilize sulfates as the final electron acceptor in anaerobic respiration, a process that is energetically more favorable than methanogenesis. In fact, as a result of the SRB activity the electron donors available in the environment are utilized and methanogenesis is slowed down (Fig. 5).

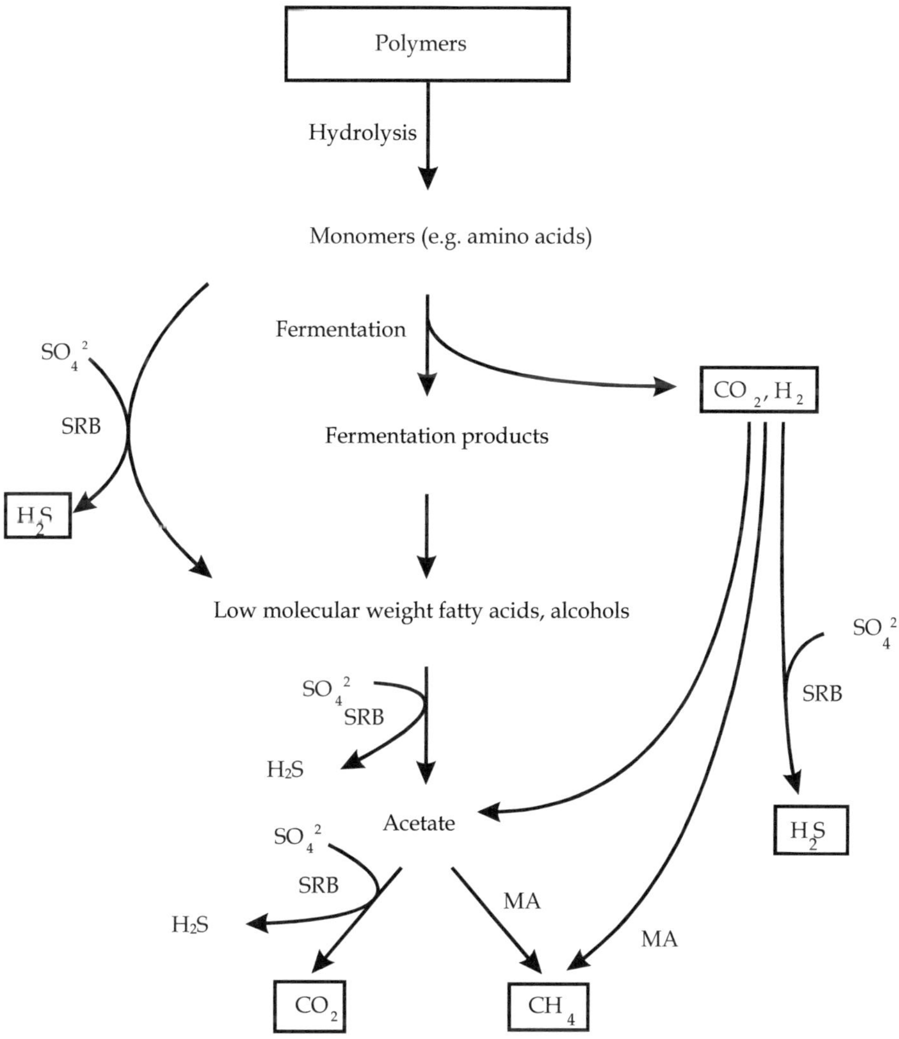

Fig. 5. Modified scheme of the biodegradation of organic compounds with the participation of various groups of microorganisms, based on Kalyuzhnyi et al. (1998); SRB - sulfate reducing bacteria, MA - methanogenic archaea.

The dominance of a group of microorganisms depends mainly on the concentration of sulfates in the environment and the organic carbon content (COD). The calculated COD/SO_4 ratio for the environment supplies information on the possible selection of a particular microorganism group (Hao et al., 1996). If the COD/SO_4 ratio reaches 0.67 or less, then according to the stoichiometry of the sulfate reduction process, the bulk of the organic compounds may be mineralized through production of CO_2 and H_2S. If the COD/SO_4 ratio is higher than 0.67, then more organic compounds are decomposed during methanogenesis. Thus based on the known concentration of both components (organic carbon [C_{org}] and S [SO_4]) it is possible to determine (or create) conditions favorable for the development of selected group of microorganisms (Oude Elferinck et al., 1998).

The most common and often prevailing group of methanogenic archaea, particularly in formation waters having low salinity concentrations, corresponds to hydrogen oxidizing methanogenic species. They include the disc-shaped *Methanococcus termolitotrophicus, Methanoplanus petrolearius,* and *Methanocalculus halotolerans;* and the rod-like *Methanobacterium thermoautotrophicum, M. bryantii,* and *M. ivanovii.* Although hydrogen oxidation during methanogenesis at surface conditions may take place only at NaCl concentrations below 9 %, the species *Methanocalculus halotolerans* are capable of carrying out this metabolic activity at conditions of high salt content up to 12 %. This capability may be an evidence of their autochthonous character related to oil reservoir ecosystems (Ollivier et al., 1998).

2.2.4 Iron III reducing bacteria

Shewanella putrefaciens is an iron reducing bacteria that also has the ability to reduce elemental sulfur, sulfites, and thiosulfates to sulfides. This bacteria is capable of withstand the harsh conditions of oil reservoirs. The electron donor may be H_2 or formate, and the acceptors - iron oxides and hydroxides. *Deferribacter thermophilus* is a bacterium that besides iron reduction also reduces manganese and nitrates using yeast extract or peptone. The source of energy is hydrogen and numerous organic acids. It is, however, not clear whether such type of metabolism occurs in conditions *in-situ* due to the lack of data on the content of iron and manganese ions (Grenne et al., 1997).

3. Microbiological biodegradation of crude oil components

The biodegradation of hydrocarbons *in-situ* (within the oil reservoir) significantly affects the composition of the crude oil. Peters & Moldowan (1993) proposed a 10-level biodegradation scale, in which 0 corresponds to unchanged material, 5 represent moderate biodegradation, and 10 corresponds to advanced biodegradation. The determination of the biodegradation stage may be also based on the relations of pristane and phytane (Pr+Ph) to nC17 + nC18, C30 alpha beta hopane to Pr+Ph, and 25-nor C29 alpha beta hopane to C30 alpha beta hopane. Depending on the stage of the process, different proportions of particular hydrocarbons are contained in the crude oil. Thus the composition of the crude oil is linked to the susceptibility of the specific crude oil to microbiological degradation. In crude oil reservoirs, a factor complicating the determination of the degree of oil biodegradation is the influx (migration) of oil from the source rock. If the initial fraction of oil was subjected to microbiological transformations and portions of unchanged oil were introduced later, then the total product will have very different composition, with biomarkers pointing to both high and low microbiological evolution of the raw material. Moreover, there will be also differences in the physical properties of the oil such as viscosity. Complete homogenization of crude oil is not possible even in the geological time scale. Due to processes of oil mixing and diffusion its composition may become uniform only locally.

Generally, biodegradation of crude oil takes place in the contact zone between crude oil and water. The resulting solution gradient causes influx of substances prone to degradation such as *n*-alkanes and isoprene alkanes into this zone and outflow of reaction products in the opposite direction. Mixing of biodegraded oil with inflowing unchanged oil takes place also in this region. At the rise of 25-nor hopane concentration, a typical feature is the removal of *n*-alkanes, followed by isoprenoid hydrocarbons with the increase of biodegradation rate. At reservoir scale, advanced biodegradation is favored by moderate temperatures in shallow reservoirs.

The influence of biodegradation on the crude oil composition is reflected in the concentration of particular compound groups. For example, in the case of PAHs, three elements determine their susceptibility to microbiological degradation: the number of rings in the compound, the number of alkyl substituents, and the location of the bonds of these substituents. The percentage content of hydrocarbons with large number of rings is inversely proportional to the rate of biodegradation (Huang et al., 2004). Similar relationship takes place in the case of the number of substituents, e.g. dimethylnaphthalene is more susceptible to biodegradation than trimethylnaphthalene. In the case of methylnaphthalenes, the most thermodynamically stable isomers are very quickly decomposed. This indicates that the stereochemical configuration of the hydrocarbon compounds is the dominant factor affecting biodegradation of crude oil rather than thermodynamic effects. In the case of methylphenanthrenes, the highest resistance to biodegradation was observed in those with methyl groups in positions 9 and 10. In the case of steroid hydrocarbons such as mono-aromatics (MSHs) and tri-aromatics (TSHs) hydrocarbons, MSHs are more resistant to microbiological degradation than the TSHs. Branched TSHs with short side chains are more effectively biodegraded than branched TSHs with long chains. In turn, short chain pregnanes are rather resistant to biodegradation in comparison to long chain typical steranes. High thermal maturity is related to high values of the concentration ratios of these compounds, while the degree of biodegradation increases when the ratio [C21/(C21+C28)] decreases and when the ratio of [C21–22-pregnanes/(C27–29-steranes)] increases. No trend has been identified for the ratios of short to long chains of MSHs. This is because thermally mature crude oils show higher concentrations of thermodynamically stable components. Thus, indicators for crude oil thermal maturity might be influenced by biodegradation because bacteria generally decompose thermodynamically stable components very quickly. In addition, thermal maturity trends and biodegradation indicators may differ between oil reservoirs due to different species within the bacterial population, red-ox conditions, crude oil composition, and nutrients availability (Huang et al., 2004).

In-situ biodegradation of the light hydrocarbon fraction, e.g. n-paraffins, iso-paraffins, cycloalkanes, as well as benzene and alkylbenzens may also take place in crude oil reservoirs (Vieth & Wilkes, 2005). All these compounds are biodegraded under anaerobic conditions that typically occur in oil reservoirs. However, these bacteria are highly selective for metabolism and microorganisms capable of simultaneously biodegrade aromatic hydrocarbons and paraffins have not been found yet.

Generally within the non-aromatic hydrocarbons, bacteria degrade first n-paraffins and then iso-paraffins therefore an increase of the i-C5/n-C5 ratio may be used as an indicator of the biodegradation progress. In the case of methylcyclohexane/n-C7 ratio, the same trend also indicates a larger degree of oil degradation. Another method that allows the determination of oil biodegradation degree is the study of the isotopic ratios of particular compounds. Microorganisms' preference toward light isotopes causes the enrichment of heavier isotopes in the crude oil. This method may be, however, applied only in the case of low molecular weight carbon compounds. The reason is that the activation of the substrate takes place in most cases on only one carbon atom, thus its share in the bulk mass of the hydrocarbon compound would be very small and the isotopic effect would be immeasurable. The removal of light oil fractions due to biodegradation is an unfavorable event because it causes the decrease of the economical value of the crude oil (Vieth & Wilkes, 2005).

The biodegradation of oil within the reservoir rock may cause the formation of an upper gas cap, which is composed mainly of CH_4 that is formed during the reduction of CO_2 with H_2 or methyl acetate. If the oil contains high quantities of sulfur, the activity of SRB would result in the formation of H_2S at concentrations that may reach over 10 % in the gas. Similarly, the activity of other microbes may result in the transformation of primarily wet gas into dry gas containing over 90 % methane. This process is linked to the preferential removal of C3 to C5 components by bacteria; thus the residual methane will be isotopically heavier and the isoalkane to *n*-alkane ratio will be increased. At the same time, the biodegradable oil fraction will become heavier; therefore the course of changes is two-directional. For instance, this type of situation has been observed in in the Troll and Frigg reservoirs in the North Sea and in other reservoirs characterized by a dominant gas cap surrounded by an oil ring. Numerical simulation indicates that these gas reservoirs could be formed by the significant biodegradation of oil, condensate, and gas in cooperation with various physical and chemical processes with the simultaneous migration of part of the gas from the hydrocarbon trap through microfractures or communicating faults (Larter & Primio, 2004).

There is also a link between the acidity of the oil, which is measured as the content of KOH required for neutralizing this acidity, and the degree of oil biodegradation. This is observed particularly in oil with high acid value (>0.5 g KOH/g oil). According to Meredith et al. (2000), this fact is connected with the production of carboxylic acids during *in-situ* microbiological degradation of hydrocarbons. However, other factors also influence the total acid value of the oil, because some non-biodegraded oils have low pH, which is probably linked with high sulfur content. In oil with advanced biodegradation, the concentration of hopane acids increases, however this increase is not enough to affect the overall acidity of the crude oil.

4. Mineral-forming processes

There is a closed relationship between crude oil, formation waters, and microorganisms (Wolicka et al., 2011). The physical and chemical conditions in oil reservoir environments selectively influence the development of particular microorganism groups; while simultaneously microorganisms may modify the chemical composition of the reservoir environment through their biological activity (Douglas & Beveridge, 1998; Onstott et al.,

2010). Nowadays, the role of microorganisms is an important subject in oil geology, due to the fact that they influence geological processes in hypergenic settings. Various mineral phases formed in biological processes may significantly influence microorganism activity and metabolism, as well as the physical and chemical properties of the environment. In many reports devoted to the crystallization of mineral phases, rock- and ore-forming processes and sedimentation in hypergenic conditions, there are evidences of the role of microorganisms in the formation of carbonates, sulfides or elemental sulfur in anaerobic conditions, however the information is insufficient and often the role of microorganisms in geological processes is just mention (Popa et al., 2004; Borkowski & Wolicka, 2007a, b; Wolicka & Borkowski, 2008b).

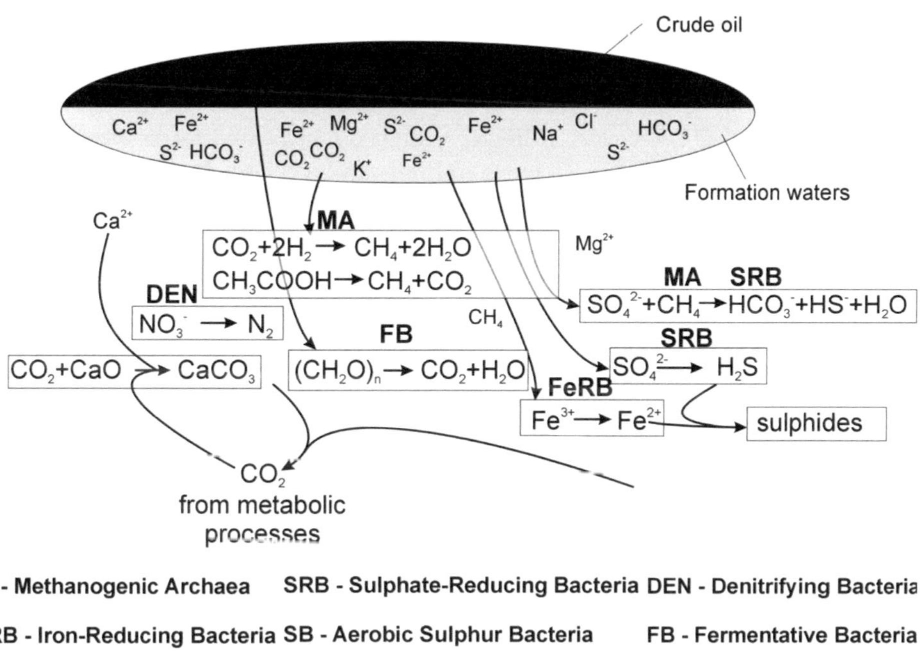

Fig. 6. Possible microbiological reactions in oil reservoirs from different groups of microorganisms.

Crude oil and oil refined products contain various organic compounds, mainly hydrocarbons, both aliphatic and aromatic. Biodegradation of crude oil and oil-derived products and formation of mineral phases are natural processes. Knowledge on these issues is, however, restricted, particularly in the case of anaerobic conditions. Particular attention is drawn to SRB, because they are the main link between the biogeochemical cycles of carbon and sulfur in anaerobic conditions. These microorganisms play an important role in the biodegradation of organic compounds and in mineral-forming processes. In some conditions, they are capable of forming mineral phases and to degrade minerals, e.g. aluminosilicates (Ehrlich, 2002). Mixed populations of SRB may begin the precipitation of many different minerals, e.g. carbonates such as calcite, dolomite, and siderite (Perry & Taylor, 2006; Wolicka & Borkowski, 2011), phosphates (e.g. apatite), elemental sulfur (Wolicka & Kowalski, 2006a, b), or sulfides (Labrenz et al., 2000). The process of dolomite, calcite, and aragonite formation in settings typical of SRB and in conditions favoring sulfate reduction is commonly described from anaerobic zones of salinated lagoons (Warthman et al. 2000), in gypsum and anhydrite deposits (Peckmann et al. 1999), as well as in salinated natural lakes rich in sulfates (Wright, 1999a, b). Figure 6 summarizes possible microbiological reactions taking place in oil reservoirs from different groups of microorganisms.

Secondary iron sulfides (pyrite, marcasite) and other metals, e.g.: galena (PbS), sfalerite (ZnS), chalkopyrite ($CuFeS_2$), chalkozine (Cu_2S), covellite (CuS), cinnabar (HgS), and realgar (AsS) are formed during dissimilative sulfate reduction.

5. Microbial enhanced oil recovery

The presence of autochthonous microorganisms, as well as their persistence and reproduction within crude oil reservoirs is advantageous for the implementation of Microbial Enhanced Oil Recovery (MEOR) methods, which are also known as Microbial Improved Oil Recovery (MIOR), allowing additional oil recovery at relatively low cost and the extension of the exploitation life of mature reservoirs, otherwise abandoned.

There are several approaches to take advantage of the natural metabolic processes of autochthonous microorganism to boost oil recovery as follows.

1. Biosurfactant sources. Based on the fact that many microorganisms synthesize biosurfactants, a microorganism community with such properties may be selected for the controlled production of biosurfactants for injection into the reservoir to decrease the interfacial tension between water and oil, thus releasing oil trapped in the rock by capillary forces. Furthermore, the presence of biosurfactants in the reservoir environment aids the biodegradation of hydrocarbons by autochthonous microorganisms. Similarly, biosurfactants may also become an easily accessible carbon source for autochthonous microorganisms, which produce biogenic gases such as CO_2, CH_4, and H_2 that would increase the reservoir pressure, which favors the displacement of crude oil towards production wells.
2. Biogenic gas production. The direct introducing of selected microorganisms that are capable of crude oil biodegradation with the ultimate goal of producing biogenic gases to increase the reservoir pressure.
3. As fluid diverting agents by producing biofilms. Microorganisms develop biofilms on the reservoir rock surfaces, which cause plugging of watered pores, thus redirecting the injected water to upswept areas of higher oil saturations making waterflooding more effective in displacing oil toward the production wells.
4. Control of H_2S production. The activity of SRBs can be hampered by the introduction of chemical compounds into the reservoir that would increase the activity of autochthonous microorganism groups to slow down the activity of the SRBs.

The main advantages of MEOR over conventional enhanced oil recovery (EOR) methods are much lower energy consumption and low or non-toxicity that makes MEOR an environmentally friendly process. Lower costs come from the fact that after the introduction of microorganisms into the reservoir their growth increases exponentially, therefore it is possible to obtain a large number of active substances from a small volume of initial organic material. After the initial injection of microorganisms, it is only necessary to supply nutrients into the reservoir rock to enable the development of the existing bacterial population that will produce the required biosurfactants. Some of the common microorganisms used in MEOR processes include: *Pseudomonas aeruginosa, Bacillus licheniformis, Xanthomonas campestris* and *Desulfovibrio desulfuricans* (Singh et al., 2007).

Biotechnological approaches use microorganisms that oxidize oil to break-up asphalthene-resin-paraffin sludges that have been accumulated in the wellbore for years affecting the oil production due to plugging of the oil production zones. In mature reservoirs, it is common to find production wells rendering water cuts of 70 to 90 % of the total production. The microbial stimulation of aged production wells to break-up unwanted paraffin/asphaltene deposits is an economical way to reactivate old production zones.

In practice, the method may be difficult to implement because uncertainty in providing suitable conditions for the development of the microorganism groups taking part in the metabolic pathway. Furthermore, after microbial application it is essential to remove the bioproducts of microorganisms' activity, as well as the bacteria themselves in order to maintain the composition of the crude oil (it is not desirable to give up light fractions of the crude oil to biodegradation). For instance, *in-situ* bacterial cultivation may cause decrease of oil production. Another disadvantage of MEOR is the unpredictability of the process (Mokhatab, 2006). Other problems linked to MEOR include plugging of the reservoir rock by the bacterial mass in undesirable locations, *in-situ* biodegradation of the applied chemical compounds, and acidification of the crude oil by the bioproduction of hydrogen sulfide in the reservoir (Almeida et al., 2004).

5.1 Examples of Microbial Enhanced Oil Recovery applications

The application a MEOR process in the Dagang oil field, in China was reported by Liu Jinfeng (2005) In this MEOR application, zymogenic organisms (*Arthrobacter, Pseudomonas* and *Bacillus*) were introduced into the reservoir. These microorganisms have the ability to oxidize crude oil. *Arthrobacter* and *Pseudomonas* are highly effective oxidizing crude oil, whereas *Bacillus* bacteria produce surfactants in crude oil fermentation and consequently reduce the oil/brine interfacial tension. The field trial was initiated with the injection of nutrients for 3 days into the reservoir, followed by the injection of the suspension of *Arthrobacter* and *Pseudomonas*. The wells were shut in for 10 days to cultivate the introduced microorganisms. After 10 days, a suspension of *Bacillus* was injected in the reservoir and the wells were shut in again for another 10 days. Follow by an additional injection of nutrients (Jinfeng et al., 2005). Positive results were observed; significant changes in crude oil and gas properties were observed. The density of the oil decreased by 0.0024 g/cm^3, oil viscosity by 4.1 mPas, paraffin content by 4.35% and asphalthenes by 2.31%. The content of methane in the accompanying gas increased from 85.4 % to 90.02 % and the content of CO_2 decreased from 5 % to 1.5 % – these favorable changes of gas composition are suggested to result from the activity of methanogenic archaea.

This field trial shows that MEOR may be successfully applied in high-temperature crude oil reservoirs. In this field trial, the reservoir temperature corresponded to the upper thermal boundary of 70°C for the development of thermophilic bacteria (Jinfeng et al., 2005).

Not all reservoir microorganisms can be applied in MEOR process; some such as the SRB may even be harmful for the exploitation and processing of crude oil. Very often applications of MEOR can result in the increase of hydrogen sulfide activity and production. This phenomenon is very unfavorable and cannot be controlled entirely due to the complex nature of biochemical and abiotic processes taking place in the reservoir and in the exploitation area. Figure 7 summarizes some of the negative and positive effects of microbial activity in the reservoir.

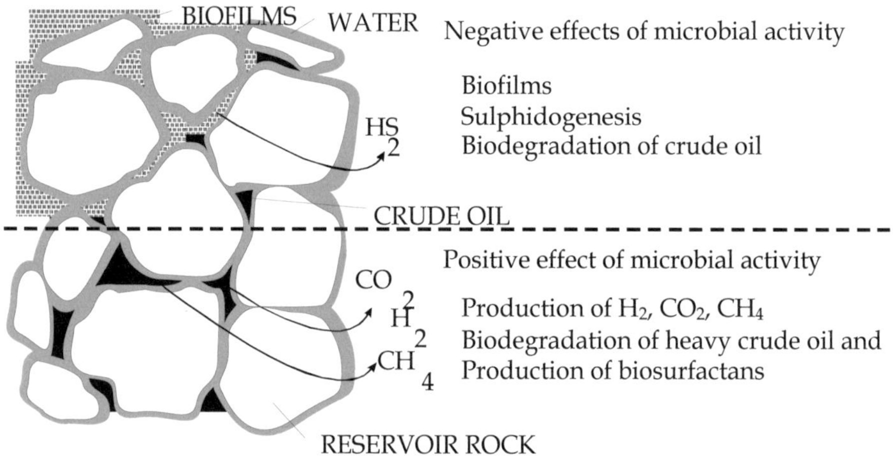

Fig. 7. Negative and positive effects of microbial activity in oil reservoirs

5.2 Changes of the reservoir rock permeability

A well-known and often studied MEOR method is the selective plugging of reservoir zones with the purpose of diverting injected water towards reservoir areas of high oil saturations. In the North Burbank in Oklahoma (1980) the growth of aerobic and anaerobic heterotrophic bacteria, SRB, and methanogenic halophilic archaea was stimulated by the addition of nutrients; the rapid growth of the bacteria population caused the decrease of permeability in the treated well by 33 %. Bacteria blocked highly permeable zones of the reservoir rock diverting water injected toward low permeability areas containing high saturation of crude oil. It is suspected that the microorganisms' mass and the products of their biological activity such as biopolymers, effectively blocked the flow of water within large pores and fractures. Biopolymers allow bacteria to connect and form biofilms, which improves nutrient gain and decreases their sensitivity to toxic substances (Mokhatab, 2006).

5.3 Biological demulsification of crude oil

MEOR applications also include the demulsification of crude oil. In crude oil processing there are two types of emulsions, e.g. oil in water and water in oil. These emulsions are formed at different stages of crude oil exploitation, production, and processing, causing significant operational problems to the industry (Singh et al., 2007). The release of oil from these emulsions is known as demulsification. Traditionally demulsification methods include: centrifuging, warming, electric current application, and the addition of chemical compounds to break up the emulsions.

However, the application of microbiological demulsification methods during drilling operations have been practiced with success, thus saving on transportation and on equipment costs (Leppchen, et.al.; 2006). There are many microorganisms with favorable demulsification properties including *Acinetobacter calcoaceticus, A. radioresistens, Aeromonas* sp., *Alcaligenes latus, Alteromonas* sp., *Bacillus subtilis, Corynebacterium petrophilium, Micrococcus* sp., *Pseudomonas aeruginosa, P. carboxydohybrogena, Rhodococcus aurantiacus, R. globerulus, R. rubropertinctus, Sphingobacterium thalophilum,* and *Torulopsis bombicola.* Microorganisms benefit from the double hydrophobic-hydrophilic nature of surfactants or the hydrophobic cell surface to remove emulsifiers from the interfacial surface between oil and water. Temperature increase is favorable for the demulsification process, because it decreases viscosity, increases the density difference between the phases, attenuates the stabilizing action of the interfacial surface, and increases the drop collision rate, which leads to coalescence (Singh et al., 2007).

6. Role of microorganisms in the biodegradation of crude oil

Microorganism activity serves to remove pollution by crude oil and oil-derived products from soil, groundwater, and seawater occurring near exploitation sites, leaking pipelines, and in dispersed locations such as petrol stations or roadsides.

The basic factors influencing the biodegradation of oil-derived products in soil include: chemical structure, concentration and toxicity of hydrocarbons in relation to the microflora, microbiological soil potential (biomass concentration, population variability, enzyme activity), physical-chemical environmental parameters (e.g. reaction, temperature, organic matter content, humidity), and availability of hydrocarbons for microorganism cells.

6.1 Biodegradation in aerobic conditions

Many soil microorganisms - bacteria and fungi - transform crude oil hydrocarbons into non-toxic compounds or conduct their complete mineralization to inorganic compounds. This natural microbiological activity is used in bioremediation to reduce the concentration and/or toxicity of various pollutants, including oil-derived substances. An unquestionable advantage of this technology is the low investment cost (complex and expensive technology is not required). Additionally, methods used in bioremediation do not involve introduction of chemical compounds that may negatively influence the biocoenosis to the soil and the soil may be used again after the remediation. The most important issue is that these processes are natural and the final products of the microbiological degradation are carbon dioxide and water.

The effectiveness of hydrocarbon degradation by bacteria does not depend on the number of carbon atoms in the compound (Klimiuk & Łebkowska, 2005; Stroud, 2007). Branched alkanes are biodegraded slower, although the process is similar to that in *n*-alkane degradation. The latter are preferred as the source of carbon by microorganisms and in their presence the degradation of branched alkanes is much slower (Klimiuk & Łebkowska, 2005). Cycloalkanes are degraded under aerobic conditions by bacterial consortia in co-metabolism as well as by pure microbial strains. However, biodegradation of cycloalkanes with high number of rings in the compound is much slower (Klimiuk & Łebkowska, 2005).

Small molecule (two or three ring) aromatic hydrocarbons are degraded by many soil bacteria as well as numerous fungi genera e.g. *Rhizopus, Aspergillus, Candida, Penicillium, Psilocybe,* and *Smittum.* In turn, the ability to degrade large-molecule (4 or more rings) of PAHs is rather rare among bacteria (e.g. *Pseudomonas putida, P. aeruginosa, P. saccharophila, Flavobacterium* sp., *Burkholderia cepacia, Rhodococcus* sp., *Stenotrophomonas* sp. and *Mycobacterium* sp.). These compounds are rather degraded by ligninolytic fungi, such as: *Phanaerochaete chrysosporium, Trametes versicolor, Bjerkandera* sp., *Pleurotus ostreatus,* and non-lygninolytic fungi, such as *Cunninghanella elegant, Penicillium janthinellum,* and *Syncephalastrum* sp. (Austin et al., 1977; Kirk & Gordon, 1988; Wolicka et al., 2009).

Degradation of compounds with five or more aromatic rings depends largely on the activity of the mixed microorganism populations. The metabolism of symbiotic systems may be the only form of metabolism of these compounds. So far, it has been reported that the biotransformation of benzo[a]pyrene by bacteria took place in co-metabolic conditions (Bogan et., al, 2003).

The microorganism's ability to grow on large-molecules of polycyclic aromatic hydrocarbons is probably a common feature of *Mycobacterium.* This property was noted in several species, e.g. *M. flavescens* and *M. vanbaalenii* sp. and strains such as: AP-1, PYR-1, BB1, KR2, GTI-23, RJGII-135, BG1, CH1. Many bacteria from the genus *Mycobacterium* that biodegrade PAHs also have the ability to degrade aliphatic hydrocarbons.

6.2 Biodegradation at anaerobic conditions

Biodegradation of hydrocarbons by microorganisms at anaerobic conditions is common knowledge since the end of the 1980s (Widdel & Rabus, 2001; Meckenstock et al., 2004). Anaerobic microorganisms utilize monocyclic aromatic hydrocarbons, such as benzene, toluene, ethylbenzene, xylene (BTEX), hexadecane, and naphthalene as the sole carbon source. Strains RCB and JJ of *Dechloromonas* (β-*Proteobacteria*) completely oxidize benzene in anaerobic conditions using nitrate as the electron acceptor. *Geobacter metallidurans* and *G. grbicium* are capable of anaerobic toluene oxidation to CO_2 with reduction of Fe(III). Some organisms were reported to link anaerobic toluene degradation with nitrate respiration (*Thauera aromatica* strains K172 and I1, *Azoarcus* sp. strain T, *A. tolulyticus* strains To14 and Td15, *Dechloromonas* strains RCB and JJ), perchlorate respiration (*Dechloromonas* strains RCB and JJ) and sulfate respiration (*Desulfobacterium cetonicum, Desulfobacula toluolica*) (Chakraborty & Coates; 2004). A methanogenic consortium composed of two archaea species related to *Methanosaeta* and *Methanospirillum* and two bacterial species, of which one is linked to *Desulfotomaculum,* were reported to degrade toluene (Beller et al., 2002). Fluorescence *in-situ* hybridization (FISH) of denitrificators degrading alkylbenzens and *n*-alkanes indicated that the bacteria belong to the *Azoarcus/Thauera* group. SRB are known to utilize benzene, toluene, ethylbenzene, and xylene as the sole carbon source (Coates et al., 2002; Kniemeyer et al., 2003; Ribeiro de Nardi et al., 2007).

Polycyclic aromatic hydrocarbons may also be degraded at anaerobic conditions. So far, only few 2-ring PAHs have studied with regard to biodegradation under sulfate reduction conditions. Metabolism studies were mainly focused on naphthalene and 2-methylnaphthalene; while less attention was given to the anaerobic metabolism of phenanthrene. Among 3-ring PAHs, phenanthrene may be degraded through an initiating reaction step that could be carboxylation. Anaerobic degradation of 3-ring PAHs also takes place through co-metabolism, for instance, benzothiophene is biodegraded in the presence of naphthalene as the subsidiary substrate. Anaerobic degradation is also possible for heterocyclic compounds such as indole or chinoline (Meckenstock et al., 2004; Widdel & Rabus, 2001).

6.3 Hydrocarbon biodegradation by psychrophilic microorganisms

Bacterial oil degradation takes place in extreme conditions e.g. in Polar Regions (Aislabie et al., 2006). Populations of microorganisms utilizing hydrocarbons are dominated in this case by psychrotolerants microorganisms rather than by psychrophiles. *Rhodococcus, Sphingomonas, Pseudomonas, Acinetobacter, Shewanella,* and *Arthrobacter* are present at conditions of low temperatures (Aislabie et al., 2006; Kato et al., 2001). Some microorganisms that biodegrade alkanes include: *Rhodococcus, Acinetobacter,* and *Pseudomonas.* It seems that alkane degradation is conducted by microorganisms of the genus *Rhodococcus,* and *Pseudomonas* have been often found in areas where elevated hydrocarbon concentration existed.

Aromatic hydrocarbons may be degraded both by *Pseudomonas* and *Sphingomonas.* Obviously, temperature is a factor hampering the bioremediation process in polar settings, because defrosting of soil as a thin layer above the permafrost takes place only for 1 to 2 months. Additionally, there are significant daily and annual temperature oscillations. The maximum temperature of the exposed soil horizon may reach 20°C during the day. Low temperature results in decrease of oil viscosity that hampers microbial metabolism. Moreover, the evaporation of light compounds with toxic properties is restricted, which prolongs the time required by microorganisms to adapt to the toxic conditions.

Low temperature unfavorably influences the degradation of mixtures comprising various oil-derived compounds, whereas the course of degradation of uniform pollution composition decreases to a lesser degree. Twenty-four hour cycles of defrosting and frosting may increase the bioavailability of the required compounds due to changes in soil structure (Aislabie et al., 2006).

In Polar Regions, the difficulties faced by microorganisms during hydrocarbon bioremediation processes in soil are not only connected with low temperature conditions, but also with the presence of relatively low concentrations of biogenic elements. The successive pollution of already polluted environments causes further impoverishment of the required elements. A factor that could also restrict the bioremediation process under these conditions is the lack of water in the soil, as well as high pH values, particularly in seaside areas (Aislabie et al., 2006). So far in the published literature, only hydrocarbon biodegradation by aerobic action of microorganisms have been described for Polar Regions.

7. Bioremediation – a natural method of removing crude oil and oil-derived products pollution from the environment

Bioremediation is a biological method that uses living organisms to reduce or completely remove pollutants (organic and inorganic compounds) from polluted areas. In the case of crude oil and oil-derived products pollution, the most effective way of biological purification of water-soil environments is the application of microbiological methods. Such approach to bioremediation of areas polluted by crude oil can be conducted by the separation of groups of microorganisms capable of the effective biodegradation of many of the organic compounds occurring in oil-derived products and crude oil. In addition, microorganisms have the most complex and comprehensive metabolism among living organisms.

Understanding the metabolic interactions between diverse hydrocarbons and microorganisms requires knowledge on the various forms of occurrence and diverse properties of hydrocarbons in the environment.

Laboratory and field studies have allowed the determination of the optimal parameters for the bioremediation of soil polluted by oil hydrocarbons: biomass content over 10^5 cells/g dry mass, relative humidity 20–30 %, temperature 20–30°C, pH 6.5–7.5, oxygen content at least 0.2 mg/l hydrocarbons, Carbon:Nitrogen:Phosporous (C:N:P) = 100:10:1 or 70:7:1 (Szlaski & Wojewódka, 2003; Sztompka 1999).

7.1 Strategies for bioremediation

Soil bioremediation may be carried out *in-situ* at the polluted site, or *ex situ* through the removal of the polluted soil from the site and placing it in a specially prepared location. Important ways of *in-situ* purification include: soil cultivation, bioventilation, bioextraction, and *in-situ* biodegradation. Bioremediation methods *ex situ* include: soil cultivation, composting, biostacks, and bioreactors.

Three types of soil remediation are distinguished: natural bioremediation (natural attenuation), biostimulation, and bioaugmentation.

Natural bioremediation is based on the process of natural biodegradation (known also as natural attenuation) carried out by microorganisms; it requires only regular monitoring of pollution concentration. The method uses microorganisms naturally occurring in the environment without stimulating their development in any way. Parallel monitoring of the polluted area is carried out to evaluate the effectiveness of decontamination and recognition of process mechanisms (Klimiuk & Łebkowska, 2005; Tiehm & Schulze, 2003). An important element in bioremediation analysis is the determination of the range and location of zones of red-ox potential. This may be carried out through hydrogeochemical methods that are based on the measurement of relative impoverishment in terminal electron acceptors and the enrichment in typical reaction products such as Fe^{2+}. The disadvantage of this method is the influence of independent abiotic processes and interactions between the resulting metabolites (e.g. Fe^{2+} and H_2S) on the target measurement. Another possible analytical technique is the electrochemical determination of the red-ox potential, despite the fact that the final measurement may be the sum of potentials from many different reactions that to not reach equilibrium state fast enough to ensure precise measurements.

In the case of native populations in a given polluted area that do not reveal the expected pollutant degradation activity, which could be the result of the toxic effect of the pollutants on the native microorganisms and if the number and quality of microorganisms is deficient, then bioaugmentation is applied. This method is most commonly used for bioremediation of soils polluted with oil-derived products. In this method, microorganism communities, which are capable of biodegradation of oil-derived products, are isolated from the polluted area, cultivated at laboratory conditions, and later introduced into the polluted environment. Previous research indicates that a 4-ply increase of microorganism activity is reached by the application of this procedure (Suthersan, 1999).

Another approach to bioaugmentation is the application of genetic engineering to optimize the biodegradation effectiveness of these microorganisms, by increasing the content of enzymes important for the biodegradation of particular substances. At present, genetically modified microbes are typically used in bioremediation (Stroud et al., 2007).

Microorganisms are introduced into soils as free cells or as immobilized form on stable media. The latter method has many advantages. In the stable media there is high concentration of bacteria, which allows obtaining high biomass density; furthermore, the method prolongs the time of biochemical activity because of the protective effect of the medium. In these conditions, the excess of substrates and products does not hamper the enzymatic processes. Additionally, the medium improves soil structure, particularly in clay soils; the soil may also be recovered and reused. Expanded clay aggregates, woodchips, volcanic rocks, and plastic materials are used in the immobilization process (Klimiuk & Łebkowska, 2005).

Another means to increase bioremediation in polluted soils is to take advantage of the fact that the pollutant (hydrocarbon) itself could be the main source of energy (carbon) for bacteria survival. Yet another method for increasing the bacteria's accessibility to substrates in the environment is the addition of surfactants or surface-active substances (SAS), which can be produced chemically or formed naturally by microorganisms. SAS increase the mobility of hydrocarbons even at lower concentrations because the decrease of interfacial tension at the oil-water boundary. However, some downsides of the use of SAS are that at high concentrations, these compounds are toxic to microorganisms and may hamper their growth. The use of natural SAS (biosurfactants) to modify bacteria's accessibility to substrates (bioavailability) is recommended. In contrast to chemically synthesized surfactants, biosurfactants properties include high surface activity, biodegradability, and low toxicity. These substances belong to different groups, including lipopeptides, glycolipids, neutral lipids, and fatty acids.

Singh et al. (2007) and Stroud et al. (2007) have shown that the application of biological SAS accelerates the time for bacteria adaptation to the polluted site and render a 5 times increase of bioremediation effectiveness in comparison to the application of chemical surfactants.

Application of beta-cyclodextrins to increase bioavailability of hydrocarbons is an interesting method. Cyclodextrins (CD) have many advantages, e.g. they may form soluble complexes with hydrophobic substances and they are non-toxic. Thus cyclodextrins may accelerate bioremediation by increasing bioavailability. In an experiment conducted by Bardi et al. (2007), a 10-ply increase on the biodegradation of PAHs was obtained after 42 days of treatment in comparison to a sample without CD.

Similarly, the stimulation of the growth of autochthonous microorganisms was more effective with the simultaneous application of biosurfactants (natural SAS) and nutrients in comparison to a case when only nutrients were introduced to the environment (Klimiuk & Łebkowska, 2005). In some situations, despite the fact that biodegradation action has previously taken place, differences in bioavailability cause that some hydrocarbons remain in the environment, while others are removed completely (Stroud et al., 2007).

7.2 Advantages and disadvantages of bioremediation

Bioremediation is an effective method for removing hydrocarbon pollution from various environments. Stimulation of microorganisms at polluted sites may solve the pollution problem at low cost. Conventional remediation methods usually generate higher costs or move the pollution problem to another location. In contrast, bioremediation is a natural and non-hazardous process to the environment. Bioremediation can be carried out at the polluted site without changing the existing land management plans (Vidali, 2001).

As every method, bioremediation has its drawbacks. Not all compounds are susceptible to relatively fast and complete degradation. In some cases, products of degradation of oil-derived products have toxic effects on microorganisms. Furthermore, sometimes bioremediation requires the application of complex techniques in the presence of mixtures of compounds that are randomly dispersed in the medium. Bioremediation lasts longer than chemical methods, usually several years; moreover, its exact duration cannot be precisely determined. *In-situ* bioremediation methods are not always easy to monitor, whereas *ex-situ* bioremediation methods are more expensive due to transportation costs and soil storage (Zieńko & Karakulski, 1997; Vidali, 2001).

8. Conclusions

The role of bacteria in crude oil reservoirs is linked to numerous interactions between oil and the environment. Bacteria influence the chemical composition of crude oil, the conditions of its *in-situ* exploitation through the decomposition of some oil fractions, through the production of metabolism products such as biopolymers, biosurfactants, organic acids, and gases (CH_4, CO_2, H_2S, H_2, among others), and by the presence of microorganism's cells that may change the properties of the reservoir rock.

The controlled influence of parameters such as oil-water interfacial tension, oil viscosity, and rock permeability through the use of microorganisms are the basis of Microbial Enhanced Oil Recovery, which has great potential as an EOR process. MEOR methods can be applied in a variety of oil reservoir conditions such as low or high temperature or at high salinity concentrations. The versatility of MEOR applications is explained by the natural adaptation of bacterial species to oil reservoir conditions. Microorganisms play also a positive role in the removal of pollutants in crude oil fields or along transportation routes. Some microorganisms can biodegrade many of the chemical compounds contained in crude oils. Bioremediations process may occur naturally (without human intervention); however natural bioremediation process may take a long time (years) to be completed.

The complex nature of biochemical processes, which take place in harsh environments such as crude oil reservoirs, causes several interpretation problems during the analysis of the models used to describe the microbiological interactions in these settings.

Although a complete description of the bio-interactions taking place in crude oil - microorganisms systems is still pending, the existing knowledge, even in a restricted mode, can be applied for the recovery of additional oil in mature oil fields (MEOR) and in bioremediation and biotechnology for the remediation of soil-water environments polluted by crude oil and oil-derived products.

9. References

Aislabie, J.; Saul, D. J. & Foght, J. M. (2006). Bioremediation of hydrocarbon-contaminated polar soils. *Extremophiles,* 10, pp. 171-179

Almeida, P. F.; Moreira, R. S. Almeida, R. C. C. Guimaraes, A. K. Carvalho, A. S., Quintella, C. Esperidia, M. C. A. & Taft, C. A. (2004). Selection and application of microorganisms to improve oil recovery. *Eng. Life Sci.* 4, pp. 319-325

Austin, B.; Calomiris, J. J. Walker, J. D. & Colwell R. R. (1977). Numerical taxonomy and ecology of petroleum-degrading bacteria. *Appl. Environ. Microbiol.* 34, pp. 60-68

Baker, B. J.; Moser, D. P. MacGregor, B. J. Fishbain, S. Wagner, M. Fry, N. K. Jackson, B. Speolstra, N. Loos, S. Takai, K. Lollar, B. Fredrickson, J. Balkwill, D. Onstott, T. C. Wimpee, C. F.& Stahl, D. A. (2003). Related assemblages of sulphate-reducing bacteria associated with ultradeep gold mines of South Africa and deep basalt aquifers of Washington State. *Environmental Microbiology,* 5, pp. 267-277.

Bardi, L.; Martini, C. Opsi, F. Certolone, E. Belviso, S. Masoro, G. Marzona, M. & Marsan F. A. (2007). Cyclodextrin-enhanced in situ bioremediation of polyaromatic hydrocarbons-contaminated soils and plant uptake. *J Incl Phenom Macrocycl Chem* 57, pp. 439-444

Bastin, E. S. (1926). The presence of sulphate reducing bacteria in oil fields water. *Science,* 63, pp.21-24

Beller, H. R.; Grbić- Galić, D. & Reinhard, M.(1992). Microbial degradation of toluene under sulfate-reducing conditions and the influence of ion on the process. Appl.Environ.Microbiol. 58, pp. 786-793

Bogan B.W.; Lahner, L.M., Sullivan, W.R. Paterek,J.R. (2003). Degradation of straight-chain aliphatic and high-molecular-weight polycyclic aromatic hydrocarbons by a strain of *Mycobacterium austroafricanum* J. App. Microbiol. 94, pp. 230-239

Borkowski, A. & Wolicka, D. (2007a). Geomicrobiological aspects of the oxidation of reduced sulfur compounds by photosynthesizing bacteria. *Polish Journal of Microbiology,* 56 (1), pp. 53-57

Borkowski, A. & Wolicka, D. (2007b). Isolation and characteristics of photosynthesizing bacteria and their utilisation in sewage treatment. *Polish Journal of Environmental Studies,* 16 (3B), pp. 38-42

Chakraborty, R. & Coates, J.D. (2004). Anaerobic degradation of monoaromatic hydrocarbons. *Appl. Microbiol. Biotechnol.* 64, pp. 437-446

Coates, J. D.; Chakraborty R.& McInerney. (2002). Anaerobic benzene biodegradation - a new era. *Research in Microbiology,* 153, 621-628.

Cohn, F. (1867). Beitrage zur Physiologie der Phycochromaceen und Floriden. *Arch. Mikroskopie Anatomie,* 3, pp. -60.

Dandie, C. E.; Thomas, S. M. Bentham, R. H. & McClure, N. C. (2004). Physiological characterization of Mycobacterium sp. strain 1B isolated from a bacterial culture able to degrade high-molecular-weight polycyclic aromatic hydrocarbons. *J. Appl. Microbiol.* 97, pp. 246-255

Davey, M. E.; Wood W.A. Key, R., Nakamura, K. & Stahl, D.A. (1993). Isolation of three species of *Geotoga* and *Petrotoga* : two new genera, representing a new lineage in the bacterial line of descent distantly related to the *Thermotogales. Systematic and Applied Microbiology,* 16, pp. 91-200

Douglas, S. & Beveridge, T. J. (1998). Mineral formation by bacteria in natural microbial communities. *FEMS Microbiology Ecology,* 26, pp. 79-88

Ehrlich H. L. (2002) . *Geomicrobiology,* Fourth Edition, Revised and Expanded. New York.

Elias, D. A; Krumholz, L. R. Tanner, R. S. & Suflita, J. M. (1999). Estimation of methanogen biomass by quantitation of coenzyme M. *Appl Environ Microbiol.* 65, pp. 5541–5545

Gałuszka, A. & Migaszewski Z. (2007). Environmenytal og geochemistry, *Wyd. Naukowo Techniczne;* Warszawa

Gibson, G. (1990). Physiology and ecology of the sulphate-reducing bacteria. *Journal of Applied Bacteriology,* 69, pp. 769-797

Gray, N. D; Sherry, A. Larter, S. R. Erdmann, M. Leyris, J. Liengen, T. Beeder, J. & Head, I.M. (2009). Biogenic methane production in formation waters from a large gas field in the North Sea, *Extremophiles,* 13, pp. 511–519

Greene, A.; Patel B.K.C. & Sheehy, A. J. (1997). *Deferribacter thermophiles* gen. nov., sp. nov., a Novel Thermophilic Manganese - and Iron - Reducing Bacterium Isolated from a Petroleum Reservoir. *International Journal of Systematic Bacteriology,* pp. 505-509

Hao, O. J.; Chen, J. M. Huang, L. & Buglass R. L. (1996). Sulfate-reducing bacteria. *Critical Reviews in Environmental Science and Technology,* 26, pp. 155-187

Huang, H.; Bowler, B. F. J. Oldenburg, T. B. P. & Larter S. R. (2004). The effect of biodegradation on polycyclic aromatic hydrocarbons in reservoired oils from the Liaohe basin, NE China. *Organic Geochemistry,* 35, pp. 1619-1634

Jeanthon, C.; L'Haridon, S. Cueff, V. Banta, A. Reysenbach, A. & Prieur, D. (2002). *Thermodesulfobacterium hydrogeniphilum* sp.nov., a thermophilic, chemolithoautotrophic, sulfate - reducing bacterium isolated from a deep-sea hydrothermal vent at Guaymas Basin, and emendation of the genus *Thermodesulfobacterium. International Journal of Systematic and Evolutionary Microbiology,* 52, pp. 765–772

Jinfeng, L.; Lijun, M. Bozhong, M. Rulin, L. Fangtian, N. & Jiaxi Z. (2005). The field pilot of microbial enhanced oil recovery in a high temperature petroleum reservoir. *Journal of Petroleum Science and Engineering,* 48, pp. 265-271

Jørgensen, B. B. (1982a). Mineralization of organic matter in the sea bed-the role of sulphate reduction. *Nature,* 296, pp. 643-645

Jørgensen, B. B. (1982b). Ecology of the bacteria of the sulphur cycle with special reference to anoxic-oxic interface environments. *Philosophical Transactions of the Royal Society,* 298, pp. 543-561

Kalyuzhnyi, S.; Fedorovich, V. Lens, P. Pol, L. H. & Lettinga, G. (1998). Mathematical modelling as a tool to study population dynamics between sulfate reducing and methanogenic bacteria. *Biodegradation,* 9, pp. 572-9729

Kato, T.; Haruki, M. Imanaka, T. Morikawa, M. & Kanaya S. (2001). Isolation and characterization of psychrotrophic bacteria from oil-reservoir water and oil sands. *Apl Microbiol Biotechnol.,* 55, pp. 794-800

Kirk, P. W. & Gordon A. S. (1988). Hydrocarbon degradation by filamentous marine higher fungi. *Mycologia,* 80, pp. 776-782

Klimiuk E. & Łebkowska M. (2005). *Biotechnology in protection of environmental* (in Polish). Wydawnictwo Naukowe PWN S.A, Warszawa

Kniemeyer, O.; Fischer, T. Wilkes, H. Glockner, F. O. & Widdel F. (2003). Anaerobic degradation of ethylbenzene by a new type of marine sulfate - reducing bacterium. *Applied and Environmental Microbiology*, 2, pp. 760-768

Kopp, R. E.; Kirschvink, J. L. Hilburn, I. A. & Nash C. Z. (2005). The paleoproterozoic snowball Earth: a climate disaster triggered by the evolution of oxygenic photosynthesis. *The Proceedings of the National Academy of Sciences*, 102, pp. 11131-11136

Kotsyurbenko, K. J.; Chin, M. V. Glagolev, S. Stubner, M. V. Simankova, A.N. Nozhevnikova & Conrad R. (2004). Acetoclastic and hydrogenotrophic methane production and methanogenic populations in an acidic West-Siberian peat bog. *Environmental Microbiology*, 6, pp. 1159–1173

Kowalewski, E.; Rueslåtten, I. Steen, K. H. Bødtker, G. & Torsæter O. (2006). Microbial improved oil recovery - bacterial induced wettability and interfacial tension effects on oil production. *Journal of Petroleum Science and Engineering*, 52, pp. 275-286

Kurr, M.; Huber, R. Konig, H. Jannasch, H. Fricke, H. Trineone, A. Kristjansson, J. K. & Stetter K.O. (1991). *Methanopyrus kandleri*, gen. and sp. nov. represents a novel group of hyperthermophilic methanogens, growing at 110 ° C, *Arch Microbiol.*, 156, pp. 239-247

Kwapisz, E. (1999). Problems of biodegradation of crude oil (in Polish). I Biotechnology Congres, Wrocław, October 23-24, 1999, pp. 227-229

Laanbroek, H. J.; Geerligs, H. J. Sijsma, L. & Veldkamp, H. (1984). Competition for sulfate and intertidal sediments. *Applied and Environmental Microbiology*, 47, pp. 329-334

Labrenz, M.; Druschel, G. K. Thomsen-Ebert, T. Gilbert, B. Welch, S. A. Kemmer, K. M. Logan, G. A., Summons R. E., De Stasio G., Bond P. L., Lai B., Kelly S. D. & Banfield J. F. (2000). Formation of sphalerite (ZnS) deposits in natural biofilms of sulfate-reducing bacteria. *Science*, 290, pp. 744-1747

Larter, S. & Primio R. (2005). Effects of biodegradation on oil and gas field PVT properties and the origin of oil rimmed gas accumulations. *Organic Geochemistry*, 36, pp. 299-310

Leppchen,K.; Daussmann, T. Curvers, S. & Bertau M. (2006). Microbial De-emulsification: A Highly Efficient

Procedure for the Extractive Workup of Whole-Cell Biotransformations Organic Process Research &

Development , 10,pp. 1119-1125Jinfeng, L.; Lijun, M. Bozhong, M. Rulin, L. Fangtian, N. & Jiaxi, Z. (2005). The field pilot of microbial enhanced oil recovery in a high temperature petroleum reservoir, *Journal of Petroleum Science and Engineering*, 48, pp. 265-271

Madigan, M. T.; Martinko J. M. & Parker J. (2006). *Biology of Microorganisms*. Southern Illinois University, Carbondal.

Magot, M.; Ollivier, B. & Patel, B. K. C. (2000). Microbiology of petroleum reservoirs. *Antonie van Leeuwenhoek*, 77, pp. 103-116

Meckenstock, R. U.; Safinowski, M. & Griebler C. (2004). Anaerobic degradation of polycyclic aromatic hydrocarbons. *FEMS Microbiol Ecol.*, 49, pp. 27-36

Meredith, W. Kelland, S. J. & Jones, D. M. (2000). Influence of biodegradation on crude oil acidity and carboxylic acid composition. *Organic Geochemistry*, 31, pp. 1059-1073

Meyer, L. (1864). Chemische Untersuchungen der Thermen zu Landeck in der Grafschaft Glatz. *Journal für Praktische Chemie*, 91, pp. 1-15

Mikesell, M. D. & Boyd, S. A. (1990). Dechlorination of chloroform by *Methanosarcina strains*. *Appl Environ Microbiol.*, 56, pp. 1198-1201

Mokhatab, S. (2006). Microbial enhanced oil recovery techniques improve production, bacteria may be valuable in offering cost-effective and environmentally beningn EOR. World Oil, 227

Nazina, T.N. & Rozanova, E. P. (1978). Thermophilic Sulfate-Reducing Bacteria from Oil Strata, *Mikrobiologiya*, 47, pp. 142–148

Nazina, T. N. Grigor'yan, A. A. Shestakova, N. M. Babich, T. L. Ivoilov, V. S., Feng, Q. Ni, F. Wang, J. She, Y. Xiang, T. Luo, Z. Belyaev, S. S. & Ivanov M. V. (2007). Microbiological investigations of high-temperature horizons of the Kongdian petroleum reservoir in connection with field trial of a biotechnology for enhancement of oil recovery. *Microbiology* 76: pp. 287-296

Nemati, M.; Mazutinec, T. J. Jenneman, G. E. & Voordouw, G. (2001). Control of biogenic H_2S production with nitrite and molybdate. *Journal of Industrial Microbiology & Biotechnology*, 26, pp. 350-355

O'Dea, V. (2007). *Understanding Biogenic Sulfide Corrosion*. NACE International 46 (11)

Ollivier, B.; Fardeau, M. L. Cayol, J. L. Magot, M. Patel, B. K. C. Prensiep, G. & Garcia, J. L. (1988). *Methanocalculus halotolerans* gen. nov., sp. nov., isolated from an oil-producing well, *International Journal of Systematic Bacteriology*, 48, pp. 821-828

Onstott, T. C.; Hinton, S. M. Silver, B. J. & King, Jr. H. E. (2010). Coupling hydrocarbon degradation to anaerobic respiration and mineral diagenesis: theoretical constraints. *Geobiology*, 8, pp. 69-88.

Ortego-Calvo, J. J. & Saiz-Jimenz, C. (1998). Effect of humic fractions and clay on biodegradation of phenanthrene by *Pseudomonas fluorescens* strain isolated from soil. *Appl. Environ. Microbiol.*, 64, pp. 3123-3126

Oude Elferinck, S. J. W. H.; Vorstman, W. J. C. Sopjes, A. & Stams A. J. M. (1998). *Desulforhabdus amnigenus* gen.sp.nov., a sulfate reducers isolated from anaerobic granular sludge. *Arch.Microbiol.*, 164, pp. 119-124

Peckmann, J.; Thiel, V. Michaelis, W. Clari, P. Gaillard, C. Martire, L. & Reitner, J. (1999). Cold seep deposits of Beauvoisin (Oxfordian; southeastern France) and Marmorito (Miocene; northern Italy): microbially induced, authigenic carbonates. *International Journal of Earth Sciences*, 88, pp. 60-75

Perry, C. T. & Taylor, K. G. (2006). Inhibition of dissolution within shallow water carbonate sediments: impacts of terrigenous sediment input on syn-depositional carbonate diagenesis. *Sedimentology*, 53, pp. 495-513

Peters, K. E. & Moldowan, J. M. (1993). *The biomarker guide: Interpreting molecular fossil in petroleum and ancient sediments.* Prentice Hall, London. p. 363.

Popa, R.; Kinkle, B. K. & Badescu, A. (2004). Pyrite framboids as biomarkers for iron-sulfur systems. *Geomicrobiology Journal*, 21, pp. 193-206

Postgate, J. R. (1984). The sulphate reducing bacteria. Cambridge University Press Cambridge, 1984.

Rabus, R. Hansen, T. & Widdel F. (2000). *Dissimilatory sulfate- and sulfur-reducing prokaryotes.* Springer-Verlag New York.

Ribeiro deNardi, I. Zaiat, M. & Foresti E. (2007). Kinetics of BTEX degradation in a packed – bed anaerobic reactor. *Biodegradation*, 18, pp. 83-90

Robertson, W. J.; Bowman, J. P. Franzmann, P. D. & Mee, B. J. (2001). *Desulfosporosinus meridiei* sp. nov., a spore-forming sulfatereducing bacterium isolated from gasolene-contaminated groundwater. *Int J Syst Evol Microbiol.*, 51, pp. 133–140

Rozanova, E. P. Nazina, T. N. & Galushko, A. S. (1988). Isolation of a new genus of sulfate-reducing bacteria and description of a new species of this genus, *Desulfomicrobium apsheronum* gen. nov., sp. nov. *Microbiology* (English translation of Mikrobiologiya) 57, pp. 514-520

Rozanova, E. P.; Borzenkov, I. A. Tarasom, A. L., Suntsova, L. A. Dong, Ch. L. Belyaev, S. S. & Ivanov, M. V. (2001). Microbiological processes in a high-temperature oil field. *Microbiology* 70, pp. 102-110

Singh, A. Van Hamme, J. D. & Ward, O. P. (2007). Surfactants in microbiology and biotechnology: Part 2. Application aspects. *Biotechnology Advances* 25, pp. 99-121

Sowers, K. R. & Ferry J.G. (1983). Isolation and characterization of a methylotrophic marine methanogen, *Methanococcoides methylutens* gen. nov., sp. nov. *Appl. Environ. Microbiol.* 45, pp. 684–690

Suthersan, S.S. (1999). In situ bioremediation. CRC Press LLCStackebrandt, E.; Sproer, C. Rainey, A. F. Burghardt, J. Pauker, O. & Hippe, H. (1997). Phylogenetic analysis of the genus Desulfotomaculum: evidence for the misclassification of *Desulfotomaculum guttoideum* and description of Desulfotomaculum orientis as Desulfosporosinus orientis gen. nov., comb. nov. *Int J Syst Bacteriol.*, 47, pp. 1134-1139

Stetter, K. O. & Hubber R. (1999). The role of hyperthermophilic prokaryotes in oil fields. Microbial ecology of oil fields? Proceedings of the 8th International Symposium on Microbial Ecology Bell CR, Brylinsky M, Johnson-Green P (ed) Atlantic Canada Society for Microbial Ecology, Halifax, Canada, 1999

Stetter, K. O.; Lauerer, G. & Thomm, M. N. (1987). Isolation of extremely thermophilic sulfate reducers: Evidence for a novel branch of archaebacteria. *Science* 236:822-824

Stroud, J. L. Paton, G. I. & Semple K. T. (2007). Microbe-aliphatic hydrocarbon interactions in soil implication for biodegradation and bioremediation; *Journal of Applied Microbiology*, 102, pp. 1239-1253

Surygała, J. (2001). *Crude oil and environment (in Polish).* Oficyna Wydawnicza Politechniki Wrocławskiej. Wrocław 2001

Swaine, D. (2000). Why trace elements are important. *Fuel Processing Technology*, 65, pp. 21-33

Szlaski, A. & Wojewódka, D. (2003). Biodegradation of spoils contaminetaed by crude oil (in Polish). *Instalator*, 61, pp. 52-53

Sztompka, E. (1999). Biodegradation of engine oil in soil. *Acta Microbiol. Pol.* 48, pp. 185-196

Tardy-Jacquenod, C. Magot, M. Patel, B. K. C. Matheron, R. & Caumette, P. (1998). *Desulfotomaculum halophilum* sp. nov., a halophilic sulfate-reducing bacterium isolated from oil production facilities. *Int J Syst Bacteriol* 48, pp. 333-338

Tiehm, A. & Schulze, S. (2003). Intrinsic aromatic hydrocarbon biodegradation for groundwater remediation. *Oil & Gas Science and Technology*, 58, pp. 449-462

Vidali, M. (2001). Bioremediation. An Overview. *Pure Applied Chemistry* 73, pp. 1163-1172

Vieth, A. & Wilkes H. (2005). Deciphering biodegradation effects on light hydrocarbons in crude oils using their stable carbon isotopic composition: A case study from the Gullfaks oil field, offshore Norway. *Geochimica* 70, pp. 651-665

Voordouw, G. (1992). Evolution of hydrogenase gene. *Advances in Inorganic Chemistry*, 38, pp. 397-423

Warthman, R. van Lith, Y. Vasconcelos, C. Mckenzie, J. A. & Karpoff A. M. (2000). Bacterially induced dolomite precipitation in anoxic culture experiments. *Geology*, 28, 1091-1094.

Widdel, F. & Rabus, R. (2001). Anaerobic biodegradation of saturated and aromatic hydrocarbons. *Biotechnology*, 12, pp. 259-276

Wolicka, D. & Borkowski A. (2008a). Geomicrobiology of crude oil and formation water (in Polish). Conference Book. Ropa i gaz a skały węglanowe południowej Polski. Czarna 16-18 kwietnia, 45.

Wolicka, D. & Borkowski A. (2008b). Participation of sulphate reducing bacteria in formation of carbonates. International Kalkowsky-Symposium Göttingen, Germany. Abstract Volume (Göttingen University Press) *Special Volume in a Geobiological Journal*, pp. 130-131

Wolicka, D.& Borkowski A. (2011). Participation of $CaCO_3$ under sulphate - reduction condition. Reitner et al., Advances in Stromatolite Geobiology, *Lecture Notes in Earth Sciences* 131, pp. 151-160

Wolicka, D. & Kowalski W. (2006a). Biotransformation of phosphogypsum on distillery decoctions (preliminary results). *Polish Journal of Microbiology*, 55, pp. 147-151

Wolicka, D. & Kowalski W. (2006b). Biotransformation of phosphogypsum in petroleum-refining wastewaters. *Polish Journal of Environmental Studies*, 15, pp. 355-360

Wolicka, D. & Borkowski A. (2007). Activity of a sulphate reducing bacteria community isolated from an acidic lake. EANA 07, 7th European Workshop on Astrobiology, Turku, Finland, October 22-24, 2007, p. 98

Wolicka D.; Borkowski, A. & Dobrzyński, D. (2010). Interactions between microorganisms, crude oil and formation waters. *Geomicrobiology Journal*, 27, pp. 430-452

Wolicka, D.; Suszek, A. Borkowski, A. & Bielecka A. (2009). Application of aerobic microorganisms in bioremediation *in situ* of soil contaminated by petroleum products, *Bioresource Technology*, pp. 3221-3227

Wolicka, D. (2008). Crude oil - environment of ocuring of microorganizsms (in Polish). Conference Book, Ropa i gaz a skały węglanowe południowej Polski. Czarna 16-18 kwietnia, 44.

Wright, D. T. (1999a). Benthic microbial communities and dolomite formation in marine and lacustrine environments - a new dolomite model. Marine Authigenesis from Global to microbial (eds. Glenn C.R., Lucas J. And revot L.), *SEPM Spec. Publ.*, 66, pp. 7-20

Wright, D. T. (1999b). The role of sulphate-reducing bacteria and cyanobacteria in dolomite formation in distal ephemeral lakes of the Coorong region, South Australia. *Sedimentary Geology*, 126, pp. 147-157

Zeikus, J.G. & Wolfe, R.S. (1972). *Methanobacterium thermoautotrophicus* sp. n., an anaerobic, autotrophic, extreme thermophile. *J. Bacteriol.*, 109, pp. 707- 713

Zieńko, J. & Karakulski, K. (1997). Hydrocarbon in environment (in Polish). Politechnika Szczecińska, Szczecin.

6

Comprehensive Perspectives in Bioremediation of Crude Oil Contaminated Environments

Chukwuma S. Ezeonu[1], Ikechukwu N.E. Onwurah[2] and Obinna A. Oje[2]
[1]*Industrial Biochemistry, Environmental Biotechnology Unit, Chemical Sciences Department, Godfrey Okoye University, Enugu,*
[2]*Pollution Control and Environmental Biotechnology Unit, Department of Biochemistry, University of Nigeria, Nsukka-Enugu, Nigeria*

1. Introduction

Bioremediation is a biotechnological approach of rehabilitating areas degraded by pollutants or otherwise damaged through mismanagement of ecosystem. It is the ability of microorganisms to degrade or detoxify organic contaminated areas by transforming undesirable and harmful substances into non-toxic compound (Bioremediation overview, 2003).

Diverse components of crude oil and petroleum such as polycyclic aromatic hydrocarbons (PAHs) have been found in waterways as a result of pollution from industrial effluents and petrochemical products (Beckles, *et al.*, 1998). Petroleum hydrocarbon pollution of the environment may arise from oil well drilling production operations, transportation and storage in the upstream industry, and refining, transportation, and marketing in the downstream industry. Petroleum hydrocarbon pollution could also be from anthropogenic sources (Oberdorster and Cheek, 2000). Some non combusted hydrocarbons escape into the environment during the process of gas flaring. Until recently, the bulk of the associated gas produced during drilling in Nigeria, was flared. Sources of petroleum and its products in the environment will also include accidental spills and from ruptured oil pipelines. Today the international oil and gas-pipelines span several million kilometers and this is growing yearly due to inter-regional trade in petroleum products. Just like any other technical appliance, pipelines are subject to ''tear and wear'', thus can fail with time (Beller, *et al.*, 1996). Spilled petroleum hydrocarbons in the environment are usually drawn into the soil due to gravity until an impervious horizon is met, for example bedrock, watertight clay or an aquifer.

Poor miscibility of crude oil accounts for accumulation of free oil on the surface of groundwater and this may migrate laterally over a wide distance to pollute other zones very far away from the point of pollution. Industrial and municipal discharges as well as urban run-offs, atmospheric deposition and natural seeps also account for petroleum hydrocarbon pollution of the environment (Baker, 1983). It is worthy of note that groundwater is one of the many media by which human beings, plants and animals come into contact with petroleum hydrocarbon pollution. In the Niger delta area of Nigeria, extensive farm land has been lost due to contamination with crude oil. Also sources of drinking water and traditional occupation such as fishing and water transportation are greatly affected by crude oil contamination.

Biotechnology is the major reliable method involved in bioremediation and is defined as a set of scientific techniques that utilize living organisms or parts of organisms to make, modify or improve products (which could be plants or animals). It is also the development of specific organisms for specific application or purpose and may include the use of novel technologies such as recombinant DNA, cell fusion and other new bioprocesses (Anon, 1991.) It is also that aspect of biotechnology, which specifically addresses issues in environmental pollution control and remediation (Onwurah, 2000). This goes to say that it involves several disciplines such as biology, agriculture, engineering, health care, economy, mathematics, chemistry and education. It is regarded today as fundamentally an engineering application of microbial ecology (Rittman *et al.*, 1990) and process design. The engineering aspect of environmental biotechnology involves the design/construction of special machines or equipment referred to as reactors or bioreactors. Environmental biotechnology also encompasses quantitative mathematical modeling whereby understanding and control of many inter-related processes become possible. Mathematical modeling technology transcends the boundary of single traditional scientific disciplines and technologies, whereby a logical framework resolves related problems (Ziegler 2005; Onwurah, 2002b). They are tools utilized economically for explaining the cost and effectiveness of different options of clean-up technology and control (Onwurah and Alumanah 2005; Ziegler 2005). One of the greatest challenges to humanity today is the endangering of biota as a result of environmental pollution from crude oil.

To estimate the biological danger of oil after a spill, knowledge of the harmful effects of the components is necessary. In order to obtain or ascertain the effects of such polluting substance, every living being and life function can be considered a potential biomarker or bio-indicator. A biomarker is an organism or part of it, which is used in soliciting the possible harmful effect of a pollutant on the environment or the biota. Biomonitoring or biological monitoring is a promising, reliable means of quantifying the negative effect of an environmental contaminant. In a broad sense, biological markers (biomarkers) are measurement in any biological specimens that will elucidate the relationship between exposure and effect such that adverse effects could be prevented (NRC, 1992). It should be instituted whenever a waste discharge or oil spill has a possible significant harm on the receiving ecosystem. It is preferred to chemical monitoring because the latter does not take into account factors of biological significance such as the combined effects of the contaminants on DNA, protein or membrane. Some of the advantages of biomonitoring include the provision of natural integrating functions in dynamic media such as water and air, possible bioaccumulation of pollutant from 10^3 to 10^6 over the ambient value and/or providing early warning signal to the human population over an impending danger due to a toxic substance. Microorganisms can be used as an indicator organism for toxicity assay or in risk assessment. Tests performed with bacteria are considered to be most reproducible, sensitive, simple, economical and rapid (Mathews, 1980). Some examples include the 'rec-assay' which utilizes *Bacillus subtilis* for detecting hydrophobic substances (hydrocarbons) that are toxic to DNA (Matsui, 1989), *Nitrobacter sp*, which is based on the effect of crude oil on oxidation of nitrite to nitrate (Okpokwasili and Odukuma, 1994), and *Azotobacter sp,* used in evaluating the effect of oil spill in aquatic environment (Onwurah, 1998). Multiple bioassays that utilize a variety of species can be applied to gain a better understanding of toxicity at a given trophic level and under field condition. Several criteria exist for selecting biomarkers of plant and animal origin. Biomarkers, as fingerprints for identifying mystery oil spills, are now in use and they include steranes, phytanes, and hopanes. Normal hexadecane, an n-alkane found in crude oil is often used because of its low volatility and high hydrophobicity (Foght, *et al.*, 1990). These markers are integral part of crude oil and are not affected or degraded easily by any biological process. Hence they remain as "skeleton" of the crude oil even after a natural degradation has taken place. Steranes in crude oil are derived from the algae or the plant from which the source rock originated, while hopanes are derived from the hopenetetrol present in bacteria (Peters and Moldowan, 1993). Hopanes can be used to determine the nature of the source rock that generated a crude oil. Biosensor is a technology that promises to be important in generating future standards regarding both bioavailability and toxicity of any pollutant being released into the environment. Biosensors are usually photo detector systems, which operate on the genes that control luminescence (King *et al.*, 1990). Most of the biosensor tests are not quantitative, but rather can detect the potential activity or presence of an environmental toxicant. Examples includes the Petro-Risk Soil Tests System (DTSC, 1996) used in detecting total petroleum hydrocarbons in a given soil after an oil spill. The test kit uses enzyme-linked immunosorbent assay (ELISA) technology. It involves an antibody with affinity to certain petroleum hydrocarbons. The antibody that does not react with the methanolic extract of the petroleum hydrocarbons or crude oil in the soil sample is detected by a color reaction. The color intensity developed decreases as the hydrocarbon concentration increases. A differential photometer is usually incorporated. Other examples include the Microtox, which utilizes the luminescent bacteria, *Vibro fischeri* (*hotobacterium phosphoreum*), in monitoring toxicity of petroleum hydrocarbon. The bacterium *Vibro fisheri* utilizes about 10% of its metabolic energy for bioluminescent activity. The luminescent pathways are linked to cellular respiration whose disruption will change the light output (Ross, 1993), or on the structure activity relationship of the individual compounds in the crude oil or petroleum (Cronin, and Schultz, 1998).

The issue of crude oil contamination of the environment should be clearly examined and proper solution proffered with respect to the level of contamination and uniqueness of the environment. Thus, this review is set to x-ray how crude oil bioremediation and bio-monitoring will benefit bioremediation scientists and experts.

2. Contamination components in crude oil

Total petroleum hydrocarbons (TPH) comprise a diverse mixture of hydrocarbons that occur at petrochemical sites and storage areas, waste disposal pits, refineries and oil spill sites. According to McElroy *et al* (1989) TPHs are considered persistent hazardous pollutants, and include compounds that can bioconcentrate and bioaccumulate in food chains. Heitkamp and Cerniglia (1988) showed that THPs are acutely toxic. Some such as benzene and benzo[a]pyrene are recognized mutagens (IARC, 2000) and carcinogens Mortelmans *et al* (1986). Since this group includes chemicals that have physical and chemical characteristics that vary in magnitude, TPHs are divided into two categories:

1. Gasoline range organics (GRO) corresponds to small chain alkanes (C6-C10) with low boiling point (60-170°C) such as isopentane, 2,3-dimethyl butane, *n*-butane and *n*-

pentane, and volatile aromatic compounds such as the monoaromatic hydrocarbons benzene, toluene, ethylbenzene, and xylenes (BTEX).

2. Diesel range organics (DRO) includes longer chain alkanes (C10–C40) and hydrophobic chemicals such as polycyclic aromatic hydrocarbons (PAH).

Whereas most of these contaminants do have natural sources, concentration and release of contaminants through anthropogenic activities has led to significant contamination of soil and groundwater. Individual contaminants behave differently. Some contaminants such as BTEX compounds are highly mobile in the environment, while others such as PAHs tend to bind strongly to soil particles near the source or remain entrapped within an organic phase, thus immobilizing them and making them difficult to move out of their percolated environments. Since hydrocarbon spills at different sites represent different mixtures, it is very difficult to find a single, efficient method of cleanup. Current treatment techniques usually involve excavation and *ex situ* treatment of the source material and the contaminated soils. However, residual contamination often exceeds regulatory limits by a relatively small margin, and occurs over extensive areas (NRC, 1994). The large volume of soil affected precludes *ex-situ* treatment due to economical constraints and requires the use of relatively inexpensive remediation schemes, such as phytoremediation.

3. Chemistry of petroleum hydrocarbon

Petroleum has been known for several years to occur in surface seepage and was first obtained in pre-Christian times by the Chinese. The modern petroleum industry had its beginnings in Romania and in a well sunk in Pennsylvania by Colonel E. A. Drake in 1859 (Alloway and Ayres, 1993). The basic use of petroleum products was for the replacement of expensive whale oil for lighting. However, today apart from its use as fuel, it also serves as sources for the production of various chemicals.

Petroleum is defined as any mixture of natural gas, condensate, and crude oil. Crude oil which is a heterogeneous liquid consisting of hydrocarbons comprised almost entirely of the elements hydrogen and carbon in the ratio of about 2 hydrogen atoms to 1 carbon atom. It also contains elements such as nitrogen, sulphur and oxygen, all of which constitute less than 3% (v/v). There are also trace constituents, comprising less than 1% (v/v), including phosphorus and heavy metals such as vanadium and nickel. Crude oil could be classified according to their respective distillation residues as paraffin, naphthenes, or aromatics and based on the relative proportions of the heavy molecular weight constituents as light medium or heavy (http/www.academicjournals.org/BMBR). Also, the composition of crude may vary with the location and age of an oil field, and may vary with the depth within an individual well. About 85% of the components of all types of crude oil can be classified as either asphalt base, paraffin base or mixed base. Asphalt base contain little paraffin wax and an asphalt residue (Atlas, 1981).

On a structural basis, the hydrocarbons in crude oil are classified as alkanes (normal or iso), cyclo-alkanes, and aromatics (fig 1). Alkenes, which are the unsaturated analogs of alkanes, are rare in crude oil but occur in many refined petroleum products as a consequence of the cracking process. Increasing carbon numbers of alkanes (homology), variations in carbon chain branching (iso-alkanes), ring condensations, and interclass combinations e.g., henylalkanes, account for the high numbers of hydrocarbons that occur in crude oil. In addition, smaller amounts of oxygen - (phenols, naphthenic acids), nitrogen- (pyridine, pyrrole, indole), and sulphur -(alkylthiol, thiophene) containing compounds, collectively designated as "resins" and partially oxygenated, highly condensed asphaltic fraction occur also in crude but not in refined petroleum (Atlas and Bartha, 1973).

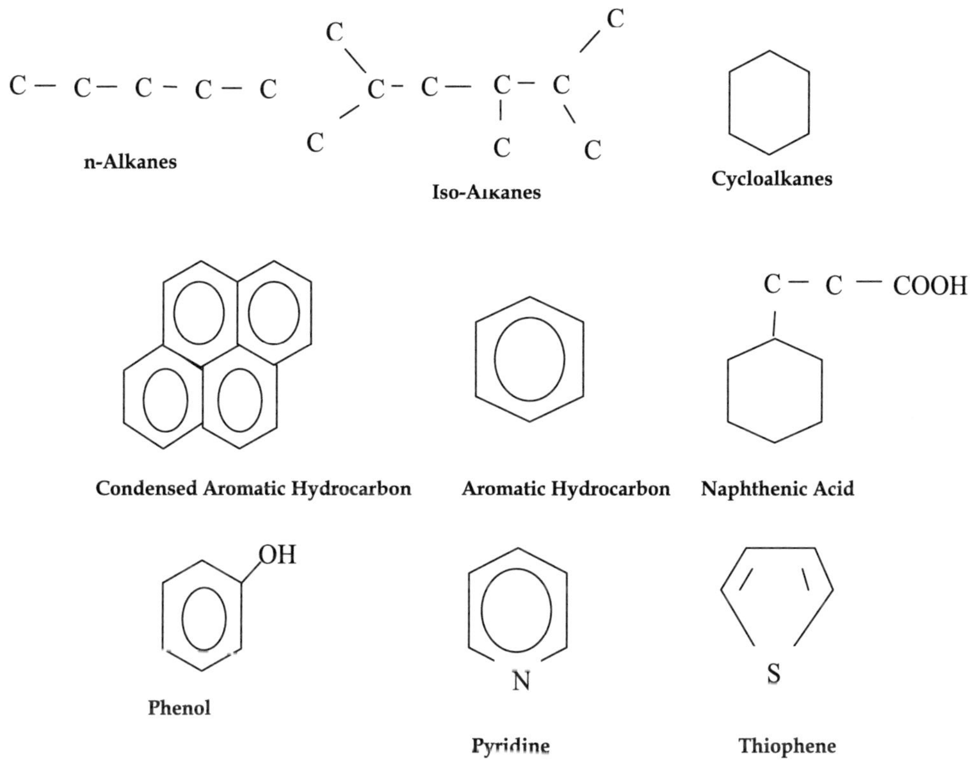

Fig. 1. Structural Classification of some Crude Oil components (Alloway and Ayres, 1993).

4. Effect of crude oil on soil nutrients

There is a direct proportionality between the quantity of oil spillage and accumulation of manganese and ferrous elements. Adam and Ellis (1960) pointed out that there was an accumulation of manganese and ferrous ions to levels which became very toxic to plants. Plant growth in crude oil contaminated soil is adversely affected due to changes in the nutrient status of the soil and disruption of microbial activities.

McGill (1976) noted that a lasting effect of crude oil in soil, eventually result in nutrient supply beneficial to crop production. According to Rowell (1977), the decomposition of spilled crude oil may result in an increased yield of soil nutrients. Plice (1948) also explained that soil polluted with crude oil remained barren for some years (i.e. seven years and above), after which the tested soil was shown to be richer in nutrients than the normal soils in the uncontaminated area. Previous research conducted by different scientists using different crops allow concluding that crude oil spillage on farm land has both positive and negative effect on soil. When a soil is polluted with crude oil, plant growth becomes adversely affected for some time and after a period the hydrocarbons become decomposed and are converted to soil organic matter which improves the nutrient content of the soil. However, this could take a long period of time under which the soil may remain unproductive for agricultural purposes.

5. Effect of crude oil spillage on germination and growth of crop

The ability of crops to germinate or grow on crude oil polluted soil is dependent on the level of crude oil spillage on soil (Odu, 1972). This means that a high level of crude oil pollution of the soil impairs germination of seedlings.

Rowell (1977) stated that at low level of spillage (e.g. one percent of oil contamination) germination may be delayed due to lack of moisture and hardening of soil structure. Moreover, at high contamination of soil, there may be no germination. Hence seed rotting will take place due to seeping of crude oil into the seeds through the outer integument. The interference of oil to soil air and water is another means of inhibiting seed germination. McGill (1976) noted that the toxic effect of crude oil coupled with poor aeration and the altered wetability of the soil due to oil spillage results in poor seed germination.

According to Baker (1970), spillage on land (soil) causes oil to enter into the leaves of plants and other economic trees through their pores and hampers the process of photosynthesis and evapo-transpiration. The pores of leaves are penetrated by films of oil, which is evidenced by the darkening of leaves as the pore becomes filled with oil. A patch of dark oil cuts sunlight from the leaves and where the shielding of sunlight becomes too much the leaves experience necrosis and the plant eventually dies (Nelson-Smith, 1979).

6. Metal accumulation abilities of plants

Soils contaminated with heavy metals cause several environmental and human health problems, which calls for an effective technological solution. Many affected sites around the world remain contaminated, because it is expensive to clean them up by available technologies. Phytoremediation is considered to be an innovative, economical, and environmentally compatible solution for remediation of heavy metal contaminated sites (Wang *et al.*, 2003; O'Connor *et al.*, 2003). Heavy metals may be bound or accumulated by particular plants, which may increase or decrease the mobility and prevent the leaching of heavy metals into ground water. Growing plants can help to reduce heavy metal pollution. The advantage of this technique is evident as the cost of phytoremediation is much less than the traditional *in situ* and *ex situ* processes; plants can be easily monitored to ensure proper growth; and valuable metals can be reclaimed and reused through phytoremediation. The metals most commonly accumulated in plants are lead, cadmium, zinc, nickel, or radioactive isotopes such as uranium or cobalt (Lombi *et al.*, 2001).

Some plants accumulate heavy metals by transporting the metals and concentrating them into their shoots for harvesting. These groups of plants are known as heavy metal hyperacccumulators. In metal accumulation it is necessary to use plants as *Polygonum hydopiper L., Rumex acetosa L.,* (Wang et al., 2003), *Thlaspi caerulescenes J. Presl, Zea mays L.* (Lombi *et al.*, 2001). Hyperacccumulators are used in the removal of metal contaminants since they take up 100 times the concentration of metals over other plants (Cunningham *et al.*, 1995). They accumulate toxic metals through their roots and transport them to the stems.

Most hyperacccumulators could be grown in contaminated soils. The metal could then be recovered and recycled when burned and the ash collected.

According to Peciulyte *et al.*, (2006), for a low metal contaminated soil, using maize and vetch plants as metal accumulator, after three weeks of growth, a negative effect on the length of shoots and roots was observed. The biomass of the plant seedlings was significantly smaller in the metal-contaminated soil in comparison to controls. This is due to metal toxicity of the plants by the metal contaminants. Metal accumulation in combination with crude oil in a contaminated site could pose a serious challenge than when contamination is by either the metal or crude oil alone.

7. Crude oil and human health

There is a direct correlation between environmental health and human health. Obviously, human health has been intensively studied for quite long, while environmental health is a recent field. Environmental health is the assessment of the health of individual organisms with a direct correlation of observable changes in the environment. Crude oil pollution has been linked to be the causative effect of many diagnosed diseases. The health problems associated with crude oil spill may be through any or combinations of the following routes: contaminated food and/or water, emission and/or vapours. Toxic components in oil may exert their effects on man through inhibition of protein synthesis, nerve synapse function, and disruption in membrane transport system and damage to plasma membrane (Prescott, *et al.,* 1996). Crude oil hydrocarbons can affect genetic integrity of mutagenesis and impairment of reproductive capacity (Short and Heintz 1997). The risk of drinking water contaminated by crude oil can be extrapolated from its effect on rats that developed hemorrhagic tendencies after exposure to water-soluble components of crude oil (Onwurah, 2002). Volatile components of crude oil after a spill have been implicated in the aggravation of asthma, bronchitis and accelerated aging of the lung (Kaladumo, 1996). Other possible health effects of oil spill can be extrapolated from rats exposed to contaminated sites and these include increased liver, kidney and spleen weights as well as lipid peroxidation and protein oxidation (Anozie and Onwurah, 2001).

8. Toxicity of Nigerian crude oils

Some of the Nigerian crude oils associated with high toxicity level are: Forcados Blend (FB), Bonny Light (BL) and Bonny Medium (BM). A work by Imevbore *et al.* (1987) aimed at determining the toxic effect of the above listed crude oil against *Desmicaris trispimosa* and *Palaemonetes africanus,* in fresh water and brackish water shrimps respectively. The work established that after 96 hours of contact between crude oil and bacteria, the concentration LC_{50} (crude oil lethal dosage) contained in FB, BL, and BM crude oils ingested by *D. trispimosa* was 2.75, 16.22 and 38.02μg/l respectively while it was 38.90 and 4.17 μg/l respectively for *P. africanus.* The toxicity could occur at even lower concentrations among species that are more sensitive than those tested such as earthworm. Another demonstration of the toxicity level of the crude oil spill was demonstrated by Ezeala (1987) with *Pistia stratoites* (a fresh water plant) abundant in most fresh water of the oil producing areas in Nigeria. In a simulated condition of very light pollution (0.5 ml m^{-2}), *Pistia stratoites* was very susceptible to the crude oil pollution. The magnitude was a reduction in leaf number of about 63%, a loss in the photosynthetic chlorophylls of about 70%, 65% decreases in leaf area and 80% loss in productivity after 7 weeks of growth period. There are several studies carried out by Nigerian scientists that support the fact that crude oil is indeed toxic. Crude oil toxicity in rats (Anozie and Onwurah, 2001), effects on chlorophyll contents (Ezeonu and Onwurah, 2009) are among many such facts.

9. Bioremediation process

Bioremediation is a natural process that can be harnessed or optimized to enhance the rate at which microbes' biodegrade organic chemicals that have been released into the environment. The rate at which biodegradation can take place may be favoured by providing an optimal living environment for the microbes. Most microbes that degrade petroleum hydrocarbon make use of appropriate levels of oxygen, water, acidity (pH), and nutrients such as nitrogen and phosphorus. When the microbial environment is optimized by having the right amounts of water, air and nutrients and by maintaining proper acidity, biodegradation rates will increase (DOE/PERF, 2002).

The rate-limiting factors that are typically assessed and which are important include oxygen, nutrients, salinity, temperature, pH, and soil moisture content or water content. In order to determine which of the rate-limiting factors may require modification, an initial assessment should be performed (DOE/PERF, 2002).

These rate-limiting factors may vary from medium to medium (i.e. soil, groundwater, fresh water, seawater, wastes, or wetlands) and they may vary from site to site depending upon climate, ecosystem, and human disturbance (DOE/PERF, 2002). Once the rate-limiting factors for biodegradation are assessed, it is important to develop a well thought out plan for correcting them in a manner that will enhance microbial degradation rates without causing more harm to the environment (DOE/PERF, 2002).

10. Degradation of petroleum hydrocarbon

The lower n-alkanes as a structural group are the most biodegradable petroleum hydrocarbons, the C_5 - C_{10} homologues have been shown to be inhibitory to the majority of hydrocarbon degraders. As solvents, these homologues tend to disrupt lipid membrane structures of microorganisms. Similar alkanes in the C_{20} -C_{40} range, often referred to as "waxes", are hydrophobic solids at physiological temperatures. Apparently, it is this physical state that strongly influences their biodegradation (Bartha and Atlas, 1977).

Zengler *et al.*, (1999) emphasized that petroleum biodegradation by bacteria can occur under both oxic and anoxic condition, albeit by the action of different consortia of organisms. According to Holba *et al.*, (1996) crude oil biodegradation at the subsurface occurs primarily under anoxic conditions, mediated by sulfate reducing bacteria or other anaerobes using a variety of other electron acceptor as the oxidant.

Most micro-organisms attack alkanes terminally whereas some perform sub-terminal oxidation. Primary attachment on intact hydrocarbons always requires the action of oxygen. In the case of alkanes, monooxygenase attack results in the production of alcohol. The alcohol product is oxidized finally into an aldehyde and finally, to a fatty acid. The latter is degraded further by beta-oxidation. Extensive methyl branching interferes with the beta-oxidation process and necessitates di-terminal attack or other bypass mechanisms. Therefore, n-alkanes are degraded more readily than iso-alkanes.

Bartha (1986) states that monocyclic compounds such as cyclopentane, cyclohexane, and cycloheptane are degraded like alkanes, since they have a strong solvent effect on lipid membranes, and are toxic to the majority of hydrocarbon degrading microorganisms. Highly condensed cycloalkane compounds resist biodegradation due to their structure and physical state.

Prokaryotes convert aromatic hydrocarbons by an initial dioxygenase attack, to trans-dihydrodiols that are further oxidized to dihydroxy products, e.g., catechol in the case of benzene (Atlas and Bartha, 1998). Eucaryotic micro-organisms use monooxygenases, producing benzene 1,2-oxide from benzene, followed by the addition of water, yielding dihydroxydihydrobenzene (cis-dihydrodiol).

In crude petroleum as well as in refined products, petroleum hydrocarbons occur in complex mixtures and influence each other's biodegradation. The effects may go in negative as well as positive direction. Some iso-alkanes are apparently spared as long as n-alkanes are available as substrates, while some condensed aromatics are metabolized only in the presence of more easily utilizable petroleum hydrocarbons, a process referred to as co-metabolism (Wackett, 1996).

11. Reduction of toxicity by biodegradation of oil

Various investigations (Stewart, 2002; Nakles and Ray, 2002 and Prince 2002), have been carried out to look at possible production of toxic substances as a result of biodegradation. There are several tests available to analyze ecosystem function analysis in terrestrial and aquatic environments such as microbial response (e.g. most probable number), Microtox ™ Solid and Liquid phase. In the case of freshwater/marine sites, some of the analysis available include: algal solid phase bioassay; daphnia survival; amphipod survival; gastropod survival; and fish bioassays (Prince, 1993). Prince (1993) reiterated that these tests pointed out that biodegradation results in a decrease, not an increase, in soil toxicity.

In addition, studies performed in Prince William Sound, Svalbard and St. Lawrence indicated that there was no evidence for any significant toxicity associated with bioremediation in marine or freshwater spills (Prince, 2002).

12. Factors affecting degradation of crude oil

12.1 Soil Type: Nakles and Ray (2002), emphatically stated that biodegradation will occur in all soil types even though some may need additives or special care and equipment. Clay soils may need to be amended with bulking agents, in order to improve oxygen transport. Sandy soils may need to be amended with organic matter to improve the soil water holding capacity. A similar study was performed for refined products (jet fuel and heating oil), and the result also indicated that biodegradation rates are lower for sandy soils than for clay and loam soils. The reasons for this may be low water holding capacity, low total organic carbon content, and/or low surface area available for microbial growth in sandy soils. Biodegradation rates for the clay soils were similar or better than those for the loam; however in both studies moisture content was maintained at optimum levels to improve soil tilt. If moisture content in clay soils is not optimized, tilt and thus aeration will be impacted, leading to slower biodegradation rates in these soils (McMillen, 2002).

12.2 Effect of brine on biodegradation: According to Sublette (2001), elevated concentrations of salt may be inhibitory or lethal to many classes of microorganisms because salts can disrupt the osmotic balance of microorganisms in the soil and interfere with their enzyme activity. As a result, soils with high electrical conductivity (EC) values could retard biodegradation rates. McMillen (1994; 1995) stated that if the soil salinity is extremely high, usually, 40mmhos, microbial activity will cease altogether. However, the same basic methodology used for in-situ bioremediation for oil spill, works well to remediate a brine spill if the salt has a pathway out of the site. It has been observed that when combined spills of oil and brine are treated for oil remediation, additional treatments are required.

12.3 Moisture Content /Metal Salt Effect: A correlation between crude oil contamination and metal salt impact on water retention ability of contaminated soil has been established, which consistently affect the activity of microorganisms in the soil microcosm (Ezeonu, 2010). Thus, a soil devoid of moisture may not tolerate the survival and growth of microorganisms. Thus, heavily contaminated soil will not permit the retention of moisture. Only lightly polluted soil will have capillarity for the movement of water within the soil and hence provide moisture for the purpose of degrading the crude oil by the microorganism. Large quantities of metal salt may also become toxic to the biotic components of the contaminated environment and become recalcitrant inhibiting biodegradation of the crude oil contaminated soil. It may also affect the survivability of the microorganisms in the soil. In such a situation additional treatment may be necessary. Combination of phytoremediation and bioremediation may be effective in the treatment of metal salt and crude oil combined contamination of soil.

12.4 Oxygen Content: Biodegradation is effective in an environment with adequate oxygen. Only on rare occasion does biodegradation take place in anoxic environments. Open marine environments have highly oxygenated surface waters offering an excellent condition for biodegradation. The presence of stagnant water along coasts decreases the amount of oxygen in the sand which decreases the rates of degradation. Biodegrading bacteria need oxygen in order to break down oil. Coincidentally, oxygen content is prevalent along marshes and beaches where oil has contaminated the soil.

12.5 pH: Mostly, life thrives at neutrality. As such if the pH is right for the microorganisms in the crude oil spill environment, it will enable them to survive and make use of the carbon source for their growth and metabolism. Thus most heterotrophic bacteria prefer a more neutral pH-value living conditions. Fungi that can also biodegrade hydrocarbons are significantly less sensitive to pH levels and are able to withstand more acidic conditions but do not contribute as much biodegradation as bacteria. As the micro-organism makes use of the carbon source of the crude oil, the quantities of crude oil spilled in such environments are drastically reduced. In order to increase the rate of biodegradation in a given environment the microorganisms must be isolated to determine the optimum pH for their survival and growth.

Sublette (2001) stated that pH is one of several environmental conditions that can serve to inactivate enzymes when levels are not optimal. This has the effect of slowing microbial metabolism (the growth rate), which in turn has a detrimental effect on biodegradation rates.

Most bacteria survive better in the pH range of 6.5 to 8.5, and yeasts and molds thrive better at pH range of 4.5 - 5.3. The optimal pH range for biodegradation is considered to be 6.0 to

8.5. Biodegradation processes may cause the soil pH to drop over time and therefore frequent monitoring of pH is important (Nackles and Ray, 2002).

Lime or limestone when added to soil can increase pH to neutral values; while aluminium sulfate and ferrous sulfate can be added to decrease high pHs (Nakles and Ray, 2002). Lime and sulfur requirements are soil type dependent. Both acidifying and neutralizing amendments should be added to topsoil gradually and thoroughly tilled in. Agricultural experts should be consulted before pH amendment is carried out.

12.6 Duration of biodegradation. The findings of a DOE/PERF (2002) study revealed that the rate at which biodegradation will occur depends on the following variables:

- The type and concentration/mass of petroleum hydrocarbons present,
- The depth of the impacted area,
- Optimization of the biodegradation environmental conditions
- The type of technology used.

12.7 Nitrogen and Phosphorus: These two elements are limiting factors for biodegradation, and their availability to bacteria can affect their ability to consume oil products. The addition of nitrogen and phosphorous increases the proliferation of biodegrading bacteria, resulting in an increase in degradation rates (Rosenberg *et al.* 1996).

12.8 Temperature: Extreme temperatures, too high or too low will naturally not allow the thriving of most microorganisms. Such temperatures will also reduce the action of enzymes in a crude oil spilled environment. As temperature decreases, the rates of degradation decrease probably because of decreased rates of enzymatic activity. However, bacteria populations in colder climates are more adapted to cold temperatures thereby increasing their capability of degradation at near freezing temperatures (Rowland *et al.* 2000). Temperature is perhaps one of the most important factors, affecting both biodegradation and the consistency of the oil spilled. Temperature is a crucial factor in the beginning stages of biodegradation. Approximately 3 months after an oil spill, the rates of degradation at different temperatures are very similar since the remaining compounds in the weathered oil are so difficult to break down. At this point, temperature becomes obsolete (Gibb et al. 2001).

Also, as temperature decreases, viscosity of oil increases, becoming thicker. This increase in viscosity leads to the clumping of oil, which facilitates the use of mechanical methods for its removal out of the water, but hinders the use of suction mechanisms. Also, as temperature decreases volatility decreases, making the oil less likely to evaporate (Rowland 2000). Therefore, biodegradation is significantly more successful at warmer temperatures as found in the tropics.

12.9 Seasonal Effects: Nakles and Ray (2002) stated that bioremediation can be initiated at any time of the year, but consideration should be given to the potential effect of climate on the results. According to Nakles and Ray (2002), rainy seasons may cause excessive runoff, while drought or seasonally dry conditions may result in the need for irrigation. In addition, temperature can have an effect, since excessively cold or hot temperatures can slow biodegradation rates. For this reason, bioremediation projects are often initiated during the "growing season" for agricultural crops.

12.10 Effect of nutrient on bioremediation-treatment: Fertilizer should be added gradually to the impacted soil zone in order to avoid excessively high pH and high concentrations of nitrogen that might be toxic to soil microbes (Prince and McMillen, 2002). Most of the time, nutrients have to be added to the impacted soil in order to enhance microbial growth. The two most important nutrients that need to be added are nitrogen and phosphorus. In slightly alkaline soils, organic nitrogen source (such as manure) can cause accumulation of nitrites that are toxic to microbes.

Research has shown that the use of specialty fertilizers such as oil soluble and slow-release fertilizers yield similar biodegradation results (Nakles and Loehr, 2002).

Prince and McMillen (2002) explained that manure can be used as a fertilizer in bioremediation, but consideration in regards to odour and nitrite accumulation should not be overlooked.

Another consideration in applying manure is that its nitrogen and phosphorus levels are not known. It is known however, that manure does supply trace nutrients and improves the soil structure, thereby stimulating microbial growth (Sublette, 2001). Manure is usually added as a bulking agent, not as the main source of fertilizer.

12.11 Molecular Weight of Oil Components: Components that are low in molecular weight such as the aliphatic hydrocarbons tend to be degraded first, leaving behind the much larger molecules (aromatic hydrocarbon) which take much longer to break down. The lighter carbon components of the crude oil are also less viscous and can easily degrade and become volatile when acted upon by weather and environmental elements. This trend indicates the presence of biodegradation by microbial bacteria who cannot break down the larger oil compounds left after the initial phases of degradation (Ezra *et al.*, 2000).

12.12 Enzymes and surfactants in bioremediation: Sublette (2001) pointed out that enzymes are protein catalysts that are responsible for driving almost all of the chemical reactions within a cell. Enzymes act as efficient catalysts and alter the reaction mechanism of a cell in such a way as to lower the energy of activation for the overall reaction (Sublette, 2001). The quantity of enzymes produced by the bacterial or fungi in a crude oil contaminated environment will determine the duration of bioremediation in such an environment. Enzymes are subject to inactivation by a variety of environmental conditions including: heat, adverse pH, salt, strong oxidizing or reducing agents, organic chemicals and detergents or surfactants (Sublette, 2001). More to this is the fact that there are commercial 'enzyme' products in the market that can be used effectively for biodegradation, but there are little of any data that prove their cost-effectiveness in bioremediation projects (Sublette, 2001).

Surfactants have a micelle action such as observed in detergents when washing out dirt and stains from fabrics. Surfactants dislodge the oil from the soil particle making it possible for the oil to be washed off by rain or acted upon easily by microorganisms. Sublette (2001) further explained that the use of surfactants in soil to break up oil is not generally recommended because they can potentially interfere with cell membranes and enzymes, which in turn can reduce biodegradation rates. Surfactants should be pre-screened for toxicity to indigenous microbes. Dispersants, a type of surfactant, has been used successfully in breaking up marine oil spills (Sublette, 2001).

12.13 Frequency of Oil Exposure: Bacterial populations that are frequently exposed to oil spills are very likely to display higher degradation rates than populations being exposed for the first time. Repetitively exposed bacterial populations become acclimated to the presence of oil and are capable of degrading it more successfully (Rowland *et al.* 2000). Exposures of the crude oil help also to increase the surface area to enable weathering and activities of microorganism. As a result of these many weathering processes, oil spilled in marine environments rapidly loses its original properties and breaks down into different hydrocarbon fractions. These fractions have different chemical compositions and exist in different forms that can be further dispersed and degraded. After their initial, rapid transformation into these fractions, the degradation rates of the oil products slow down. In the final phase of the transformation, the compounds are completely converted into carbon dioxide and water. It is possible then for marine environments to naturally cleanse themselves in time, as long as the amount of toxic chemicals does not exceed a certain threshold (Patin, 1999). Figure 2 presents a summary of the routes of crude oil degradation in marine environments.

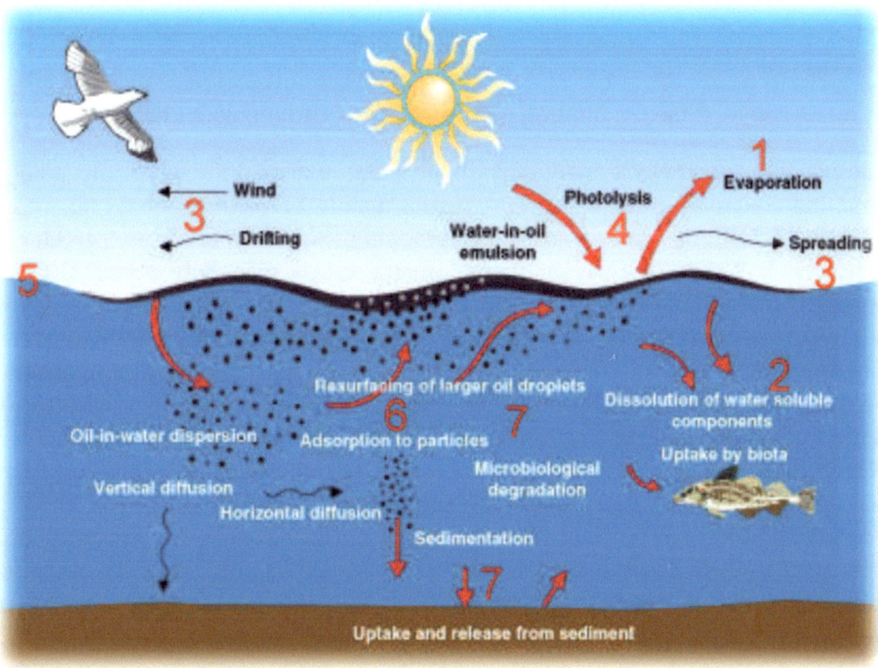

Fig. 2. Routes of Crude oil degradation. Adapted from Patin, 1999.

13. Factors limiting petroleum hydrocarbon biodegradation

Most crude oil polluted sites are not successfully treated at a given period of time due to environmental as well as inherent negative quality of the crude oil itself. The toxicity of the crude oil is an inherent factor that limits it breakdown since it inhibits the thriving of biodegraders. The aggregation of crude oil in piles, blocks or slurry also limits the action of environmental elements in its biodegradation. Successful application of bioremediation technology to contaminated systems requires knowledge of the characteristics of the site and the parameters that affect the microbial biodegradation of pollutants (Sabate *et al.*, 2004). A number of limiting factors however, have been identified to affect the biodegradation of petroleum hydrocarbons. Some of these factors are listed in Table 1.

Limiting Factor	**Explanations or Examples**
Petroleum hydrocarbon composition (PHC)	Structure, amount, toxicity
Physical State	Aggregation, Spreading, dispersion, adsorption
Weathering	Evaporation, Photo-oxidation
Water Potential	Osmotic and matrix forces, exclusion of water from hydrophobic aggregates.
Temperature	Influence on evaporation and degradation rate.
Oxidant	O_2 required to initiate oxidation, NO_3 or SO_4^{2-} to sustain, PHC biodegradation.
Mineral nutrients	N, P, Fe may be limiting.
Reaction	Low pH may be limiting.
Microorganisms	PHC degraders may be absent or low in numbers.

Source: Bartha 1986.

Table 1. Some common factors limiting petroleum hydrocarbon biodegradation.

The composition of the oil pollutant to a great extent influences the rate of its biodegradation. Kerosene, for example, consists mainly of medium chain alkanes which under suitable condition, achieves total biodegradation. Similarly, crude oil is biodegradable quantitatively, but for heavy asphaltic-naphthenic crude oils, only about 11% may be biodegradable within a reasonable time period, even if the conditions are favourable (Bartha, 1986). Okoh (2002) explained that the composition and inherent biodegradability of the petroleum hydrocarbon pollutant, therefore, is the first and most important to be evaluated. Hence the heavier crude oils generally contain more degraders; therefore more petroleum hydrocarbon degraders could be isolated in heavier crude oils than lighter ones. Also, Okoh *et al.*, (2002) noted that the amount of heavy crude oil metabolized by some bacterial species increased with increasing concentration of starter oil up to 0.6% (w/v), while degradation rates appeared to be more pronounced between the concentration of 0.4 and 0.6% (w/v) oil. In another report, Rahman *et al.* (2002) noted that the percentage of degradation by the mixed bacterial consortium decreased from 78% to 52% as the concentration of crude oil was increased from 1 to 10%. Thouand *et al.*, (1999) also stated that in some situations microbial biomass will be required only to a particular threshold, enough to produce the appropriate enzyme system that carry through the degradation process even when biomass production had ceased.

Recent studies have reported that photo-oxidation increases the biodegradability of petroleum hydrocarbon by increasing its bioavailability and thus enhancing microbial activities (Maki *et al.*, 2005). In a related study as reported by Trindade *et al.*, (2005) on assessing the bioremediation efficiency of a weathered and recently contaminated soil in Brazil, the authors reported low biodegradation efficiencies in the weathered soil contaminated with a high crude oil concentration compared to recently contaminated soil. Also, both soils (weathered and recently contaminated) submitted to bioaugmentation and biostimulation techniques presented biodegradation efficiencies approximately twice higher than the ones without natural attenuation.

Temperature plays very important roles in biodegradation of petroleum hydrocarbons, firstly by its direct effect on the chemistry of the pollutants, and secondly on its effect on the physiology and diversity of the microbial milieu. Ambient temperature of an environment affects both the properties of the spilled oil and the activity or population of microorganisms (Venosa and Zhu, 2003). According to Atlas (1981), at low temperatures, the viscosity of the oil increases, while the volatility of toxic low-molecular-weight hydrocarbons is reduced, delaying the onset of biodegradation. Temperature also affects the solubility of hydrocarbons (Foght *et al.*, 1996). Although hydrocarbon biodegradation can occur over a wide range of temperatures, the rate of biodegradation generally decreases with decreasing temperature. Bossert and Bartha (1984), Cooney (1984), variously observed that the highest degradation rates generally occur in the range of 30-40°C in soil environments, 20-30°C in some freshwater environments, and 15-20°C in marine environments.

Inadequate mineral nutrient, especially nitrogen, and phosphorus, often limits the growth of hydrocarbon utilisers in water and soils. Iron was reported to be limiting only in clean, offshore seawater (Swannell *et al.*, 1996). Cooney (1984) observed that nutrients are very important ingredients for successful, biodegradation of hydrocarbon pollutants, especially nitrogen, phosphorus and in some cases iron. Depending on the nature of the impacted environment, some of these nutrients could become limiting thus affecting the biodegradation processes. When a major oil spill occurs in marine and freshwater environments, the supply of carbon is dramatically increased and the availability of nitrogen and phosphorus generally becomes the limiting factor for oil degradation (Atlas, 1984). According to Okolo *et al.* (2005), using poultry manure as organic fertilizer in contaminated soil increases biodegradation. Biodegradation was enhanced in the presence of poultry manure alone, but the extent of biodegradation was influenced by the incorporation of alternate carbon substrates or surfactants. Hence the addition of nutrients is necessary to enhance the biodegradation of oil pollutants (Choi *et al.*, 2002; Kim *et al.*, 2005). However, excessive nutrient concentrations can inhibit the biodegradation activity (Challain *et al.*, 2006), and several authors have reported the negative effect of high Nitrogen, Phosphorus and Potassium (NPK) on the biodegradation of hydrocarbons (Oudot *et al.*, 1998; Chaineau *et al.*, 2005) and more especially the degradation of the aromatic compounds.

14. Fate of contaminants in soil systems

Organic contaminants can reach the groundwater zone in dissolved form mixed with water or as organic liquid phases that may be immiscible in water. According to Hall and Quam (1976), contaminants travel with the soil moisture and are retarded in their migration by various factors. One of the most important factors in contamination of water by petroleum products is the extremely low concentration of the product that can give rise to objectionable tastes and odour. The major aspects of contamination can be broadly classified into:

1. The formation of surface films and emulsions and
2. The solubility in water of certain petroleum products.

The problems associated with surface films are minimized by the ability of aquifers to absorb much of the product. However, this phenomenon magnifies the problems associated with the solubility components of the product, since hydrocarbons held in this manner are subject to leaching as water passes over them. Surface films may affect the aesthetics and interfere with treatment or industrial processes. They may also be toxic to animal or plant life if they emerge into surface waters.

Hall and Quam (1976) further stated that, the water soluble components of petroleum products that give rise to taste and odour problems are the aromatic and aliphatic hydrocarbons. Phenols and cresols are examples of these compounds and generate taste and odours at concentrations as low as 0.01 mg/L. Hence, when chlorine is added to drinking water, as in most municipal water supplies, it reacts with phenols to form chlorophenols, which have objectionable taste and odours at concentrations as low as 0.001mg/L. Therefore, very small quantities of hydrocarbons can cause widespread contamination of water resources.

15. Transport and distribution of soil contaminants

The properties that enhance the transport and transformation process for organic dissolved contaminants such as crude oil can be classified into:

a. **Physical Processes** e.g. advection, dispersion and volatilization, adsorption and ion-cation exchange.
b. **Chemical Processes** e.g. ionization, hydrolysis, oxidation-reduction and complexation.
c. **Biological processes** e.g. bioaccumulation and biodegradation (Mackay and Roberts, 1985).

However, Mackay and Roberts (1985) explained that the migration of an immiscible organic liquid phase is governed largely by its density, viscosity, and surface-wetting properties. Density differences of about 1% are known to influence fluid movement significantly. Organic liquids less dense than water floats and spreads across the water table, and organic liquids more dense than water (e.g. halogenated hydrocarbon) sink through water and plummet through sand and gravel aquifers to the underlying aquitard (relatively impermeable layer) if present. Method of movements of crude oil and other liquid contaminants are as follows:

15.1 Advection, dispersion, and volatilization

According to Mackay and Roberts (1985), in sand and gravel aquifers, the dominant factor in the migration of a dissolved contaminant is advection, the process by which solutes are transported by the bulk motion of flowing groundwater. Groundwater velocity ranges between 1 and 1000 m/year. So that dissolved contaminants spread as they move with the ground water. Dispersion results from two basic processes: molecular diffusion in solution and mechanical mixing. The process results in an overall net flux of solutes from zone of high concentration to a zone of lower concentration. Diffusion in solution is the process whereby ionic or molecular constituents move under the influence of their kinetic activity in the direction of their concentration gradient.

Volatilization refers to the process of pollution transfer from soil to air. It is a form of diffusion that takes place by the movement of molecules or ions from a region of high concentration to a region of low concentration. Volatilization is an extremely important pathway for many organic chemicals; while most ionic substances are usually considered to be non volatile.

15.2 Adsorption and non-cation exchange

Adsorption is a common phenomenon in all viscous substances. Crude oil is highly viscous with a little percentage of volatiles. Adsorption of crude oil has to do with adhesion of pollutant ions or molecules of crude oil to the surface or soil solids, causing an increase in the crude oil concentrations on the soil surface over the concentration present in the soil moisture. Adsorption occurs as a result of a variety of processes with a variety of mechanisms, and some processes may cause an increase of pollutant concentration within the soil solids not merely on the soil surface. Mackay and Roberts (1985) illustrated that adsorption can drastically retard the migration of pollutants in soils; therefore, knowledge of this process is of importance when dealing with contaminant transport in soil and groundwater. For organic compounds, it appears that partitioning between water and the organic compound content of soil is the most important adsorption mechanism.

16. Biomonitoring in crude oil pollution

In a broad sense, biological monitoring involve any component that make use of living organisms, whole or part as well as biological systems to detect any harmful, toxic or deleterious change in the environment especially with the aim of detecting deleterious, harmful or toxic substances in that environment. There are various components employed in biomonitoring of contaminants in the environment. They include biomarkers (biological markers) biosensors and many others. Biomonitoring or biological monitoring is a promising, reliable means of quantifying the negative effect of an environmental contaminant.

16.1 Biological Markers: A biomarker is an organism or part of it, which is used to establish the possible harmful effect of a pollutant on the environment or the biota (Onwurah, *et al.*, 2007). Biological markers (biomarkers) are measurements taken from any biological specimen that will elucidate the relationship between exposure and effect such that adverse effects could be prevented (NRC, 1992). The use of chlorophyll production in *Zea mays* to estimate deleterious effect of crude oil contaminants on soils is a typical plant biomarker of crude oil pollution (Ezeonu and Onwurah, 2009). When a contaminant interacts with an organism, substances like enzymes are generated as a response. Thus, measuring such substances in fluids and tissue can provide an indication or "marker" of contaminant exposure and biological effects resulting from the exposure.

The term biomarker includes any such measurement that indicates an interaction between an environmental hazard and biological system (NRC, 1989). It should be instituted whenever a waste discharge has a possible significant harm on the receiving ecosystem. It is preferred to chemical monitoring because the latter does not take into account factors of biological significance such as combined effects of the contaminants on DNA, protein or membrane. Onwurah *et al.* (2007) stated that some of the advantages of biomonitoring include the provision of natural integrating functions in dynamic media such as water and air, possible bioaccumulation of pollutant from values of 10^3 to 10^6 over the ambient value and/or providing early warning signal to the human population over an impending danger due to a toxic substance. Microorganisms can be used as an indicator organism for toxicity assay or in risk assessment. Tests performed with bacteria are considered to be most reproducible, sensitive, simple, economic and rapid (Matthew, 1980). Table 2 provides an outline of common biomarkers and their application.

Biomarker type	Uses	Reference
'Rec-assay' utilizes *Bacillus subtilis*	Detection of hydrophobic substances (hydrocarbons) toxic to DNA	Matsui, 1989
Chlorophyll content *Zea mays* L	Detection of level of hydrocarbon contamination of agricultural soil	Ezeonu and Onwurah, 2009
Sensitivity of *Nitrobacter sp*	Based on the effect of crude oil on oxidation of nitrite to nitrate	Okpokwasili and Odukuma, 1994
Azotobacter sp	Used in evaluating the effect of oil spill in aquatic environment	Onwurah, 1998
Algae/plant steranes and Bacteria hopanes	Steranes formed as components of crude oil and hopanes used to determine the source rock that generated a crude oil	Peters and Moldown, 1993
Ethoxyresorufin-O-deethylase (EROD) in fish in-vivo	Indicates exposure of fish to planar halogenated hydrocarbons (PAHs) by receptor-mediated induction of cytochrome P-450 dependent monooxygenase exposed to PAHs and similar contaminants	Bucheli and Fent, 1995 Stegeman and Hahn, 1994

Table 2. Biomarkers and their applications

16.2 Biosensor

A biosensor is an analytical device consisting of a biocatalyst (enzyme, cell or tissue) and a transducer, which can convert a biological or biochemical signal or response into a quantifiable electrical signal (Wilson and Walker, 1994). A biosensor could be divided into two component analytical devices comprising of a biological recognition element that outputs a measurable signal to an interfaced transducer (Ripp *et al.*, 2010). Biorecognition typically relies on enzymes, whole cells, antibodies, or nucleic acids, whereas signal transduction exploits electrochemical (amperometric, chronoamperometric, potentiometric, field-effect transistors, conductometric, capacitative), optical (absorbance, reflectance, luminescence, chemiluminiscense, bioluminescence, fluorescence, refractive index, light scattering), piezoelectric (mass sensitive quartz crystal microbalance), magnetic, or thermal (thermistor, pyroelectric) inferfaces (Ripp *et al.*, 2010). The biocatalyst component of most biosensors is immobilized on to a membrane or within a gel, such that the biocatalyst is held in intimate contact with the transducer, and may be reused. Biosensors are already of major commercial importance and their significance is likely to increase as the technology develops (Wilson and Walker, 1994). Biosensors are still emerging biotechnology for the future in environmental biomonitoring since they have specific limitations. Biosensors on a general sense are often employed for continuous monitoring of environmental contamination or as bioremediation process monitoring and biocontrol tools to provide informational data on what contaminants are present, where they are located, and a very sensitive and accurate evaluation of their concentrations in terms of bioavailability. Ripp *et al.* (2010) explained that bioavailability measurements are central to environmental monitoring as well as risk assessment because they indicate the biological effect of the chemical, whether toxic, cytotoxic, genotoxic, mutagenic, carcinogenic, or endocrine disrupting, rather than mere chemical presence as is achieved with analytical instruments. As the name suggest they are biological instruments that detect and signal the presence of harmful contaminants in the environment. There are different types based on the biological components on which their sensitivities are based. Some of them, though not exhaustive are presented in Figure 3.

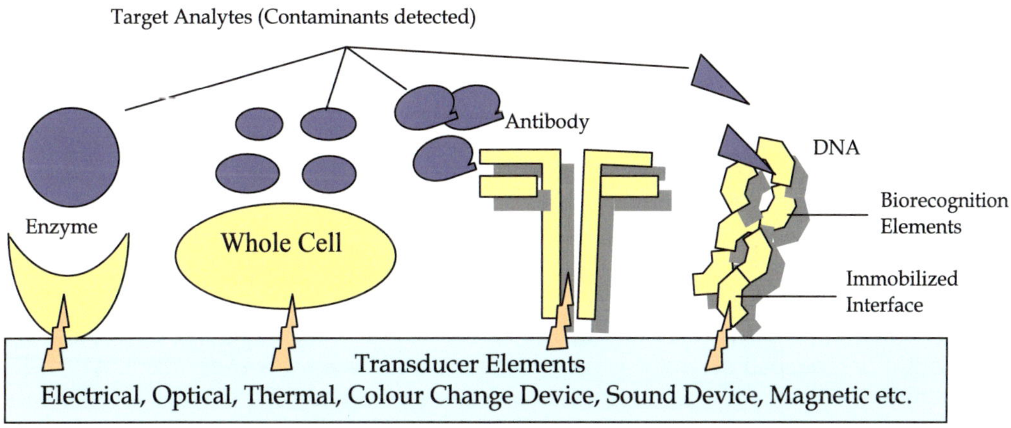

Fig. 3. Anatomy of a Biosensor. The interaction between the target analyte and the biorecorgnition element creates a signaling event detectable by the interfaced transducer element. Modified from source: Ripp *et al.* (2010).

16.3 Enzyme-based biosensors

Leyland Clark in the 1960s used an enzyme biosensor which consists of glucose oxidase enzyme immobilized on an oxygen electrode for blood glucose sensing. This historical application of enzyme based biosensor has found a world wide lucrative application in especially medical diagnosis. Nevertheless, enzymes based biosensor gradually gained application in environmental monitoring. According to Ripp *et al.* (2010) enzymes act as organic catalysts, mediating the reactions that convert substrate into product. Since enzymes are highly specific for their particular substrate, the simplest and most selective enzyme based biosensors merely monitor enzyme activity directly in the presence of the substrate. A good example of enzyme base biosensor in the oil industry is Ethoxyresorufin-O-deethylase (EROD) in fish in-vivo. This enzyme based biosensor (Bucheli and Fent, 1995, Stegeman and Hahn, 1994) indicates exposure of fish to planar halogenated hydrocarbons (PAHs) by receptor-mediated induction of cytochrome P-450 dependent monooxygenase exposed to PAHs and similar contaminants.

Various immobilization techniques are adopted in the attachment of the enzyme to the transducing element (Lojou and Bianco, 2006) they include: adsorption, covalent attachment, entrapment in polymeric matrices such as sol-gels or Langmuir-Blodgett films, or direct cross-linking using polymer networks or antibody/enzyme conjugates. Immobilization provides the biosensor longevity and with recent integration of redox active carbon-based nanomaterials (nanofibers, nanotubes, nanowires, and nanoparticles) as transducers and their unique ability to interact with biological material, a promising advancement in enzyme biosensor design and sensitivity is in sight. When devices such as these are developed specifically for crude oil detection, it will be easier to monitor crude oil pollutants in drinking water and agricultural products.

Optical transducers (absorption, reflectance, luminescence, chemiluminiscence, evanescent wave, surface plasma resonance) are also commonly employed in enzyme based biosensor (Ripp *et al.*, 2010).

16.4 Antibody-based biosensors (immunosensors)

These types of biosensors make use of antibodies as recognition elements (immunosensors). They are used widely as environmental monitors because antibodies are highly specific, versatile, and bind stably and strongly to target analytes (antigens) (Ripp *et al.*, 2010). Antibodies can be highly effective detectors for environmental contaminants, and advancements in techniques such as phage display for the preparation and selection of recombinant antibodies with novel binding properties assures their continued environmental application. Perhaps the best introduction to antibody-based biosensing is the Automated Water Analyzer Computer Supported System (AWACSS) which is an environmental monitoring system developed for remote, unattended, and continuous detection of organic pollutants for water quality control (Tschmelak *et al.*, 2005). AWACSS uses an optical evanescent wave transducer and fluorescently labelled polyclonal antibodies for multiplexed detection of targeted groups of contaminants, including endocrine disruptors, pesticides, industrial chemicals, pharmaceuticals, polycyclic aromatic hydrocarbons (PAHs), polychlorinated biphenyls (PCBs) and other priority pollutants, without requiring sample pre-processing. Antibody binding to a target sample analyte occurs in a short 5 minute preincubation step, followed by microfluidic pumping of the sample over the transducer element, which consists of an optical waveguide chip impregnated with 32 separate wells of immobilized antigen derivatives (Ripp *et al.*, 2010). As the antibody/analyte complexes flow through these wells, only antibodies with free binding sites can attach to the well surface (in what is referred to as a binding inhibition assay). Thus, antibodies with both of their binding sites bound with analyte will not attach to the surface and will pass through the detector. A semiconductor laser then excites the fluorophore label of bound antibodies, allowing for their quantification, with high fluorescence signals indicating high analyte concentrations. A fibre optic array tied to each well permits separation and identification of signals by the well, thereby yielding a simultaneous measurement of up to 32 different sample contaminants. The instrument has been used for groundwater, wastewater, surface water, and sediment sample testing with detection limits for most analyte in the ng/L range within assay times of approximately 18 minutes (Ripp *et al.*, 2010). Another design by Glass *et al.*, (2004), similar to the above but less refined benchtop flow-through immunosensor (KinExA) demonstrated to detect analytes successively based on a replaceable flow cell containing fluorescently labeled antibody. Their time of assay was approximately 26 minutes, with detection limits at picomolar concentrations.

Although not as elaborate as the AWACSS, a multitude of other antibody-based biosensors have been applied as environmental monitors, traditionally serving as biosensors for pesticides and herbicides, but their target analytes have broadened considerably over the past several years to include heavy metals, polycyclic aromatic hydrocarbons (PAHs), polychlorinated biphenyls (PCBs), explosives (TNT and RDX), phenols, toxins such as microcystin, pharmaceutical compounds, and endocrine disruptors (Ferre *et al.*, 2007).

16.5 DNA-based biosensors

The principle underlying the DNA based biosensor is the ability of a transducer to monitor a change in the nucleic acid's structure occurring after exposure to a target chemical. These structural changes are brought on either by the mutagenic nature of the chemical, resulting in mutations, intercalations, and/or strand breaks, or by the chemical's ability to covalently or non-covalently attach to the nucleic acid (Ripp *et al.*, 2010). Immobilizing the nucleic acid as a recognition layer on the transducer surface forms the biosensor, and detection of the chemically induced nucleic acid conformational change is then typically achieved electrochemically (ie., a change in the current) or less so through optical or other means (Fojta, 2002).

Nucleic acid biosensors are generally nonselective and provide an overall indication of a potentially harmful (genotoxic, carcinogenic, cytotoxic) chemical or chemical mix in the test environment and, depending on the biosensor format, an estimate of concentration. Bagni *et al.* (2005) illustrated a conventional DNA biosensor which was used to screen soil samples for genotoxic compounds, using benzene, naphthalene, and anthracene derivatives as model targets. Double-stranded DNA was immobilized on a single-use disposable screen-printed electrochemical cell operating off a handheld battery-powered potentiostat (Sassolas *et al.*, 2008). A 10 μL drop of a preprocessed and preextracted contaminated soil sample was placed onto the working electrode for 2 minutes, and resulting electrochemical scans, based on the chemical's propensity to oxidize DNA guanine residues, were measured. The magnitude of these "guanine peaks" in relation to a reference electrode was linearly related to their concentration in solution (ie., the higher the concentration of the target chemical, the more damage is imposed on the DNA, and the lower the electrochemical measurement of the oxidation signal). In a very discrete application of this DNA biosensor, the authors also applied it to the detection of this DNA biosensor and also to the detection of PAHs in fish bile, using the accumulation of PAH compounds in live fish to monitor for water contamination events (Lucarelli *et al.*, 2003). Nucleic acid can be manipulated similarly to create target specific aptamers using a process called SELEX (systematic evolution of ligands by exponential enrichment) (Ripp, 2010). By iteratively incubating nucleic acid with the desired target, one can select for oligonucleotide sequences (or aptamers) with the greatest affinity for the target.

Selectivity, though, has been demonstrated by several groups using deoxyribozymes (DNAzymes) or ribozymes (RNAzymes). These engineered catalytic oligonucleotides can mediate nucleic acid cleavages or ligation, phosphorylation, or other reactions. For example, DNAzyme biosensor for lead uses a single stranded DNAzyme absorbed to a gold electrode (Xiao *et al.*, 2007). The DNAzyme incorporates a methylene blue tag at concentrations as low as 62 ppb, the DNAzyme strand is cleaved, allowing the methylene blue tag to approach the transducer and transfer electrons, thereby instigating an electrochemical signal (Ripp *et al.*, 2010).Advantages of this technique are promptness (only a few minutes to detect sample processing is often necessary), sensitivity (typically down to low part-par billion levels), ease of use, and cost-effectiveness screening environmental sites for toxic chemical intrusions or monitoring operational endpoints of bioremediation efforts. A calourimetric DNAzyme-based biosensor for lead has also been demonstrated (Wei *et al.*, 2008).

Currently, perhaps, one of the greatest applications of DNA for the detection of crude oil pollutants is the use of 'Rec-assay' which utilizes *Bacillus subtilis* to detect hydrophobic substances (hydrocarbons) (Matsui, 1989) toxic to DNA.

17. Bioremediation of crude oil polluted environments

Crude oil pollutants introduction into the environment occurs through three major routes: land, aquatic environments and the atmosphere. Each of these environmental components is a habitat for biotic existence. In the refining of crude oil most components are broken down into highly volatile fragments and are disposed in gaseous form. Thus gas flaring is a major source of environmental pollution that is introduced into the atmosphere and has direct effect in Ozone layer perforation as well as increase in green house gases emission. Acid rain and petrochemical smog are other environmental hazards resulting from crude oil processing. Both the land and the aquatic environments are affected due to the impact of the non-volatile petroleum hydrocarbon (polycyclic aromatic hydrocarbons). In the course of petroleum exploration, drilling fluids are also other sources of pollution normally neglected as emphasis is shifted to crude oil pollutants. Aquatic and terrestrial environments polluted by crude oil results from man's carelessness and negligence of his environment. Crude spills are the major source of contamination of land and marine bodies. During drilling, uncontrolled spill occurs in offshore environments thereby damaging the aquatic environment. In Nigeria, over 20 percent of crude oil spill results from vandalism of pipelines. Transportation spill accounts for over 10 percent pollution in both marine and soil pollution. Automobile spent engine oil and exhaust release of gases such as carbon dioxide and carbon monoxide are other ways of polluting the environment in cities. Some of the bioremediation methods used in Nigeria and elsewhere are clean up procedures which use bio-microbs. Some of these clean up procedures include: introduction of organic manure, phytoremediation, and optimizing the factors that enhance the thriving of microorganisms.

Some of these standard eco-friendly methods used for crude oil bioremediation will be emphasized in this chapter.

18. Natural attenuation

Natural attenuation refers to decrease in concentrations of chemicals in the environment due to natural phenomena such as microbial degradation, evaporation, and adsorption (where chemicals adsorb unto solids). Prince (2002) explained that as the age of an oil spill increases, there will be more opportunity for the oil to weather and for its constituents to attenuate in the environment. Generally, the more weathering that take place consequently reduces the biodegradation that occurred.

19. Bioremediation of soil

Bioremediation can be a cost effective technology often used to treat oil spills in all types of environmental media including: soils, ground-water, and surface water (both fresh and marine) (DOE/PERF, 2002). Oily wastes are also treated using bioremediation processes.

Spilt oil is subjected to a wide range of physical, biological and chemical processes that actually "weather" the oil, and attenuate it in the environment (DOE/PERF, 2002). Biodegradation is one of the weathering processes that are unique because it is the primary process by which the oil is actually removed from the environment. Most of the other weathering processes transfer the oil from one medium to another (as in the case of volatilization where certain oil constituents evaporate into the air), or dilute it (such as from wave action, which may disperse oil throughout the water column in a marine environment) (DOE/PERF, 2002).

Methodologies for bioremediation of terrestrial oil spills are well developed. Research on bioremediation in surface soils is currently focused on optimizing bioremediation rates, developing appropriate treatment goals, and site restoration following remediation (DOE/PERF, 2002). Crude oil and refined oil products are frequently stored and transported on or over land, and as a result, oil spills that impact soil and groundwater tend to be quite common although usually smaller in volume than marine or freshwater spills (DOE/PERF,2002).

Bioremediation uses microscopic organisms (primarily bacteria) that live on soil and 'eat' chemicals, such as petroleum hydrocarbons. They use certain components of the petroleum hydrocarbons as their food source; leaving other chemicals behind. The waste products of the process are generated in form of water and carbon dioxide a process known as mineralization that characterizes the complete removal of crude oil from the environment.

20. Use of bulking agents/tilling for bioremediation

Nackles and Ray (2002), clearly elaborated that in composting applications, organic bulking agents allow for successful treatment of higher oil concentrations, increase the biodegradation rate and also increase the temperature. Locally available bulking agents can be cost-effective, and can include palm husks, wood chips, saw dust, rice hulls, manure, straw and hay. Of these, wood chips, palm husks, straw and hay can serve as structural bulking agents. The source of bulking agents should always be carefully checked to ensure that there is no potential for residual substance (like pesticides) to be present that could be toxic to microbes. Bulking agents are added to soil or wastes to improve permeability and water holding capacity, which in turn increases biodegradation rates (Sublette, 2001). Bulking agents fall into two general categories: structural and organic (Sublette, 2001). Structural bulking agents improve the porosity and permeability of the soil by creating larger and more numerous pore spaces. Organic bulking agents initially act as structural bulking agents but also biodegrade themselves, producing degradation products that build soil structure on a long term basis. Sublette (2001), insisted that bulking agents improve the water holding capacity which is especially important in sandy soils, but they can also increase fertilizer demand.

For soil treatment, organic bulking agents should be blended into the soil until a porous soil structure is obtained with no visual evidence of oil. The amount of bulking agent to be used in composting depends upon the original texture of the soil or waste (DOE/PERF, 2002).

Prince and McMillen (2002) stated that tilling the active biological zone should be performed regularly to overcome any oxygen deficiencies and to mix the soil with the nutrients and bulking agent. Prince and McMillen (2002) further stated that mixing also helps to optimize contact among the microorganisms, hydrocarbons, moisture and nutrients to enable maximum degradation rates. Furthermore, according to Nakles and Ray (2002), in farming operations, tilling is recommended at an interval of twice per month. When composting is used, tilling can be performed or a force or passive aeration system can be used to supply oxygen. Generally, the aeration method is determined by equipment availability and engineering considerations such as the compost pile configuration. Rocky soils may not be ideal for bioremediation because of lower microbial populations, the inability to retain moisture, and difficulty in tilling. For rocky soils, one needs to consult with a bioremediation expert.

21. Microbial bioremediation

Bioremediation is the use of biological systems for the reduction of pollution from air, aquatic or terrestrial systems (EFB, 1999), it also involve extracting a microbe from the environment and exposing it to a target contaminant so as to lessen the toxic component (Vallero, 2010). Thus, the goal of bioremediation is the employment of bio-systems such as microbes, higher organisms like plants (phytoremediation), and animals to reduce the potential toxicity of chemical contaminants in the environment by degrading, transforming, and immobilizing these undesirable compounds.

Biodegradation is the use of living organisms to enzymatically and otherwise attack numerous organic chemicals and break them down to lesser toxic chemical species. Biotechnologists and bio-engineers classify pollutants with respect to the ease of degradation and types of processes that are responsible for their degradation, sometimes referred to as treatability (Vallero, 2010).

Biodegradation with micro-organisms is the most frequently occurring bioremediation option. Micro-organisms can break down most compounds for their growth and/or energy needs. These biodegradation processes may or may not need air. In some cases, metabolic pathways where organisms normally use for growth and energy supply may also be used to break down pollutant molecules. In these cases known as co-metabolisms, the micro-organism does not benefit directly. Researchers have taken advantage of this phenomenon and used it for bioremediation purposes (EFB, 1999).

A complete biodegradation results in detoxification by mineralising pollutants to carbondioxide (CO_2), water (H_2O) and harmless inorganic salts (EFB, 1999). Incomplete biodegradation (i.e mineralization) will produce compounds that are usually simpler (e.g cleared rings, removal of halogens), but with physical and chemical characteristics different from the parent compound. In addition, side reactions can produce compounds with varying levels of toxicity and mobility in the environment (Vallero, 2010).

Biodegradation may occur spontaneously, in which case the expressions "intrinsic bioremediation" or "natural attenuation" are often used (EFB, 1999). In many cases the natural circumstances may not be favourable enough for natural attenuation to take place due to inadequate nutrients, oxygen or suitable bacteria. Such situations may be improved by supplying one or more of the missing/inadequate environmental factors. For instance extra nutrients (EFB, 1999) were disseminated to speed up the break down of the oil spilled on 1000 miles of Alaskan shoreline by the super tanker Exxon Valdez in 1989.

According to Vallero (2010), there are millions of indigenous species of microbes living at any given time within many soil environments. The bioengineer simply needs to create an environment where those microbes are able to use a particular compound as their energy source. Biodegradation processes had been observed empirically for centuries, but putting them to use as a distinct field of bioremediation began with the work of Raymond *et al.* (1975). This seminal study found that the addition of nutrients to soil increases the abundance of bacteria that was associated with a proportional degradation of hydrocarbons, in this case petroleum by-products (Raymond *et al.*, 1975).

Vallero (2010) indicates that the success of bioremediation depends on the following factors:

1. The growth and survival of microbial populations
2. The ability of these organisms to come into contact with the substances that needs to be degraded into less toxic compounds.
3. The size of the micro-organisms population.
4. The appropriate microbial environment that must be habitable for the microbes to thrive.

Sometimes, concentrations of compounds can be so high that the environment is toxic to microbial populations. Therefore, the bioengineer must either use a method other than bioremediation or modify the environment (e.g dilution, change of pH, oxygen pump, adding organic matter, etc) to make it habitable. An important modification is the removal of non-aqueous phase liquids (NAPLs) since the microbes' biofilm and other mechanisms usually work best when the microbe is attached to a particle, thus, most of the NAPLs need to be removed, by vapour extraction (Vallero, 2010). Thus, low permeability soils, like clays, are difficult to treat, since liquids (water, solutes and nutrients) are difficult to pump through these systems. Usually bioremediation works best in soils that are relatively sandy, allowing mobility and greater probability of contact between the microbes and the contaminant (Vallero, 2010). Therefore, an understanding of the environmental conditions sets the stage for problem formulation (i.e identification of the factors at work and the resulting threats to health and environmental quality) and risk management (i.e what are the various options available to address these factors and how difficult it will be to overcome obstacles or to enhance those factors that make remediation successful). In other words, bioremediation is a process of optimization that involves the selection options among a number of biological, chemical and physical factors. This process includes the correct match of the degrading microbes to the conditions of the contaminated soil, the understanding and controlling of the movement of the contaminant (microbial food) so as to come into contact with microbes, and the characterization of the abiotic conditions controlling both of these factors (Vallero, 2010). Optimization can vary among options, such as artificially adding microbial populations known to break down the compounds of concern. Only a few species can break down certain organic compounds (Vallero, 2010). Two major limiting factors of any biodegradation process are toxicity to the microbial population and inherent biodegradability of the compound. Numerous bioremediation projects include in-situ (field treatment) and ex-situ (sample/laboratory treatment) waste treatment using bio-systems (Vallero, 2010).

21.1 A practical application of microorganism in crude oil bioremediation: According to (Onwurah, 2003) many microorganisms can adapt their catabolic machinery to utilize certain environmental pollutants as growth substrates, thereby bioremediating the environment. Some microorganisms in carrying out their normal metabolic function may fortuitously degrade certain pollutants as well. This process termed cometabolism obviously requires adequate growth substrates. Diazotrophs, such as *Azotobacter vinelandii,* beyond their ability to fix atmospheric nitrogen also have the capacity in some case, to cometabolise petroleum hydrocarbons (Onwurah, 1999).

Onwurah (2003) carried out a bioremediation study that involved two bacteria, a hydrocarbonoclastic and diazotrophic bacteria. The hydrocarbonoclastic was tentatively identified as *Pseudomonas sp* and designated as $NS_{50}C_{10}$ by the Department of Microbiology, University of Nigeria, Nsukka. The diazotrophic bacteria were *Azotobacter vinelandii,* which was isolated from previously crude oil contaminated soil (Onwurah, 1999). This study describes the mineral media and procedure for isolation and multiplication of the bacteria to the required cell density. Crude oil spill was simulated by thoroughly mixing 50, 100, 150 mg fraction of crude oil with 100g batches of a composite soil sample in beakers. The soil samples were taken from a depth of 0 - 50cm from the Zoological garden, University of Nigeria, Nsukka. The mixing was conducted using a horizontal arm shaker adjusted to a speed of 120rpm for 30 minutes. The contaminated soil samples, in beakers, were inoculated with optimal combinations (cell density) of $NS_{50}C_{10}$ and *A.vinelandii.* Water was added to the crude oil-contaminated soil samples (both inoculated and those not inoculated to a saturation point but not in excess) and then the samples were left to stand undisturbed for seven days. $NS_{50}C_{10}$ was applied first, followed by A.*vinelandii,* 12 hours later. At the seventh day of soil treatment, 20 sorghum grains (previously soaked overnight in distilled water) were planted in each soil sample followed by irrigation to aid germination. Seven days after the planting of the sorghum grains, the soil from each beaker was carefully removed. The number of germinated seed per batch of soil sample was noted and the length of radicule was measured and the mean length was taken from each batch.

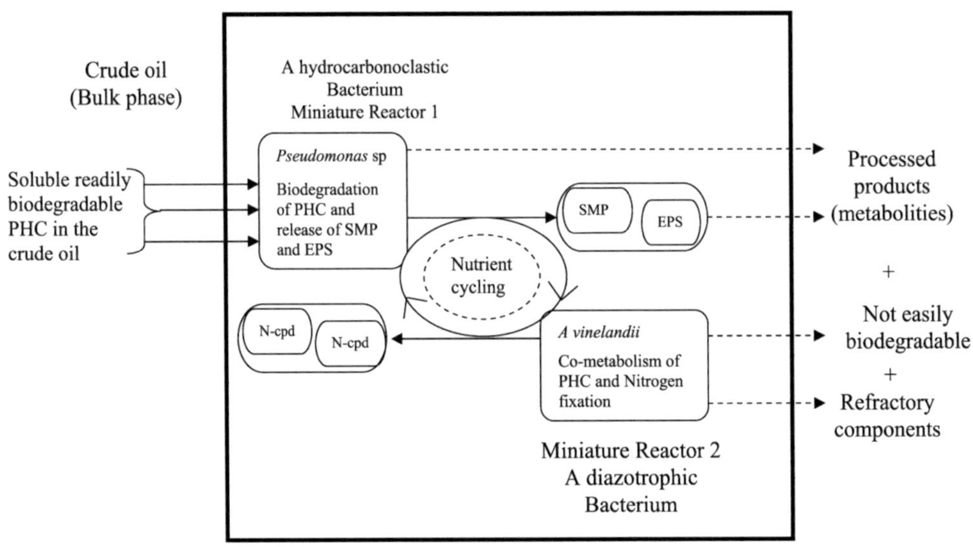

Fig. 4. Simplified bioremediation conceptual model of *Pseudomonas sp* and *A vinelandii* operating as a unit of two miniature sequencing bioreactors, in situ, (SMP= Soluble microbial products; N-cpd= fixed nitrogen compounds; EPS= Exopolysaccharide; PHC=petroleum hydrocarbons) (Onwurah, 2003).

The results of this experiment showed that *Pseudomonas sp.* grew well on agar plates containing a thin film of crude oil as the only carbon source while *A. vinelandii* did not. However, cell free extract of *Azotobacter vinelandii*, fixed atmospheric nitrogen as ammonium ion (NH^{4+}) under appropriate condition. The specific growth rate values in contaminated soil samples inoculated with both normal $NS_{50}C_{10}$ and *A.vinelandii* (consortium) were highest in all cases. By adding an aerobic, free living diazotroph *A.vinelandii* with the *Pseudomonas sp.* ($NS_{50}C_{10}$), an improvement on bioremediation of soil over that of the pure $NS_{50}C_{10}$ alone was achieved to the order of 51.96 to 82.55%. This innovative application that uses the synergetic action of several microorganisms to clean up oil polluted soil has potential application for the bioremediation of oil contaminated soil in the Niger delta region.

The method described above is the biotechnological application known as **Bioaugmentation** which is the addition of selected organisms to contaminated soils (sites) in order to supplement the indigenous microbial population and speed up degradation. Figure 4 presents a model of the process involved in this bioremediation technique.

22. Phytoremediation techniques

Phytoremediation is a biological technology process that utilizes natural plant processes to enhance degradation and removal of contaminants in contaminated soil or groundwater (Schooner, *et al.*, 1995). Phytoremediation utilizes physical, chemical, and biological processes to remove, degrade, transform, or stabilize contaminants within soil and groundwater. Hydraulic control, uptake, transformation, volatilization, and rhizodegradation are important processes used during phytoremediation. These processes are briefly in the subsequent paragraphs.

22.1 Hydraulic Control: Phytoremediation applications can be designed to capture contaminated groundwater plumes to prevent off-site migration and/or decrease downward migration of contaminants. Trees and grasses act as a solar "pump" removing water from soils and aquifers through transpiration. Contaminant plume capture relies on the formation of a cone of depression within an aquifer due to uptake of water by plants and subsequent transpiration. Downward migration of contaminants due to percolation of rainwater can also be controlled with phytoremediation. Within the upper region of an aquifer, grasses with dense, fibrous root systems are used to transpire water and limit percolation of contaminants through the vadose zone and to intercept rainwater that may discourage tree root penetration through the water table. However, plume capture is not limited to shallow aquifers, as poplar trees planted in well casings have been used to tap water tables at a depth of 10-m (Gatliff, 1994).

22.2 Phytovolatilization: Volatile pollutants diffuse from the plant into the atmosphere through open stomata in leaves. Radial diffusion through stem tissues has also been reported (Zhang, *et al.*, 2001; Narayanan, *et al.*, 1999; Davis *et al.*, 1999). For example, methyl-tert-butyl ether (MTBE) can escape through leaves, stems, and the bark to the atmosphere (Hong, *et al.*, 2001; Trapp and McFarlane, 1994).

The natural ability of a plant to volatilize a contaminant that has been taken up through its roots can be exploited as a natural air-stripping pump system. Phytovolatilization is most applicable to those contaminants that are treated by conventional air-stripping.

Tree core samples of hybrid poplars exposed to TCE (Trichloroethylene) also showed radial diffusion from the stem (Ma, and Burken, 2003) rather than transpiration from leaves (Ma, and Burken, 2003; Newman *et al.*, 1998) as the main dissipation mechanism. Generally, the concentration of volatile organic components (VOCs) in the xylem decreases with increasing distance from the roots (Ma, and Burken, 2003). Once released into the atmosphere, compounds with double-bonds such as TCE and perchloroethylene (PCE) could be rapidly oxidized in the atmosphere by hydroxyl radicals. However, under certain circumstances (e.g., poor air circulation) phytovolatilization may not provide a terminal solution. For example, MTBE is long lived in the atmosphere and can pose a risk to shallow groundwater during precipitation (Penkow *et al.*, 1997). In such cases, simple mass balance models can be utilized to determine if phytovolatilization poses a significant risk to humans and/or the environment (Narayan *et al.*, 1999; Ma, and Burken, 2003; Aitchison, *et al.*, 2000).

Nevertheless, the rate of release of VOCs from plant tissues is generally small relative to other emissions (Aitchison, *et al.*, 2000). Thus, phytovolatilization is a potentially viable remediation strategy for many volatile organic chemicals.

22.3 Rhizodegradation: Microbial degradation in the rhizosphere might be the most significant mechanism for removal of diesel range organics in vegetated contaminated soils (Aprill, *et al.*, 1990; Banks, *et al.*, 1999; Binet, et al., 2000; Liste and Alexander, 2000; Reilley, et al., 1996; Miya and Firestone, 2000; Miya and Firestone, 2001) Rhizodegradation occurs because contaminants such as PAHs are highly hydrophobic and their sorption to soil decreases their bioavailability for plant uptake and phytotransformation. Briggs *et al.* (1982) first demonstrated that the lipophilicity of a pesticide determines its fate in a barley plant. High Kow values (an indicator of hydrophobicity) corresponds to a greater possibility that the compound would be retained in the roots. Burken and Schnoor (1998) published similar results for the sorption of a wide range of organic contaminants to roots of hybrid poplar plants grown hydroponically. Where the Root Concentration Factor (RCF) (L/kg dry roots) is the ratio of organic chemical sorbed on the root (mg/kg of fresh root tissue) to that in hydroponic solution (mg/L). This equilibrium partitioning coefficient has generally proved to be a good indicator of whether a plant retains a contaminant in the root, which increases the probability of microbial degradation (not withstanding significant bioavailability limitations). However, a few exceptions exist such as phenol and aniline, which bind irreversibly to the root (especially aniline) and are chemically transformed.

23. Factors enhancing phytoremediation site treatability

23.1 Source removal: For phytoremediation to succeed, it is very important to physically remove the source of contamination (e.g., excavation of highly-contaminated soil and/or extraction of free phase). The presence of a continuous source can be detrimental to the health of the plants and can extend the life of the phytoremediation project indefinitely. Perhaps pretreatment methods that reduces the bulk quantity of crude oil such as application of cow dung, poultry wastes as well as extended time between contamination and phytoremediation can also help to reduce the bulk accumulation of crude oil in surface soil so that the plant is able to grow in such environments. The crude oil source could also be dispersed over a large surface area by mixing with bulk substances such a rice husk and other lignocelluloses and exposed to weathering and microbial degradation before being made available to the plants as source of nutrient.

23.2 Depth of Contamination: Dept of crude oil contamination of soil is also monitored before application of phytoremediation. Phytoremediation is most effective at sites with shallow (i.e., root accessible) contaminated soils where contaminants can be treated in the rhizosphere and/or by plant uptake. Roots of phreatophytic trees can be expected to grow at least 3 meters into a soil profile, and it is possible to encourage rooting to a depth of 5 meters or more using the tree-in-a-well concept (Gatliff, 1994). On the other hand, roots of some grasses (alfalfa, switchgrass, and tall fescue) can reach soil depths of only 0.25- 0.4 m. Buffel-grass roots to a depth of 0.75 m but has been observed to have dense rooting pattern within 0.3 m from the topsoil layer. Hawaiian plants, Milo and Kou were used to remediate saline soils contaminated with TPHs (Total Petroleum Hydrocarbons), had roots which grew to a depth of more than 1.5 m by growing through the brackish water table into a zone of concentrated contaminants (USACE, 2003). Optimizing irrigation patterns can also facilitate biodegradation of contaminants by creating an "expanded rhizosphere" due to translocation of organic root exudates and inorganic nutrients to relatively deep soil layers. Phytoremediation can therefore influence soils to the depth where irrigation water reaches, even if the roots are sparse in the contaminated zone.

23.3 Soil composition and quality: Soil quality is another important factor for determining successful germination, growth and health of plants. Heavily contaminated soils have a tendency towards poor physical conditioning which is unsuitable for vigorous growth of vegetation and rhizosphere bacteria. It is therefore critical to use amendments to improve the quality of soil before planting. Common limitations are poor moisture-holding capacity, insufficient aeration, low permeability and nutrient deficiencies. Agronomic soil analysis and preliminary greenhouse or pilot scale experiments can help identify these constraints. For example, nutrient analysis of contaminated soils from a site at the Unocal Bulk Storage Terminal at Superior, Wisconsin (Rentz, *et al.*, 2004) indicated general deficiencies in nitrogen, phosphorus, potassium, and zinc. To decrease the soil pH addition of sulfur was recommended. This information was subsequently used in greenhouse treatability studies, from which a formula of 50 lb/ac phosphorus, 225 lb/ac zinc, and 50 lb/ac potassium was identified as optimum for growth of native grasses. Organic amendments such as aged manure, sewage sludge, compost, straw, or mulch can be used to increase the water-holding capacity of a contaminated soil. Soil pH can be increased and decreased by the addition of lime and sulphur respectively.

23.4 Weather: Phytoremediation might be best suited for tropical countries such as Nigeria, Africa and most part of Asia where plant growth occurs all year round. Plants are known to take up numerous inorganic and organic contaminants and store them in various plant organs when they can utilize them effectively for their benefits. In temperate climates, the active contribution of phytoremediation is restricted to the growing period only. Winter operations may pose problems for phytoremediation when deciduous vegetation loses its leaves, transformation and uptake cease, and soil water is no longer transpired due to continues humid nature of the environment. However, a combination of grasses can be used to prolong the growing period.

23.5 Plant Selection Criteria: Plants should be selected according to the needs of the application, the contaminants of concern, and their potential to thrive on contaminated soil. Design requirements should include the use of native plants, to avoid introduction of invasive species. Apart from this, vegetation should be fast growing, hardy, easy to plant and maintain. The main aim is to ensure that roots expand throughout the entire contaminated zone. In temperate climates with shallow contaminated aquifers, phreatophytes, such as *Populus* sp. (hybrid poplar, cottonwood, aspen) and *Salix* sp. (willow) are often selected because of fast growth, deep rooting ability down to the surface of groundwater, large transpiration rates, and the fact that they are native throughout most of the country. Among tropical plants tested for use in the Pacific Islands, three coastal trees, kou (*Cordia subcordata*), milo (*Thespesia populnea*), and kiawe (*Prosopis pallida*) and the native shrub beach naupaka (*Scaevola serica*) tolerated field conditions and facilitated clean-up of soils contaminated with diesel fuel (Hetch and Badiene, 1998). Wang and Meresz (1981) assessed onions, beets, tomatoes, and soil for 17 PAHs including Barium Phosphate. They found most of the PAH contamination localized in the onion 'peels'. Other factors influencing the localization of PAH in plants include: the rate of PAH uptake by plant species, the nature of the substrate that the plant is growing in, PAH solubility, PAH phase (vapor or particulate), and molecular weight (Edwards, 1983). These latter findings are of potential significance since in an oil spill the PAH compounds would be present along with benzene. Grasses are often planted in tandem with trees at sites with organic contaminants as the primary remediation method. They provide a tremendous amount of fine roots in the surface soil, which is effective at binding and transforming hydrophobic contaminants such as TPH, BTEX, and PAHs. Grasses are often planted between rows of trees to provide for soil stabilization and protection against wind-blown dust that can move contaminants off-site. Legumes such as alfalfa (*Medicago sativa*), alsike clover (*Trifolium hybridum*), and peas (can be used to restore nitrogen to poor soils. Fescue (*Vulpia myuros*), rye (*Elymus sp.*), clover (*Trifolium* sp.) and reed canary grass (*Phalaris arundinacea*) have been used successfully at several sites, especially petrochemical wastes. Once harvested, the grasses can be disposed off as compost or burned. Plant tolerance to high contaminant concentrations is also a very important factor to keep in mind. The phytotoxicity of petroleum hydrocarbons is a function of the specific contaminant composition, its concentration, and the plant species used. Major adverse effects typically include reduced germination and growth if contaminant concentrations are sufficiently high. In general, TPH values of 15 percent or greater can result in significant reductions in plant growth and in some cases mortality. Compared with uncontaminated soil, soils with 2% TPH reduced alfalfa yields by 32 percent (Wiltse *et al.*, 1998). Production of biomass by ryegrass was reduced 46 percent at a soil concentration of 0.5 percent (5000 mg/kg) hydrocarbons (Gunther *et al.*, 1996). It was found that plants pre-grown in clean soil and subsequently transplanted to the contaminated soil grew nearly as well as the control, showing that toxicity was associated with germination and/or early plant growth. Similarly, poor rooting of ryegrass compared to legumes appeared to adversely affect the removal of TPH from Gulf War-contaminated soils (Yateem, 1999).

Also, although the germination of sunflower seeds and beans was greater than that of maize, vegetative growth was greater for maize than beans, demonstrating that germination and later plant growth may be affected differently (Chaineau *et al.*, 1997). Aged spills tend to be much less phytotoxic than fresh ones, possibly because of the lower bioavailability of toxic compounds in the aged spills. However, the speciation of petroleum hydrocarbons is also very important in determining phytotoxicity. A fuel oil with 30 percent aromatics resulted in LC50 germination (oil concentration lethal to 50 percent of test plants) values of 7 percent (70,000 mg/kg) for sunflower seeds. The volatile fraction can prove most toxic to plants. Aromatic volatile petroleum hydrocarbons such as benzene have been used as herbicides in the past years, illustrating their phytotoxicity when applied to plant leaves (Baker, 1970). In contrast, no phytotoxic effects were observed in hybrid poplar trees exposed to a simulated groundwater containing a mixture of VOCs including BTEX, chlorinated aliphatics, and alcohols at a total concentration of 169 mg/L (Ferro *et al.*, 1999). Reduction of the volatile fraction may be accomplished through management, such as tillage of the soil. If initial efforts at plant establishment at a site fail, replanting the area may ultimately lead to success as concentrations or bioavailability of the more phytotoxic components decline. Solution-phase concentrations of hydrocarbons are also important, particularly for aquifer remediation applications of phytoremediation. Additional components with phytotoxic effects include various unsaturated hydrocarbons and acidic hydrocarbons such as alicyclics with carboxylic acid groups (naphthenic acids) (Baker, 1970). A screening test and knowledge from the literature of plant attributes is essential for selection of plants. Most experts recommend a mixture of grasses or legumes to address surface soils contaminated with petroleum hydrocarbons. However, design engineers should work in interdisciplinary teams that include a botanist and/or agricultural specialist to identify and select plants that will grow well at the site. Preliminary greenhouse studies should also be used to identify plants that can thrive and enhance transformation of contaminants of concern to non-toxic or less toxic products.

23.6 Time scale of clean-up: Degradation of organics may be limited by mass transfer, i.e., desorption, and mass transport of chemicals from soil particles to the aqueous phase may become the rate determining step. Therefore, phytoremediation may require more time to achieve clean-up standards than other more costly alternatives such as excavation or ex-situ treatment, especially for hydrophobic pollutants that are tightly bound to soil particles. In many cases, phytoremediation may serve as a final "polishing step" to close sites after more aggressive clean-up technologies have been used to treat the hot spots.

24. Agronomic inputs of phytoremediation

24.1 Irrigation: Irrigation is a common practice in Nigeria and other tropical region especially during the dry seasons of the year. The dry season period is between November and April. Exposure of crude oil to elements such as sunshine, aeration and organic fertilizer followed by tillage and irrigation is bound to give the best result. For terrestrial phytoremediation applications, it is often desirable to include irrigation costs on the order of 10-20 inches of water per year, in the design. Spray irrigation is less efficient than drip irrigation as it encourages the growth of weeds that compete for nutrients with plants and hinder their delivery to the contaminated zone. Results suggest that irrigation can enhance bioremediation of certain diesel components. Irrigation of the plants is especially important during the start of the project. However, after the first year, hydrologic modeling can be used to estimate the rate of percolation to groundwater under irrigation conditions. Over time, irrigation can be withdrawn from the site, provided the area receives sufficient rainfall to sustain the plants.

24.2 Fertilizer Requirements: Contaminated soils are usually deficient in macro- and micro-nutrients necessary for establishing healthy vigorously growing plants and stimulating microbial contaminant degradation. Nitrogen fertilization of motor oil-contaminated soils was found to increase the growth of corn and reduce what appeared to be nitrogen-deficient yellowing of the leaves (Giddens, 1970). The source of nutrients also appeared to affect the germination and growth of plants. Organic sources of nitrogen are better than inorganic sources. This is probably because organic nitrogen sources provide a slow release source of nitrogen, and also help to improve soil structure and soil water relationships for plant growth. It was found that poultry manure increased the growth of corn in a soil containing 3 percent weight per volume crude oil more than an inorganic fertilizer containing nitrogen, phosphorus, and potassium (Amadi *et al.*, 1993). The addition of sawdust alone improved germination by decreasing oil contact with seeds, but accentuated the adverse effect of the oil on later growth, apparently by further widening the carbon-to-nitrogen ratio (Amadi *et al.*, 1993). With respect to TPH degradation, nutrient addition during phytoremediation has yielded mixed results. Hutchinson *et al.* (2001) observed better degradation of TPH using grasses with N/P amendments than without inorganic amendments. Joner *et al.* (2002) reported improved degradation of 3 and 4 ringed PAHs with the addition of N/P, but diminished degradation of 5 and 6 ringed PAHs. Finally, Palmroth *et al.* (2002) observed no improved degradation of diesel fuel with nutrient amendments during phytoremediation with pine, poplar, or grasses. Microbial bioremediation of TPH contaminants with nutrient addition also produced widely varying results. Diesel fuel degradation was stimulated with the addition of N/P using cold region soils (Walworth *et al.*, 2003) and potassium amendments stimulated creosote degradation (Phillips *et al.*, 2000). There was an observed improved degradation of 4 ringed PAHs with N/P addition, but no increased degradation of 3 ringed PAHs (Breedveld and Sparrevik, 2000). However, Graham *et al.* (1999) assessed an array of N/P amendments for hexadecane biodegradation and suggested that amendments above stoichiometric requirements can lead to diminished rates of degradation. This potentially occurs because addition of excessive nitrogen results in an increase in soil salinity and this increases the osmotic stress and suppresses the activity of hydrocarbon-degrading organisms (Walworth *et al.*, 2003). Carmichael and Pfander (1997) observed slower degradation of 3 and 4 ringed PAHs with N addition and no effects for P addition. Johnson and Scow (1999) reported similar results indicating N/P addition inhibited or did not change phenanthrene degradation (3 ringed PAH). Their results showed that soil with initial low concentrations of nitrogen or phosphorus is more likely to show decreased degradation with N/P addition. Many PAH-degrading organisms are adapted to low nutrient conditions and activity may decrease with the addition of soil amendments. Thus, addition of nutrients should be considered on a site-by-site basis and a balance should be considered between biodegradation and plant growth. Application of amendments exclusively for plant growth may result in diminished contaminant degradation, the ultimate goal of phytoremediation.

24.3 Oxygen requirements: Soil oxygen is required for optimal aerobic microbial degradation of petroleum hydrocarbon contaminants. Similar to nutrient deficiencies, oxygen depletion is caused by natural microbial respiration of contaminants. Within phytoremediation, plants may provide a net positive or negative oxygen source (Lee, 2000).

Plants may improve soil oxygen through two mechanisms. First, specially adapted plants use parenchyma, channels of reduced air resistance, to transport oxygen to the root zone, enhancing aerobic biological degradation (Erickson *et al.*, 1993). Second, soil dewatering and fracturing increases soil porosity, allowing increased diffusion of atmospheric oxygen (EPA, 2001). Plant roots can also be a net oxygen sink within petroleum-contaminated soils. Rentz *et* al., (2003) observed stimulation of hybrid poplar growth and increased poplar root density with the addition of Oxygen Release Compound (ORC) when plants were grown in petroleum smear zone soils (high biochemical oxygen demand). Flux of oxygen into soil by plants could be offset by root turnover and root exudation that provides microbial populations with simple carbon sources that could deplete soil oxygen when metabolized (Lynch, 1990). Furthermore, plant roots are known to require oxygen (Neuman *et al.,* 1996). For soils with a high biochemical oxygen demand, oxygen addition may be required to promote plant growth and stimulate microbial degradation. Passive methods of oxygen delivery are suggested to keep costs of phytoremediation low. These methods include the use of include the use of perforated aeration tubes placed next to cuttings that can supply oxygen to roots along a vertical axis (Ferro et al., 2001). Perforated ADS (Adsorption) tubing that are placed at depth prior backfilling the planting trench provides oxygen on a horizontal plane. Gravel used to backfill planting trenches allows permeation of oxygen on vertical and horizontal axis. Finally, the use of solid peroxides (e.g. Oxygen Release Compounds) can provide oxygen to soils when in contact with water (Koenigsberg and Norris, 1999).

25. Biodegradation end points

It is important to monitor the rate at which biodegradation takes place so as to adjust the important parameters that will enhance the biodegrading process. The endpoint is however used to measure the completion of the biodegradation.

Endpoint criteria are typically concentrations of specific components or chemicals that are measured in the impacted soil. Accordingly, endpoints can be the bulk hydrocarbon content or the contraction of specific petroleum hydrocarbons (DOE/PERF, 2002). Desired end points should be considered when selecting a treatment technology, since certain technologies may be capable of achieving the end points in less time and with less money spent. The end-points that can be achieved by bioremediation are related to how much oil you start with, the composition of the oil, and the age of the spill (Nakles and Loehr, 2002).

If regulatory treatment goals have been established prior to initiating bioremediation treatment, then bioremediation treatment will be complete when the goal has been achieved. There may be regulatory constraints concerning whether bioremediation is permissible, and which biotreatment methods are approved. Therefore, experts should always consult with all the local and state regulatory agencies prior to the application of biotreatment (DOE/PERF, 2002). Life cycle assessment can also be used to investigate the viability of the biodegradation process by looking at the crude oil remaining at each stage of the bioremediation processes as well as gasses evolved and toxic level at each phase of the bioremediation project.

26. Conclusion

Crude oil spill in our environments have been a consistent challenge and as long as crude oil exploitation takes place spillage is bound to occur. This chapter is review of bioremediation techniques for the reclamation of oil contaminated lands and water bodies. Some of these bioremediation techniques include: bio-monitoring, microbial bioremediation and phytoremediation methods. Important factors that limit bioremediation were also described.

It is the wish of the authors that this review would be useful to those who are looking at practicable ways of bioremediating crude oil polluted environments.

27. Acknowledgement

To all who gave encouragement to go ahead with this work! All the authors whose works are hereby cited are duly acknowledged for their contributions to the body of knowledge to solve the crude oil pollution challenges in our respective environments. The greatest acknowledgement of all is to God the source of wisdom, strength, and knowledge.

28. References

Adams, C. and Ellis, A.(1961). Oil spillage. A problem in the oil producing Areas. *Journal of environmental quality*. 3:936-984.

Aitchison, E. W. Kelley, S. L. Alvarez, P. J. J and Schnoor, J. L. (2000) *Wat. Environ. Res.*, 72:313.

Alloway, B.J and Ayres, D.C. (1993). Organic pollutants. In: chemical principles of Environmental Pollution. 1st edition Chapman and Hall, India, Publishers, pp. 201.

Amadi, A. A. Dickson, and Maate, G.O. (1993) *Wat. Air Soil Pollut.*, 66:59.

Anon.(1991) Report on National Biotechnology Policy. Whitehouse Council on competitiveness Washington DC,.

Anozie, O. and Onwurah, I.N.E (2001). Toxic effects of Bonny light crude oil in rats after ingestion of contaminated diet. *Nigerian J. Biotechnology and Molecular Biology* (Proceedings supplement). 16 (3):1035-1085.

Aprill, W. and Sims, R. C. (1990) *Chemosphere*, 20 :253.

Atlas, R.M and Bartha, R (1998). Fundamentals and Applications. In: microbial Ecology. 4th edition. Benjamin/Cumming Publishing Company, Inc. California, USA, pp 523 - 530.

Atlas, R.M and Bartha, R. (1973). Simulated biodegradation of oil slicks using oleophilic fertilizers. *Environ. Sci. Tech.* 7:538-541.

Atlas, R.M. (1981) Microbial degradation of petroleum hydrocarbons: an environmental perspective. *Microb. Rev.* 45:180-209.

Atlas, R.M.,(1984). Petroleum Microbiology. Macmillan Publishing Company, New York.

Bagni, G., Hernendez, S., Mascini, M., Sturchio, E., Boccia, P., and Marconi, S., (2005). DNA biosensor for rapid detection of genotoxic compounds in soil samples. *Sensors.*, 5:394-410.

Baker, J. M. (1970). The Effects of oil on plants. *Environmental pollution* . 1:27-44.

Baker, J. M., (1983). Impact of Oil Pollution on Living Resources: Commission on ecology Papers No 4 International Union for Conservation of Nature and natural Resources , Gland, Switzerland.

Banks, M. K. Lee, E. Schwab, A. P. (1999) J. Environ. Qual., 28 :294.

Bartha, R (1986). Microbial ecology: fundamentals and applications. Addisson_Wesley Publ., Reading, Mass.

Beckles, M. D., Ward, C. H., Hughes, J. E., (1998). Effect of mixtures of polycyclic aromatic hydrocarbons and sediments of fluoromethane biodegradation pattern Environ. Toxicol. Chem., 17, 1246-1257.

Beller, M., Schoenmaker, H., Huuskonen, E., (1996).Pipeline inspection environmental protection throughon-line inspection, Proceeding of the NNPC seminar: Oil industry and the Nigerian Environment, PortHarcourt, Nigeria. 233-241.

Binet, P. Portal, J. M. and Leyval, C. (2000) Soil Biol. Biochem., 32 :2011.

Bioremediation overview (2003). http://www.integraenvironmental.com/bior.overviewew.htm.

Bossert, I., and Bartha, R. (1984). The Fate of Petrolelum in soil Ecosystems. In Petroleum Microbiology, R.M.Atlas (ed.), Macmillan, New York, pp 453-473.

Breeveld, D. and Sparrevik, M. (2000) Biodegradation, 11 :391.

Briggs, G. G. Bromilow, R. H. and Evans, A. A. (1982) *Pesticide Sci.*, 13: 495.

Bucheli, T., and Fent, K., (1995). Induction of Cytochrome P450 as a biomarker for environmental contamination in aquatic ecosystems. *Critical Reviews in Environmental Science and Technology*. 25:201-268.

Burken, J. G. and Schnoor, J. L. (1998) *Environ. Sci. Technol.*, 32 :3379.

Carmichael, L. M. and Pfaender, F. K. (1997) Biodegradation, 8 :1.

Chaillan, F., Chaineau, C.H., Point, V., Saliot, A. and Oudot, J. (2006). Factors inhibiting bioremediation of soil contaminated with weathered oils and drill cuttings. *Environ. Pollut.* In Press.

Chaineau, C.H., Rougeux, G., Yepremian, C. and Oudot, J. (2005). Effect of Crude Oil Concentration on the biodegradation of crude oil and associated microbial populations in the soil. *Soil. Biol. Biochem.* 37:1490- 1497.

Choi, S. C., Kwon, K.K., Sohn, J.H., Kim, S.J.,(2002). Evaluation of fertilizer additions to stimulate oil biodegradation in sand seashore mesocosms. *J. Microbiol. Biotechnol.* 12:431-436.

Cooney, J.J.(1984). The fate of petroleum pollutants in fresh water ecosystems. K 03099 *Pollution*; J 02905 Water; P 2000 *Freshwater Pollut.* 1:95-98.

Cronin, M. T. D., Schultz, T. W., (1998). Structure toxicity relationships of three mechanism of actions of toxicity to *Vibrio fischeri*. *Ecotoxicol. Environ. Saf.*, 39:65-69.

Cunningham, S. D., Berti, W.R. and Huang, J. W. (1995). Phytoremediation of contaminated soil. *TIBECH*. 13 : 393-397.

Davis, L. C. Lupher, D. Hu, J. and Erickson, L. E. (1999) Proceedings of the 1999 Conference on Hazardous Waste Research, Kansas State University, Manhattan, (eds. L. E. Erickson and M. M.Rankin), Kansas, pp. 219- 223. http://www.engg.ksu.edu/HSRC.

Department of Toxic Substances Control (DTSC), (1996). State of California EPA Petro Rise Soil Test System. A rapid immuno- assay screen for total petroleum hydrocarbons in soil. Ensys. Inc. Research Triangle Park, N.C 27709.

DOE/PERF (2002). A Summary of the DOE/PERF Bioremediation Workshop.

Environmental Protection Agency (EPA) (2001) Treatment Technologies for Site Cleanup: Annual Status Report (Tenth Edition).

Erickson, L. E. Banks, M. K. Davis, L. C. Schwab, A. P. Muralidharan, N. Reilley, K and Tracy, J.C. (1993) *Environ. Progress*, 13: 226.

European Federation of Biotechnology (EFB) (1999). Environmental Biotechnology. European Federation of Biotechnology. Task group on public perceptions of Biotechnology Briefing paper 4, Second Edition January 1999. Availableon: http://www.kluyver.stm.tudelft.nl/efb/home.htm

Ezeala, D. O. (1987). The Sensitivity of Pistia stratoites (A fresh water plant) to crude oil pollution. The petroleum Industry and the Nigerian Environment: Proceeding of 1987 by The Nigerian National Petroleum Corporation.

Ezeonu, C. S. (2010). Impact of soluble metal salts and crude oil contaminants on water retention capacity of soil and protein content of *zea mays*. *EJEAFChe*, 9(5):885-891.

Ezeonu, C.S., and Onwurah, I.N.E.,(2009). Effect of crude oil contamination on Chlorophyll content in *Zea mays L. International Journal of Biology and Biotechnology (IJBB)*, 6(4):299-301.

Ezra, S., Feinstein, S., Pelly, I., Bauman, D. and Miloslavsky, I. (2000). Weathering of fuel oil spill on the east Mediterranean coast, Ashdod, Israel.*Organic Geochemistry*. 31:1733-1741

Ferro, J. K., Kjelgren, R., Rieder, J., and Perrin, S. (1999) *Int. J. Phytoremediation*, 1:9.

Ferro, J., Chard, R., Kjelgren, B., Chard, D., Turner, and Montague, T. (2001) *Int. J. Phytoremed.*, 3:105.

Fojta, M., (2002) Electrochemical sensors for DNA interactions and damage. *Electroanalysis*, 14:1449-1463.

Forght, J. M., Fedorak, P. M., Westlake, D. W. S., (1990). Mineralization of [^{14}C] hexadecane and [^{14}C] phenanthrene in crude oil. Specificity among bacterial isolates. *Can J Microbiol.*, 36:169-175.

Forght, J.M., Westlake, D., Johnson, W.M., and Ridgway, H.F. (1996). Environmental gasoline-utilizing isolates and clinical isolates of *Pseudomonas aeroginosa* are taxonomically indistinguishable by chemotaxonomic and molecular techniques. *Microbial*.142:1333-2340.

Gatliff, E. G. (1994). *Remediation*, 8:343-352.

Gibb, A., A. Chu, Ron Chik Kwong, R.H. Goodman. 2001 Bioremediation kinetics of Crude Oil at 5 degree C. *Journal of Environmental Engineering*. 127:818-824.

Giddens, J. (1970) *J. Environ. Qual.*, 5 :179.

Graham, D. W. Smith, V. H. Cleland, D. L. and Law, K. P. (1999) *Water., Air, and Soil Poll.*, 111:1.

Gunther, T. Dornberger, U. and Fritsche, W. (1996) *Chemosphere*, 33 :203.

Hall, P. L. and Quam, H. (1976). Counter measures to control oil spill in Western Canada, *Groundwater*. 14(3):163.

Hecht, D. and Badiane, G. (1998) *New Internationalist*, June 12.

Heitkamp, M. A. and. Cerniglia, C. E (1988) *Appl. Environ. Microbiol.*, 54:1612.

Holba, A.G., Dzou, I.L., Hickey, J.J., Franks, S.J. and Lenney, T. (1996) Reservoir Geochemistry of South Pass 61 field, Gulf of Mexico: Compositional heterogeneities reflecting filling History and biodegradation: *Org. Geochem*. 24: 1179-1198.

Hong, M. S. Farmayan, W. F. Dortch, I. J. Chiang, C. Y. McMillan, S. K. and Schnoor, J. L. (2001) *Environ. Sci. Technol.*, 35 :1231

Hutchinson, S. L. Banks, N. K. and Schwab, A. P. (2001) *J. Environ. Qual.*, 30:395.

Imevbore, A.M.A., Adeyemi, S.A., and Afolabi, O.A.(1987) The toxicity of Nigerian crude oils to aquatic organisms. The petroleum Industry and the Nigerian Environment: Proceeding of 1987 by The Nigerian National Petroleum Corporation, pp 171 - 176.

International Agency for Research on Cancer (IARC) (2000) Monographs on the valuation of the Carcinogenic Risk of Some Industrial Chemicals to Humans, Lyon, France.

Johnson, C.R. and Scow, K.M. (1999) *Biodegradation,* 10:43.

Joner, E. J. Corgie, S. C. Amellal, N. and Leyval, C. (2002) *Soil Biol. Biochem.*, 34 :859.

Kaladumo, C.O.K.(1996). The implications of gas flaring in the Niger Delta Environment. Proceedings of the 8th Biennial International NNPC Seminar. In: The petroleum Industry and the Nigerian Environment, Portharcourt, Nigeria, pp 277-290.

King, J. M. H., DiGrazia, P. M., Applegate, B., Bulage, R., Sanseverio, J., Dunbar, P., Larimwer, F., Sayler, G. S (1990). Rapid sensitive bioluminescent reporter technology for naphthalene exposure and biodegradation. *Science,* 249: 778.

Koenigsberg, S. S. and Norris, R. D. Accelerated Bioremediation Using Slow Release Compounds, Regenesis Bioremediation Products, San Clemente, CA, 1999.

Lee, R.W. Jones, S.A. Kuniansky, E.L. Harvey, G. Lollar, B.S. and Slater, G.F. (2000) *Int. J. Phytoremed.*, 2:193.

Liste, H. H and Alexander, M. (2000). *Chemosphere,* 40:11.

Lojou, E., and Bianco, P., (2006). Application of the electrochemical concepts and techniques to amperometric biosensor devices. *J. Electroceram.,* 16:79-91.

Lombi, E., Zhao, F. J., Dunham, S. J., McGrath, S. P. (2001). Phytoremediation of heavymetal-contaminated soils: natural hyper-accumulation versus chemically enhanced phyto-extraction. *J. Environ. Qual.* 30(6): 1919- 1926.

Lucarelli, F., Authier, L., and Bagni, G. (2003). DNA biosensor investigations in fish bile for use as a biomonitoring tool. *Analytical Letters.,* 36:1887-1901.

Lynch, J. M. (1990). *The Rhizosphere,* New York, Wiley.

Ma, X. and Burken, J. G. (2003) *Environ. Sci. Technol.,* 37:2534

Mackay, D.M. and Roberts, P.V.(1985). Transport of organic contaminants in groundwater. *Environ. Sci. Technol.* 19(5):384.

Maki, H., Sasaki, T. and Haramaya, S. (2005). Photooxidation of biodegradable crude oil and toxicity of the photooxidized products. *Chemosph.* 44:1145-1151.

Martin, A (1983). Microbial Technologies to overcome environmental problems of persistent pollutants Ithaca NY USA.

Mathews, P. J., (1980). Toxicology for water scientists. *Journal of Environmental Management,* 11(1):1-14.

Matsui, S., (1989). Molecular Biology and Industrial Application. In: *Bacillus subtilis.* Maruo and Yosikawa H. eds. Elsevier, Kodansha Ltd. Tokyo. 241-260.

McElroy, E. Farrington, J. W. and Teal, J. M. (1989.) In:U Varanasi (ed.) Metabolism of PolycyclicAromatic Hydrocarbons in the Aquatic Environment, CRC Press inc, Boca Raton, Florida.

McGill, W.B.(1976). An introduction to field personnel act of oil spills on soil and some general restoration and clean up procedures. Albert Institute of Petrology. AIP publisher, No. C-76-1.

McMillen, S. (1994). Society of Petroleum Engineers distinguished lecture, "Biotreatment at E & P Sites."

McMillen, S. (2002). Bioremediatiion Overview - Chevron Texaco. Presentation at DOE/PERF Bioremediation Workshop.

Miya R. K. and. Firestone, M. K (2000) *J. Env. Qual.*, 29 :584.

Miya, R. K. and Firestone, M. K. (2001) *J. Env. Qual.* 30 :1191.

Mortelmans, K. Harworth, S. Lawlor, T. Speck, W. Tainer, B. and, and Zeiger, E. (1986) *Environ. Mutagen*, 8:1

Nakles, D and Ray, L.(2002) Overview of Bioremediation Research of Texas and Gas Research Institute. Presentation at DOE/PERF Bioremediation Workshop.

Nakles, D. and Loehr, R. (2002) Overview of bioremediation research of University of Texas and Gas Research Institute. Presentation at DOE/PERF bioremediation workshop.

Narayanan, M. Erickson, L. E. and Davis, L. C. (1999) *Environ. Progress,* 18:231.

National Research Council (NRC) (1989). Biologic Markers in Reproductive Toxicology. National Academic Press, Washington, DC.

National Research Council (NRC) (1994) Report of the National Research Council Committee on Groundwater Cleanup Alternatives, National Academy Press, Washington, DC, 1994.

National Research Council (NRC), (1992). *Environmental Neurotoxicology.* National Academy Press, Washington DC.

Nelson-smith, A.(1979). Effect of crude oil spill on land. J: Wardley-smith Graham and Tortemen. London. *Pro.* 17-18.

Neuman, D. S. Wagner, M. Braatne, J. H. and Howe, J. (1996) Stress Physiology–abiotic, In:Biology of Populus and its implications for management and conservation, (eds. R. F. Stettler, H. D. Bradshaw, Jr., P. E. Heilman, and T. M.Hinckley), NRC Research Press, National Research Council of Canada, Ottawa, ON,pp. 423-458.

Newman, L. A. Doty, S. L. Gery, K. L Heilman, P. E. Muiznieks, I. Shang, T. Q. Siemieniec, S. T. Strand, S. E. Wang, X. P. Wilson, A. M and Gordon, M. P. J.(1998) *Soil Contamination*, 7:531.

O'Connor, C.S., Leppi, N.W., Edwards, R. and Sunderland, G.(2003). The combined use of electro-kinetic remediation and phytoremediation to decontaminate metal-polluted soils: laboratory-scale feasibility study. *Environ. Monit. Assess.* 84(1-2): 141-158.

Oberdorster, E., Cheek, A. O., (2000). Gender benders at the beach, endocrine disruption in marine and estuarine organisms. Environ. Toxicol. Chem., 20(4),23-36.

Odu, C.T.T (1972). Microbiology of soils contaminated with petroleum hydrocarbon. Extent of contamination and some soil and microbial properties after contamination. *J. Institute of Petroleum*, pp 58, 201-204.

Odu, E.A (1987) Impact of Pollution on Biological Resources within the Niger Delta Basin of Nigeria, *Environmental Consultancy Group*, University of Ife. 6:69-121.

Okoh, A. I. (2002). Assessment of the potentials of some bacterial isolate for application in the bioremediation of petroleum hydrocarbon polluted soil. Ph.D. Thesis, Obafemi Awolowo University, Ile-Ife, Nigeria.

Okoh, A.I., Ajisebutu, S., Babalola, G. O. and Trejo-Hemandez, M.R. (2002). "Biodegradation of Mexican heavy crude oil (maya) by *Pseudomonas aeruginosa*" *J. Trop. Biosci.* 2(1):12-24.

Okolo, J.C., Amadi, E.N. and Odu, C.T.I. (2005). Effects of soil treatments containing poultry manure on crude oil degradation in a sandy loam soil. *Appl. Ecol. Environ. Res.* 3(1):47-53.

Okpokwasili, G.C. and Odukoma, L.O.,(1994). Tolerance of Nitrobacter to toxicity of some Nigerian crude oil. *Bull. Environ. Contan. Toxicol.* 52:388-395.

Onwurah, I. N. E., (2000). A Perspective of Industrial and Environmental Biotechnology. Snaap Press/ Publishers Enugu, Nigeria, pp 148.

Onwurah, I.N.E (1999) Role of diazotrophic bacteria in the bioremediation of crude oil polluted soil. *J. Chem. Tech. Biotech.* 74: 957-964.

Onwurah, I.N.E. (2002). Anticoagulant potency of water soluble fractions of Bonny light oil and enzyme induction in rats. *Biomed. Res.* 13(1): 33-37.

Onwurah, I.N.E. (2003) An Integrated Environmental Biotechnology for Enhanced Bioremediation of Crude Oil Contaminated Agricultural Land. *J. Bio. Research. & Biotech.* 1 (2): 51 - 60.

Onwurah, I.N.E. and Alumanah, E. A., (2005). Integration of biodegradation half-life model and oil toxicity model into a diagnostic tool for assessing bioremediation technology, *Industrial Biotechnology* 1(4), 292 - 296.

Onwurah, I.N.E., Ogugua, V.N., Onyike, N.B., Ochonogor, A.E., and Otitoju, O.F.,(2007). Crude oil spills in the Environment, Effects and some Innovative Clean up Biotechnologies. *International Journal of Environmental Research.*, 1(4):307-320.

Oudot, J., Merlin, F.X., and Pinvidic, P. (1998) Weathering rates of oil components in a bioremediation experiment in estuarine sediments. *Mar. Environ. Res.* 45:113-125.

Palmroth, M. R. T. Pichtel, J. and Puhakka, J. A. (2002) *Bioresource Technol.*, 84 :221.

Pankow, J. F. Thompson, N. R. Johnson, R. L. Baehr, A. L. and Zogorski, J. S. (1997) *Environ.Sci. Technol.*, 31:2821.

Peciulyte, D., Repeckiene, J., Levvinskaite, L. Lugauskas, A., Motuzas, A. and Prosycevas, I.(2006).Growth and metal accumulation ability of plants in soil polluted with Cu, Zn and Pb., *EKOLOGIJA* pp. 48-52.

Peters, K. E., Moldown, J. M. (1993). The Biomarker guide, Interpreting molecular fossils in petroleum and ancient sediments. Prentice- Hall, Englewood Cliffs, NJ. USA.

Plice, M.J (1948) 'Some effects of Crude petroleum on soil fertility'. *Soil Sci. Amer. Proc.* 13:413-416.

Prescott, M.L., Harley, J.P. and Klan, A.D.(1996). Industrial Microbiology and Biotechnology.In: Microbiology. 3rd Ed. Wim C Brown Publishers, Chicago, pp 923-927.

Prince, R. and McMillen, S. (2002). Summary of PERF bioremediation project. Presentation at DOE/PERF bioremediation workshop.

Prince, R.(2002). Bioremediation effectiveness: removing hydrocarbons while minimizing environmental impact. ExxonMobil Research and Engineering. Hand-out at DOE/PERF - Bioremediation Workshop.

Rahman, K.S.M., Thahira-Rahman, J., Lakshmanaperumalsamy, P. and Banat, I.M. (2002). Towards efficient crude oil degradation by a mixed bacterium consortium. *Biores. Tech.* 85: 257-261.

Raymond, R. L., Jamisen, V. W., and Hudson Jr. J. O., (1975). Final Report on Beneficial simulation of Bacterial activity in groundwater containing petroleum products. *American Petroleum Institute*. Washington, DC.

Reilley, K. A. . Banks, M. K and Schwab, A. P. (1996) *J. Environ. Qual.*, 25 :212.

Rentz, J. A. Chapman, B. Alvarez, P. J. J. and Schnoor, J. L. (2003) Int. J.Phytoremed., 5:57.

Ripp, S., Diclaudio, M.L., and Sayler, G.S.,(2010). Biosensors as Environmental Monitors: In Environmental Microbiology. Second Edition (Mitchell, R., and Gu, J. ed) Wiley-Blackwell. New Jersey, pp 213-233.

Rittmann, B. E., Smets, B. F., Stahl, D. A., (1990). Genetic capabilities of biological processes Part 1. *Environ. Sci. Technol.*, 24, 23-30.

Rosenberg, E., Legman, R., Kushmaro, A., Adler, E., Abir, H. and Ron, E.Z. (1996) Oil bioremediation using insoluble nitrogen source. *Journal of Biotechnology.* pp 273-278.

Ross, P., (1993). The use of bacterial luminescence systems in aquatic toxicity testing. In: Richardson M. ed, *Ecotoxicology Monitoring* VCH, New York, USA. 155-194.

Rowell, M.J. (1977). The effect of crude oil spills on soils. In: The reclamation of agricultural soils after oil spills, J.A.Toogood (editor). Department of Soil Science, University of Alberta, Edmonton. pp 1-33.

Rowland, A.P., Lindley, D.K., Hall, G.H., Rossal, M.J., Wilson, D.R., Benhan, D.G.,Harrison, A.F. and Daniels, R.E., (2000). Effects of beach and sand properties,temperature and rainfall on the degradation rates of oil buried in oil/beach sand mixtures. *Environmental Pollution.* 109:109-118.

Sabate, J., Vinas, M. and Solanas, A. M (2004). Laboratory-Scale bioremediation experiments on hydrocarbon-contaminated soils. *Internl. Biodeter. Biodegrad,* pp 54:19-25.

Sassolas, A., Leca-Bouvier, B.D., and Blum, L.J., (2008) DNA biosensors and microarrays. *Chemical Reviews.,* 108:109-139.

Schooner, J. L. Licht, L. A. McCutcheon, S. C. Wolfe, N. L. and Carriera, L. H. (1995) *Environ.Sci. Technol.,* 29:318A.

Shimp, J.F. Tracy, J.C. Davis, L.C. Huang, W. Erickson, L.E. and Schnoor, J.L. (1993) *Crit. Rev. Environ. Sci. Technol.,* 23 :41.

Short, J.W. and Heintz, R. A. (1997). Identification of Exxon Valdez oil in sediments and tissue from Prince William Sound and the North Western Gulf of William based in a PAH weathering model. *Environ. Sci. Technol.,* 31: 2375-2384.

Stegeman, J., and Hahn, M., (1994). Biochemistry and Molecular biology of monooxygenases: Current perspectives on forms, functions, and regulation of cytochrome P450 in aquatic species. In: D.Malins and G.Ostrander (Eds), *Aquatic Toxicology: Molecular, Biochemical, and Cellular Perspectives.* CRC Press, Boca Raton, FL.

Stewart, A. (2002). Toxicity after Bioremediation - Oak Ridge National Laboratory. Presentation at DOE/PERF Bioremediation Workshop.

Sublette, K.L. (2001). Fundamentals of Bioremediation of Hydrocarbon Contaminated Soils. The University of Tulsa, Continuing Engineering and Science Education. Houston, TX.

Swannell, R.P.J., Lee, K. and McDonagh, M. (1996).Field evaluation of marine oil spill bioremediation. *Microb. Rev.*, pp 342-365.

Thouand, G., Bauda, P., Oudot, J., Kirsch, G., Sutton, C. and Vidallie, J. F. (1999). Laboratory evaluation of crude oil biodegradation with commercial or natural microbial inocula. *Can. J.Microb.* 45(2):106-115.

Trapp, S. McFarlane J. C (1994).Plant Contamination: Modeling and Simulation of Organic Chemical Processes, Lewis Publishers, Boca Raton, Florida.

Trindade, P.V.O., Sobrah, L.G., Rizzo, A.C.L., Leite, S.G.F. and Soriano, A.U. (2005). Bioremediation of a weathered and a recently oil-contaminated soils from Brazil: a comparison study. *Chemosph.* 58:515-522.

Tschmelak, J., Proll, G., and Riedt, J., (2005). Automated water analyser computer supported system (AWACSS): I. Project objectives, basic technology, immunoassay development, software design and networking. *Biosensors and Bioelectron.,* 20:1499-1508.

U.S. Army Corps of Engineers, (USACE) (2003).Agriculturally Based. Bioremediation of Petroleum- Contaminated Soils and Shallow Groundwater in Pacific Island Ecosystems.

Vallero, A. D.,(2010). Environmental Biotechnology: A Biosystems Approach, 1st Edition. Elsevier Academic Press, Burlington, MA.

Venosa, A.D. and Zhu, X. (2003). Biodegradation of crude oil contaminating marine shorelines and Freshwater Wetlands. *Spill Sci. Tech. Bull.* 8(2):163-197.

Wackett, T.M (1996). Co-metabolism: is the emperor wearing any clothes? *Curr. Opin. Biotechn.* 7:311-316.

Walworth, J. L. Woolard, C. R. and Harris, K. C. (2003) Cold Regions Sci. Technol.,37 :81.

Wang, D.T., and Meresz, O. (1981). Occurrence and potential uptake of polynuclear aromatic hydrocarbons of highway traffic origin by proximally grown food crops (abst.). Sixth Int. Symp on PAH, Batlelle Columbus Lab, Columbus, Ohio. *Zea mays* Sourced from: http://www.tropicalforages.info/key/forages/media/Html/Zea-mays.htm.

Wang, Q.R., Cui, Y.S., Liu, X.M., Dong, Y.T. and Christine, P. (2003). Soil contamination and plant uptake of heavy metals at polluted sites in China.*J.Environ. Sci. Health.* 38(5): 823-838.

Wei, H., Li, B.L., Li, J., Dong, S.J., and Wang, E.K.,(2008). DNAzyme-based colorimetric sensing of lead (Pb^{2+}) using unmodified gold nanoparticle probes. *Nanotechnology,* 19. Article 095501.

Wilson, K., and Walker, J.M., (1994). Principles and Techniques of Practical Biochemistry, (4th ed). Cambridge University Press. Pp 565-573.

Wiltse, C.C. Rooney, W. L. Chen, Z. Schwab, A. P. and Banks, M. K. (1998) *J.Environ. Qual.,* 27 :169.

Wu, H. Haig, T. Pratley, J. Lemerle, D. and An, M. (2000) J. Agric. Food Chem., 48:5321.

Xiao, Y., Rene, A.A., and Plaxco, K.W.,(2007). Electrochemical detection of parts-per-billion lead via an electrode-bound DNAzyme assembly. *J.Am.Chem.Soc.,* 129:262-263.

Yateem, A., A.S. El-Nawawy, and N. Al-Awadhi,. Soil and Groundwater Cleanup, 1999, pp. 31-33.

Zengler, K., Richnow, H. H., Rossello-Mora, R., Michaelis, W. and Widdel, F (1999). Methane formation from long-chain alkanes by anaerobic microorganisms: *Nature.* (401): 266-269.

Zhang, Q. Davis, L. C. and Erickson, L. E. (2001) *Environ. Sci. Technol.,* 35 :725.

7

Hydrocarbon Pollution: Effects on Living Organisms, Remediation of Contaminated Environments, and Effects of Heavy Metals Co-Contamination on Bioremediation

Shukla Abha and Cameotra Swaranjit Singh
Institute of Microbial Technology, Chandigarh, India

1. Introduction

Hydrocarbon contamination in the environment is a very serious problem whether it comes from petroleum, pesticides or other toxic organic matter. Environmental pollution caused by petroleum is of great concern because petroleum hydrocarbons are toxic to all forms of life. Environmental contamination by crude oil is relatively common because of its widespread use and its associated disposal operations and accidental spills. The term petroleum is referred to an extremely complex mixture of a wide variety of low and high molecular weight hydrocarbons. This complex mixture contains saturated alkanes, branched alkanes, alkenes, napthenes (homo-cyclics and hetero-cyclics), aromatics (including aromatics containing hetero atoms like sulfur, oxygen, nitrogen, and other heavy metal complexes), naptheno-aromatics, large aromatic molecules like resins, asphaltenes, and hydrocarbon containing different functional groups like carboxylic acids, ethers, etc. Crude oil also contains heavy metals and much of the heavy metal content of crude oil is associated with pyrrolic structures known as porphyrins. Petroleum is refined into various fractions such as light oil, naphtha, kerosene, diesel, lube oil waxes, and asphaltenes, etc. The light fractions, which are distilled at atmospheric pressure, are commonly known as light ends and the heavy fractions like lube oil and asphaltenes are known as the heavy ends. The light and the heavy ends of petroleum have different hydrocarbons composition, the light ends contain low molecular weight saturated hydrocarbons, unsaturated hydrocarbons, naphthenes, and low percentage of aromatic compounds; while the heavier ends consist of high molecular weight alkanes, alkenes, organometallic compounds, and high molecular weight aromatic compounds. This portion is comparatively rich in metals and N,S,O containing compounds. Figure 1 shows some of the chemical structures of common hydrocarbons compounds. These hydrocarbon molecules are widespread in the environment due to the wide range of petroleum uses, which are presented elsewhere (http://www.dep.state.fl.us/waste/quick_topics/publications/pss/pcp/PetroleumProductDescriptions.pdf).

Heavy metals are naturally present in soils; however due to human activity, the concentration of heavy metals in soil is increasing. Some areas contain such a high concentrations of heavy metals and metalloids that they are affecting the natural ecosystem. It has been observed that hydrocarbon contaminated sites co-contaminated with heavy metals are difficult to bioremediate. The reason is that heavy metals and metalloids restrict microbe's activity rendering it unable to degrade hydrocarbon or reducing its efficiency.

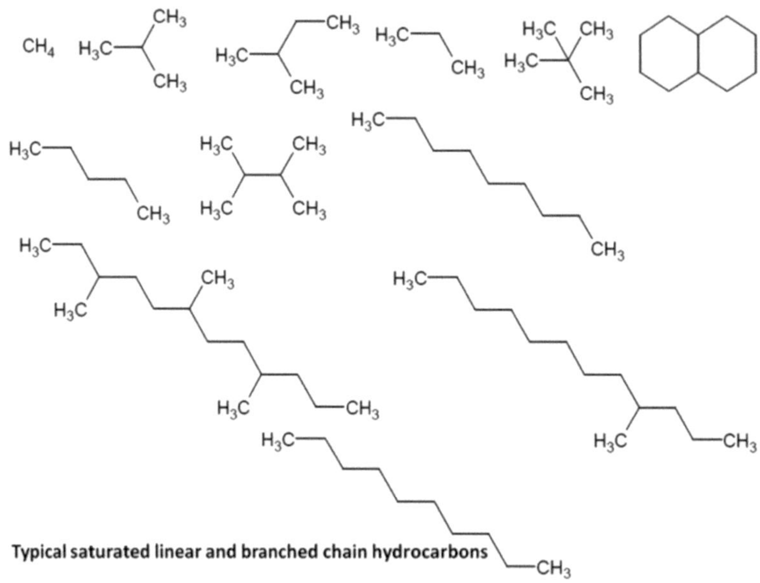

Typical saturated linear and branched chain hydrocarbons

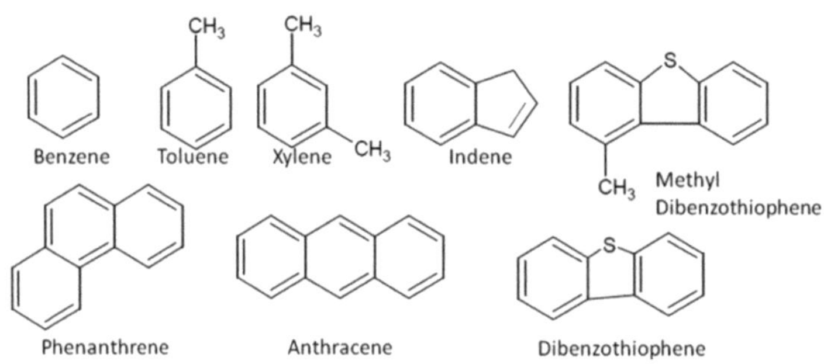

Typical smaller aromatics

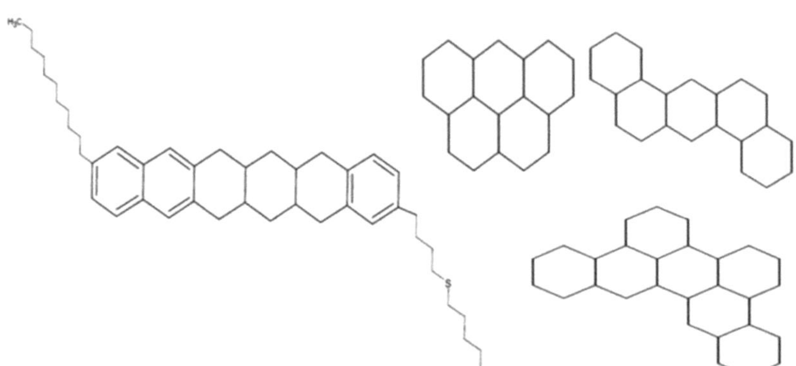

Polynuclear model structures

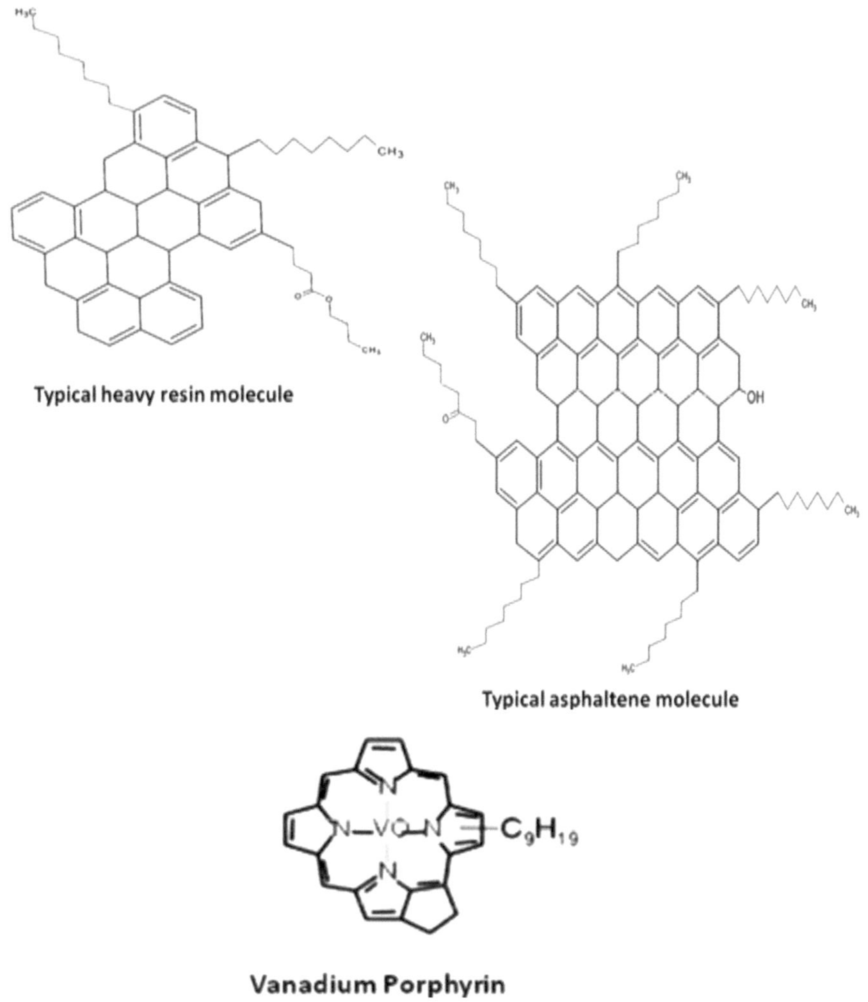

Fig. 1. Some of the representative molecules in the light and heavy fractions found in petroleum (http://nepis.epa.gov/Adobe/PDF../P1000AE6.pdf; Wenpo et al., 2011;, Dmitriev and Golovko, 2010, Qian et al., 2008).

1.1 Effects of petroleum hydrocarbons on living matter

Living matter is exposed to petroleum in many ways directly or indirectly. Some byproducts, formed during petroleum refining and processing which are used for the manufacturing of other products are highly toxic. Constantly, these toxic compounds are inadvertently released into the environment and if this effect is connected to the effect of accidental crude oil spills worldwide, then these combined sources of unrestricted hydrocarbons constitute the major cause of environmental pollution. Despite the large number of hydrocarbons found in petroleum products, only a relatively small number of the compounds are well characterized for toxicity. Petroleum hydrocarbon molecules, which have a wide distribution of molecular weights and boiling points, cause diverse levels of toxicity to the environment. The toxicity of the hydrocarbon molecules and their availability for microbial metabolism depend on their chemical and physical nature. Petroleum is toxic and can be lethal depending upon the nature of the petroleum fraction, the way of exposure to it, and the time of exposure. Chemicals and dispersants in crude oil can cause a wide range of health effects in people and wildlife, depending on the level of exposure and susceptibility. The highly toxic chemicals contained in crude oil can damage any organ system in the human body like the nervous system, respiratory system, circulatory system, immune system, reproductive system, sensory system, endocrine system, liver, kidney, etc. and consequently can cause a wide range of diseases and disorders (Costello, 1979).

Individuals more susceptible to harm by the toxic effects of crude oil are as follows.

1. Infants, children, and unborn babies.
2. Pregnant women.
3. People with pre-existing serious health problems.
4. People living in conditions that impose health stress.

The damage caused by the toxicity of crude oil to organ systems may be immediate or it may take months or years. Singh et al.(2004) studied the toxicity of fuels with different chemical composition on CD-1 mice (A swiss mice strain that is used as a general purpose stock and an oncological and pharaceutical research. This is a vigorous outbred stock. These mice are fairly docile and easy to handle). The objective of the study was to establish a correlation between the physico- chemical properties of the fuel and their biologic effects on mice. The results of the study demonstrated that the automobile derived diesel exhaust particles were more toxic than the exhaust generated by forklift engines. It was also found that the diesel exhaust particles contain ten times more extractable organic matter than the standard exhaust material generated by forklift engines. A similar type of study conducted by Kinawy (2009) revealed that the inhalation of leaded or unleaded (containing aromatics and oxygenated compounds) gasoline vapors by rats impaired the levels of monoamine neurotransmitters and other biochemical parameters in different areas of the rats' brains. Likewise, several behavioral changes causing aggression in rats were observed. Menkes and Fawcett (1997) discussed the toxicities of lead and manganese added gasoline and the public health hazards due to aromatic and oxygenated compounds in gasoline. The extent of absorption of petroleum components by inhalation, oral, and dermal routes varies significantly because of the wide range of physicochemical properties of these components. The incorporation of crude oil into the body may affect the reproductive health of humans and to other lives. Obidike et al. (2007) observed that when the male rats were given an oral crude oil treatment using a drenching tube, degeneration and necrosis of interstitial cell occurred followed by the exudation into the interstices in the testes of rats. The study concluded that exposure of rats to crude oil induces reproductive cytotoxicity confined to the differentiating spermatogonia compartment, likewise it may also harm human reproductive cells. The extent of absorption through the various routes depends on the volatility, solubility, and other properties of the specific component or mixture. The more volatile and soluble the oil fractions (low molecular weight aliphatics and light aromatic compounds) are the faster they can leak into groundwater or vaporize into the air. Therefore, living matter may be easily exposed to these crude oil fractions by breathing the contaminated air and by drinking the contaminated water (Welch et al., 1999). As reported by Knox and Gilman (1997), petroleum derived volatiles are one of the causes for the geographically associated childhood cancers.

On the other hand, the non-volatile heavy fractions of crude oil tend to be absorbed by the soil and persist at the site of release, which may harm living beings by skin contact, by intake of contaminated water or food. The heavy fraction of crude oil consists mainly of napthene-aromatics and poly-aromatic compounds that are carcinogenic and long exposure to these compounds often leads to tumors, cancer, and failure of the nervous system. The aromatic compounds found in petroleum are an important group of environmental pollutants. These aromatic compounds are introduced into the environment from various sources such as natural oil seeps, refinery waste products and emissions, oil storage wastes, accidental spills from oil tankers, petrochemical industrial effluents and emissions, and coal tar processing wastes, etc. Petroleum hydrocarbons can rapidly migrate from the site of contamination and adversely affect terrestrial and aquatic ecosystems and humans. Crude oils also contain polar organic compounds that contain N, S, and O atoms in various functional groups. The chemical properties of these NSO compounds, particularly their solubility and toxicity, are of environmental concern. It has been documented, that at the sites where oil spills have occurred a portion of the polar organic compounds present in the oil had partitioned into the groundwater rendering high concentration of total petroleum hydrocarbons in drinkable water sources (Mahatnirunkul et al. 2002, Delin et al., 1998, Oudot, 1990). Therefore, drinkable water sources turn out to be unsafe for the human population as well as for the aquatic animals, aquatic plants, and microbes (Griffin and Calder, 1976). Fortunately a diverse group of microorganisms like bacteria, fungi, and yeast may efficiently breakdown crude oil fractions into nontoxic components. There are a large number of studies available on the microbial degradation of crude oil or hydrocarbons.

2. Remediation

At present, it is widely recognized that contaminated land or water systems are a potential threat to human health. Awareness of this reality has led to international efforts to remediate many of these sites, either as a response to the health risks or to control the detrimental effects on the environment caused by contamination aiming the recovery of the contaminated sites. For decades efforts have been directed toward the evaluation of cost effective methods to cleanup oil contaminated soils. Over the years, many cleanup methods have been developed and applied. However, the remediation of oil contaminated environments is difficult because petroleum is a complex mixture of chemical compounds, and their degradation whether chemical or biological is not easy as different class of compounds needs different treatments. Furthermore, crude oil composition is reservoir dependent; therefore it is of great importance to know first the composition of the oil and the physico-chemical nature of the contaminated site before deciding the remediation strategies. Crude oil degradation in the natural environment depends on several factors such as pH, chemical composition, and physical properties of the contaminated soil and/or water, among others.

More than 50% of the crude oil produced worldwide comes from the Arabian Gulf area (onshore and offshore oil reservoirs). In the period from 1995 to 1999, 550 oil spill incidents for a total volume of 14,000 barrels of oil were reported in the Arabian Gul; while 11,000 barrels of oil were spilled in the period from 2000 to 2003. Oil spills affect mostly water systems during oil production operations, as many oil production wells are located near sea shore.

There exixts some chemical and biological methods to control oil spills, but most frequently applied are chemical methods because the chemical remediation of oil spills is faster if compared to bioremediation. However, bioremediation is getting worldwide attention. In general the remediation of oil contaminated sites can be performed by two basic processes: in-situ and ex-situ treatment using different cleaning technologies, such as biological treatment, thermal treatment, chemical extraction and soil washing, and through aerated accumulation techniques. However, at the present there is not a fully effective or universal remediation method for the cleaning up of oil contaminated sites. The subsequent paragraphs summarize some existing chemical and biological methods for the remediation of crude oil spills in soil and water systems such as bioventing, biosparging, in-situ biodegradation, and bioaugmentation.

2.1 Chemical and mechanical remediation methods

2.1.1 Oil spilled on the sea surface. There are different methods to remove oil from the sea surface and to prevent the oil from reaching the shoreline. Mechanical recovery and the application of dispersants are methods widely used. At sea, crude oil, which is usually lighter than water, spreads over the water surface and in a short time the thickness of the oil film on the water surface becomes very thin. The oil spreading velocity on the water surface depends on the type of oil, water temperature, atmospheric temperature, wind, and tide. The evaporation of the oil light components take place immediately and up to around 40 % of crude oil might evaporate during a short period of time. If accidental oil spills take place, booms (floating barriers placed around the oil or around the oil spill source) and skimmers (boats, vacuum machines, and oil-absorbent plastic ropes) are used to prevent the spreading of oil and to restrict the spill to a concise area, after which the oil is collected into a container by the skimmers. Chemical dispersants are frequently used to render the oil spill harmless for the aquatic life and other living organisms by, reducing the oil slick to droplets that can be degraded by naturally occurring bacteria. Dispersants can be applied to a large area but sometimes its hazards to the environment are of great concern.

Biosurfactants can also be applied to solubilise the oil, with the advantage that biosurfactants are generally not harmful to the environment. Sometimes in-situ burning of oil is also used as an optional method for oil spill remediation; however this method is applied only when the oil spill is fresh and the spilled oil layer is floating on the water surface or after the oil has been concentrated into a small area by the booms. This method has the downside that oil burning generates smoke and other by products, which pollutes the aquatic and atmospheric systems.

Every method has its advantages and disadvantages, and the effectiveness of any cleaning method depends on ocean currents and tides, as well as the weather. If the oil reaches the shoreline, then other methods are applied to clean up the sand and gravels. Sometimes sponge like oil sorbents are used to absorb oil. The oiled birds are washed manually with a diluted liquid soap solution, however sometimes in spite of an efficient manual cleaning the affected birds cannot be saved. Oiled vegetation can be flushed with water to remove the oil, but severely damaged plants will need to be destroyed and removed entirely.

The ex-situ remediation involves the pumping of contaminated water to a processing site; this process is applied only when the volume of contaminated water is low. The shore sand

and gravels are removed from the contaminated site and cleaned elsewhere. The remediated sand and gravel are then replaced on the shoreline.

2.1.2 Oil Spilled on Soil. Soil contamination occurs during offshore drilling and oil production operations due to leakages from well heads and pipelines, overflow from gathering stations, petroleum products, improper disposal of petroleum wastes and leakage from underground petroleum storage tanks. Conventional soil remediation techniques consist of the excavation and removal of the contaminated soil to a landfill, or to cap and contain the contaminated areas at the site. This method may create significant risks during the excavation, handling, and transport of contaminated material. Additionally, it is very difficult and increasingly expensive to find new landfill sites for the final disposal of the material. The cap and containment method is only an interim solution since the contamination remains on the site, requiring continuous monitoring and maintenance of the separation barriers long into the future, with all the associated costs and potential liability. Contaminated soil can be treated by two ways: in-situ or ex-situ. In-situ method of contaminated soil cleaning means that the treatment is applied to the soil at the site through physico-chemical processes such as ignition, air sparging, and soil air extraction or by combinations of these two processes, depending on the nature of the contaminants.

The application of in-situ treatments make use of technology and equipment from oil and gas drilling operations including vertical and horizontal drilling. In-situ techniques are more effective on sandy soils than in clay soils. Air sparging, also referred as soil venting (volatilisation) can be applied to take out the contaminants from soil and from groundwater-saturated soil by mobilization of the volatile compounds. This method also accelerates the growth of aerobic bacteria in the contaminated area by oxygen feeding. However, if oil contamination occurs in the groundwater, air sparging can be also conducted below the water table to extract the volatile compounds to unsaturated zones or to the surface by gravity segregation or through extraction wells. Another process known as slurping uses vacuum to extract oil from the boundary of the groundwater oil saturated or partially saturated soil. Steam injection into the contaminated oil is very effective in extracting volatile components trapped in the soil.

Ex-situ remediation of oil contaminated soil is usually applied to a small volume of soil. Ex-situ soil remediation involves the removal and transportation of the soil from the contaminated area to an off site remediation facility. The ex-situ remediation can perform through different processes as follows. Land farming in which the oil contaminated soil is excavated and spread over a bed where it is periodically tilled until the degradation of the contaminants takes place. This technique is usually limited to the treatment of 15 to 35 cm of surface soil. Composting which involves mixing of contaminated soil with nonhazardous organic matter to enhance the development of microbial species. Biopiles is a hybrid process that involves land farming and composting, in which engineered cells are constructed to stack the contaminated soil that can be aerated. The bio-processing of contaminated soil, sediment, and water is sometimes carried out in slurry or aqueous reactors known as bioreactors where the three phases solid (soil), liquid, and gas are continually mixed which enhances the biodegradation rate. The contaminated soil requires pre-treatment before loading into the bioreactors. The contaminants undergo chemical oxidation reactions converting the toxic compounds into less toxic or harmless compounds. The oxidation reactions can be catalyzed through dechlorination or UV. Generally, these methods present some drawbacks such as elevated costs due to the complexity of the technology required; while bioremediation is an option that offers the possibility to degrade hydrocarbon contaminants using natural biological activity.

2.2 Bioremediation treatments

Bioremediation of oil contaminated sites is an efficient alternative to the conventional mechanical and chemical processes that is gaining worldwide attention. This technique utilizes the natural biological activity of microorganisms or enzymes to transform the toxic petroleum components into less toxic or harmless metabolites. Bioremediation requires the addition of nitrate or sulphate fertilizers to aid the decomposition of hydrocarbon compounds. Advantages of bioremediation are: it can be applied directly at the site, low cost, the degradation of pollutants do not produce side effects, and that it is a technology worldwide acceptable. Downsides of of bioremediation are: (1) its application is restricted to petroleum components that are microbially degradable and (2) the bioremediation process take long times for completion. Hydrocarbon compounds having high aromaticity, chlorinated compounds, or having other resistant toxic functional groups are resistant to microbial attack and are almost impossible to degrade at aerobic conditions. However, these compounds can be degraded under anaerobic conditions. In this technique, the contaminated site is supplied with an oil degrading microorganism or a consortium of microorganisms that may be indigenous or isolated from elsewhere and inoculated into the contaminated site, which is known as bioaugmentation. For bioremediation to be effective, microorganisms must enzymatically attack the pollutants during their metabolic processes. Microorganisms synthesize surfactants known as biosurfactants, which have both hydrophobic and hydrophilic domains and are capable of lowering the surface tension and the interfacial tension of the growth medium. Biosurfactants are synthesized during the bacterial growth period on water-immiscible substrates, providing an alternative to chemically prepared conventional surfactants. In the case of the bacterial degradation of hydrocarbons, microbial cells synthesize biosurfactants that solublizes oil droplets into the aqueous phase making easier the oil uptake by microbial cells as shown in Fig. 2 (Cameotra et al., 1983, Ganesh and Lin, 2009).

Biosurfactants show different chemical structures: lipopeptides, glycolipids, neutral lipids,and fatty acids (Cameotra et al. 2010, Muthusamy et al.,2008). Biosurfactants are nontoxic, biodegradable biomolecules that have gained importance in the fields of enhanced oil recovery, environmental bioremediation, food processing, and pharmaceutical due to their unique properties. The economics of biosurfactants production is a bottle neck because of the low production efficiency and the high costs associated to the recovery and purification of biosurfactants (Cameotra and Makkar, 1998). To overcome these problems, the use of unconventional substrates together with biosurfactant producing bacterial strains or genetically modified strains have been suggested, which could make possible for biosurfactants to compete with synthetic surfactants (Soumen et al., 1997, Makkar and Cameotra, 2002). Similarly, biosurfactants production costs could be reduced by using different production routes (development of economical engineering processes) aiming the increased biosurfactant production yieldand the use of cost free or cost credit feed stock for microorganism growth and biosurfactant production(Gautam and Tyagi, 2006).

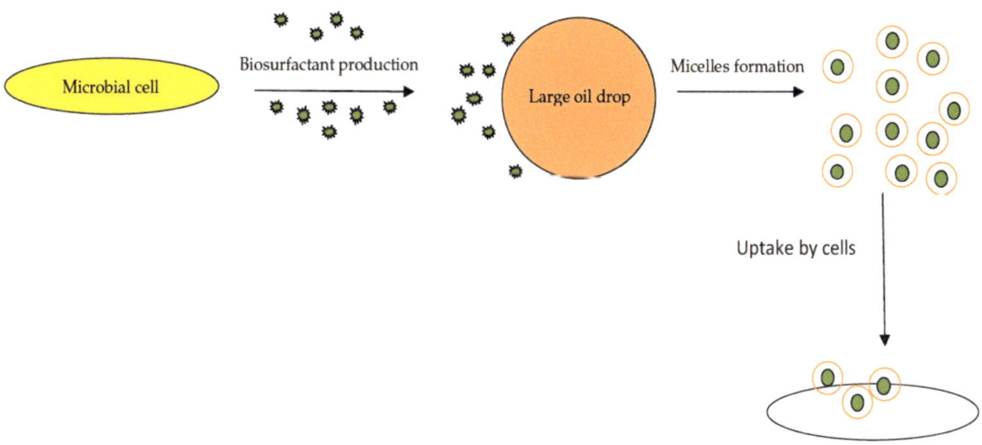

Fig. 2. Mechanism of oil solubilization by biosurfactants and uptake of solubilized oil by microbial cells.

The efficient biodegradation of petroleum hydrocarbon often involves the manipulation of the environmental parameters to allow microbial growth and degradation to proceed at a faster rate, which include the availability of sufficient amount of oxygen or other electron acceptors, essential nutrients, the penetration depth of the hydrocarbon pollutants into the soil, and the nature of the soil.

Usually, the application of bioremediation of hydrocarbon contaminated sites involves the injection of sufficient amounts of oxygen and nutrients to stimulate the oil degrading bacteria.

Figure 3 shows general pathways for the metabolism of different hydrocarbons. It is well known that alkanes (short or long chains) are oxidized mono-terminally to the corresponding alcohols, aldehydes, and mono basic fatty acids followed by beta oxidation (Nieder and Shapiro, 1975) however, further observations suggest di-terminal and sub-terminal n- alkane or fatty acid oxidation that is frequently found in branched alkanes (Forney and Markovetz,1970, Cong et al.,2009). The enzymology of alkane oxidation is not clear yet, however common accepted mechanisms are dehydrogenation, hydroxylation, and hydroperoxidation. n-alkanes are the most easily degradable components of petroleum, followed by cyclo alkanes; while the aromatic fraction is the most resistant to microbial attack. A review of microbial petroleum hydrocarbon degradation is presented elsewhere (Atlas 1981).

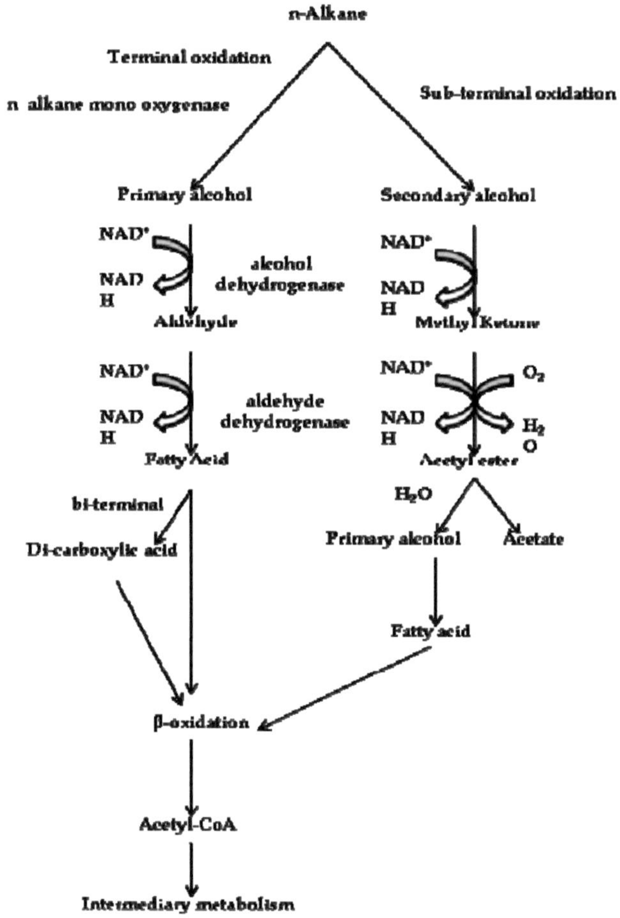

Fig. 3. General pathways for the metabolism of n-alkanes, cyclo alkanes, aromatics and other compounds (Fritsche & Hofrichter, http://www.wiley-vch.de/books/biotech/pdf/v11b_aero.pdf

3. Effect of heavy metals on bioremediation

A number of microbial strains have been isolated which are very efficient in degrading petroleum hydrocarbons (Sandrin and Maier, 2003,). The field application of bioremediation needs to consider several variables such as soil pH, temperature, salt concentration (salinity), pressure, presence of other co-contaminants as PCBs, pesticides, dyes, and heavy metal ions (Atlas, 1981). Likewise, the presence of recalcitrant hydrocarbons as poly aromatic/substituted aromatic can affect the degradation of other degradable hydrocarbons. These aromatic compounds can be toxic to individual microorganisms, thus inhibiting the microbial degradation capability. For instance, alkyl- and nitro-PAHs are highly toxics among common substituted PAHs. Substitution on aromatic rings produces several problems in their degradation. The ability of PAH dioxygenase to remove the substitutions is currently the subject of debate. The presence of alkyl branches may inhibit the proper orientation and accessibility of the PAHs into dioxygenases. Transition metals, some metalloids, lanthanides and actinides can be referred as heavy metals. Heavy metal ions are naturally found in soil but human activities are increasing unexpectedly the concentration of heavy metals in the soil, which is of great concern. Heavy metals affect the ecosystem adversely for example mercury at low concentrations represents a major hazard to microorganisms and other biota. Inorganic mercury has been reported to produce harmful effects at concentrations of 5 μg/l in a culture medium (Boening, 1999, Orct et al., 2006). Similarly other metals also have an adverse effect on microorganisms and ecosystems (Babich and Stotzky, 1983; Sobolev and Begonia,2008). Heavy metals not only contaminate the soil but also restrict the microbial activity at the polluted site consequently affecting the degradation of other organic pollutants (Mittal and Ratra, 2000).

Divalent heavy-metal cations are structurally very similar as is the case of the following divalent cations: manganese, ferrous, cobalt, nickel, copper and zinc. Oxyanions like chromate, arsenate, phosphate with four tetrahedrally arranged oxygen atoms and two negative charges, differ mostly in the size of the central ion, so the structure of chromate resembles that of sulfate. Thus, cells uptake systems for heavy-metal ions have to bind those ions tightly. Most cells solve this problem by using two types of uptake systems for heavy-metal ions: one is the fast uptake and the second uses high substrates specificity. These fast systems are usually driven only by the chemiosmotic gradient across the cytoplasmic membrane of bacteria. The second type of uptake system has high substrate specificity; it is slow and often uses ATP hydrolysis as the energy source. Microbes are very susceptible to metal toxic environments when exposed to high concentrations of heavy metals (Rajapaksha et al., 2004; Utgikar et al.,2001). Different microbes interact with metals in a different enzymatically or non-enzymatically way depending upon the microbial ability, gene regulation, protein expression, the type of metal, and the metal's toxicity (Ehrlich, 1997).

The expression of the gene for the fast and for the unspecific transporter may be diminished by mutation and the resulting mutants might be metal-tolerant. The protein families involved in heavy metal transport are as follows (Nies, 1999).

1. ABC transporters which are involved in both uptake and efflux process.
2. P-type proteins that are also involved in both functions.
3. A-type proteins involve in the efflux process.
4. RND family that is involved in the efflux process.
5. HoxN takes part in the uptake process driven by chemiosmotic energy.
6. CHR that is an integral protein membrane, which is involved in chromate transport.
7. MIT is also an integral protein membrane that is involved the uptake of most cations.
8. CDF is involved in the chemiosmotically efflux of zinc, cadmium, cobalt, and iron.

Table 1 summarizes a list of microbes capable of degrading petroleum hydrocarbon components and hydrocarbon derived components in the presence of metal contaminants (Sandrin and Maier, 2003).

Generally, heavy metals exert an inhibitory action on microorganisms by blocking essential functional groups, dislodging essential metal ions or transforming the active conformations of bio-molecules (Saurav and Kannabiran, 2011; Nies et al., 1999). After heavy metals ions entered into the cell, toxicity occurs through numerous biochemical pathways, which can be divided into five categories as follows (Harrison et al., 2007).

Metal	Organic	Lowest metal concentration reported to reduce biodegradation	Microbe studied	Environment	pH	Reference
Cd^{2+}	2,4-DME	0.100 mg/L[a]	Indigenous community	Sediment (microcosm)	6.5	Said and Lewis 1991
Cd^{2+}	2,4-DME	0.629 mg/L[a]	Indigenous community	Aufwuchs[c] (microcosm)	5.6	Said and Lewis 1991
Cd^{2+}	4CP, 3CB, 2,4-D, XYL, IPB, NAPH, BP	< 25.3-50.6 mg/L[a,b]	Alcaligenes spp., Pseudomonas spp., Moraxella sp.	Tris-buffered minimal medium plates	7.0	Springael et al. 1993
Cd^{2+}	2,4-D	> 3 mg/L[a]	Alcaligenes eutrophus JMP134	Mineral salts medium	6.0	Roane et al. 2001
Cd^{2+}	2,4-D	24 mg/L[a]	Alcaligenes eutrophus JMP134	Mineral salts medium containing cadmium-resistant isolate	6.0	Roane et al. 2001
Cd^{2+}	2,4-D	0.060 mg/g[a]	Alcaligenes eutrophus JMP134	Soil microcosms	8.2	Roane et al. 2001
Cd^{2+}	2,4-D	0.060 mg/g[a]	Alcaligenes eutrophus JMP134	Field- scale bioreactors	8.2	Roane et al. 2001
Cd^{2+}	PHEN	1 mg/L[d]	Indigenous community	Soil microcosms	7.6	Maslin and Maier 2000
Cd^{2+}	NAPH	1 mg/L[d]	Burkholderia sp.	Dilute mineral salts medium containing 1.4 mM phosphate	6.5	Sandrin et al. 2000
Cd^{2+}	TOL	37 mg/L[a]	Bacillus sp.	Mineral salts medium containing 36 mM phosphate	5.9	Amor et al. 2001
Cd^{2+}	Naphthalene	500 μM	Comamonas testosteroni	Tris buffered MSM	6.5	Douglas et al.
		100μM		PIPES buffered MSM	6.5	
		500 μM(not complete inhibition)		Bushnell-Hass medium	6.5	
Co^{2+}	4CP, 3CB, 2,4-D, XYL, IPB, NAPH, BP	< 13.3-1,330 mg/L[a,b]	Alcaligenes spp., Pseudomonas spp., Moraxella sp.	Tris-buffered minimal medium plates	7.0	Springael et al. 1993
Cr^{3+}	2,4-DME	0.177 mg/L[a]	Indigenous community	Aufwuchs[c] (microcosm)	6.1	Said and Lewis 1991
Cr^{6+}	4CP, 3CB, 2,4-D, XYL, IPB, NAPH, BP	< 131 mg/L[a,b]	Alcaligenes spp., Pseudomonas spp., Moraxella sp.	Tris-buffered minimal medium plates	7.0	Springael et al. 1993
Cr^{6+}	Phenol	200mg/ L[a]	Burkholderia cepacia	Minimal medium	7.0	Silva et al.2007

Cu^{2+}	2,4-DME	0.076 mg/L[a]	Indigenous community	Sediment (microcosm)	6.1	Said and Lewis 1991
Cu^{2+}	2,4-DME	0.027 mg/L[a]	Indigenous community	Aufwuchs[c] (microcosm)	5.0	Said and Lewis 1991
Cu^{2+}	4CP, 3CB, 2,4-D, XYL, IPB, NAPH, BP	< 14.3-71.6 mg/L[a,b]	Alcaligenes spp., Pseudomonas spp., Moraxella sp.	Tris-buffered minimal medium plates	7.0	Springael et al. 1993
Cu^{2+}	PHB	8 mg/L[d]	Acidovorax delafieldii	Agar plates containing 4.70 mM phosphate	6.9	Birch and Brandl 1996
Cu^{2+}	Crude oil	6.30 mg/L[a]	Pseudomonas sp.	Mineral salts medium containing 31 mM phosphate	7.2	Benka- Coker and Ekundayo 1998
Cu^{2+}	Crude oil	11.25 mg/L[a]	Micrococcus sp.	Mineral salts medium containing 31 mM phosphate	7.2	Benka- Coker and Ekundayo 1998
Cu^{2+}	PH	0.01 mg/L[a]	Acinetobacter calcoaceticus, AH strain	Bioreactor medium containing 0.15 mM phosphate	7.8	Nakamura and Sawada 2000
Hg^{2+}	2,4- DME	0.002 mg/L[a]	Indigenous community	Aufwuchs[c] (microcosm)	6.8	Said and Lewis 1991
Hg^{2+}	4CP, 3CB, 2,4-D, XYL, IPB, NAPH, BP	< 45.2-226 mg/L[a,b]	Alcaligenes spp., Pseudomonas spp., Moraxella sp.	Tris-buffered minimal medium plates	7.0	Springael et al. 1993
Mn^{2+}	Crude oil	317.0 mg/L[a]	Pseudomonas sp.	Mineral salts medium containing 31 mM phosphate	7.2	Benka- Coker and Ekundayo 1998
Mn^{2+}	Crude oil	28.2 mg/L[a]	Micrococcus sp.	Mineral salts medium containing 31 mM phosphate	7.2	Benka- Coker and Ekundayo 1998
Ni^{2+}	4CP, 3CB, 2,4-D, XYL, IPB, NAPH, BP	5.18- 10.3 mg/L[a,b]	Alcaligenes spp., Pseudomonas spp., Moraxella sp.	Tris-buffered minimal medium plates	7.0	Springael et al. 1993
Ni^{2+}	TOL	20 mg/L[a]	Bacillus sp.	Mineral salts medium containing 36 mM phosphate	5.9	Amor et al. 2001
Pb^{2+}	Crude oil	2.80 mg/L[a]	Pseudomonas sp.	Mineral salts medium containing 31 mM phosphate	7.2	Benka- Coker and Ekundayo 1998
Pb^{2+}	Crude oil	1.41 mg/L[a]	Micrococcus sp.	Mineral salts medium containing 31 mM phosphate	7.2	Benka- Coker and Ekundayo 1998

Zn^{2+}	2,4- DME	0.006 mg/L[a]	Indigenous community	Sediment (microcosm)	6.4	Said and Lewis 1991
Zn^{2+}	2,4- DME	0.041 mg/L[a]	Indigenous community	Aufwuchs[c] (microcosm)	5.6	Said and Lewis 1991
Zn^{2+}	4CP, 3CB, 2,4-D, XYL, IPB, NAPH, BP	< 29.5- 736 mg/L[a,b]	Alcaligenes spp., Pseudomonas spp., Moraxella sp.	Tris-buffered minimal medium plates	7.0	Springael et al. 1993
Zn^{2+}	PH	10 mg/L[a]	Acinetobacter calcoaceticus, AH strain	Bioreactor medium containing 0.15 mM phosphate	7.8	Nakamura and Sawada 2000
Zn^{2+}	Crude oil	0.43 mg/L[a]	Pseudomonas sp.	Mineral salts medium containing 31 mM phosphate	7.2	Benka- Coker and Ekundayo 1998
Zn^{2+}	Crude oil	0.46 mg/L[a]	Micrococcus sp.	Mineral salts medium containing 31 mM phosphate	7.2	Benka- Coker and Ekundayo 1998
Zn^{2+}	TOL	2.8 mg/L[a]	Bacillus sp.	Mineral salts medium containing 36 mM phosphate	5.9	Amor et al. 2001

BP:biphenyl; IPB: isopropylbenzene; MTC: maximum tolerated concentration; TOL: toluene; XYL: xylene. [a]Value represents total metal added to system. [b]Value represents minimum inhibitory concentration (MIC) calculated by multiplying MTC by a factor of 2.25. MIC = MTC × 2.25. [c]Floating algal mats. [d]Value represents solution phase concentration of metal present in system.

Table 1. Microbes capable of degrading petroleum hydrocarbon components in the presence of metal contaminants (Sandrin and Maier, 2003).

1. Toxic metal species can bind to proteins in lieu of essential inorganic ions, thereby altering the biological function of the target molecule.
2. Heavy metals tend to react with thiol (SH) or disulfide groups destroying the biological function of proteins that contain sensitive S groups. By binding to these groups, metals can inhibit the functions or activity of many sensitive enzymes.
3. Certain transition metals can participate in catalytic reactions, known as Fenton-type reactions that produce reactive oxygen species (ROS). Collectively, these reactions place the cell in a state of oxidative stress, and increased levels of ROS damage DNA, lipids, and proteins through a range of biochemical routes,
4. The transporter-mediated uptake of toxic metal species might interfere with the normal transport of essential substrates owing to competitive inhibition.
5. Some metal oxyanions are reduced by the oxidoreductase DsbB, which draws electrons from the bacterial transport chain through the quinone pool. In fact, certain toxic metal species starve microbial cells by indirectly siphoning electrons from the respiratory chain.

It has been found that heavy metals greatly affect the biotransformation or biodegradation of organic pollutants by interacting with microbial enzymes or their cell walls, by interfering with the microbial general metabolism or by interrupting the functioning of the enzymes participating in the degradation of hydrocarbons. (Beveridge and Murray, 1976; Sobolev and Begonia, 2008; Merroun , 2007; Haferburg and Kothe, 2007; Kim, 1985; Sandri and Maier, 2003; Wataha et al., 1994). Mergeay et al. (1985) found a plasmid bound metal resistance in Alcaligenes eutrophus that is a gram negative bacterium, which was considered to be the first gram negative bacterium showing plasmid bound resistance against cadmium and zinc. Said and Lewis (1991) claimed that heavy metal ions exert a stronger inhibitory effect on biodegradation than the inhibitory effect caused by high concentrations of toxic organic compounds. Therefore, it is very important to include the effect of heavy metal concentration in bioremediation studies. For instance, during the evaluation of metal impacted biological systems, is critical to monitor the metal speciation to understand the microbial response to heavy metal stress (Sandrin and Maier, 2003).

The extent of the metal stress on the microbes can only be determined if chemical speciation analysis is performed during the degradation studies. Values of the minimum inhibitory concentration (MIC) can be misinterpreted for a metal if the reported metal concentration is not in the solution phase. In most standard media, it is found that the metal ions chelate or bind with organic entities of the media leading to the false interpretation of the MICs (Angle and Chaney, 1989). Silva et al. (2007) reported the high metal tolerance of the strain *B. cepacia*-JT50 for chromium and mercury, however a reduction in the degradation efficiency was observed at high metal concentrations. The study was carried out using minimal medium at pH 7.0 in which different chromium and mercury concentrations were incorporated; however it is not clear from this study the actual metal speciation of the mercury and chromium ions during testing.. Kuo and Genthner (1996) showed that the anaerobic degradation or biotransformation of phenol, benzoate and their choloro derivatives 2-chlorophenol and 3-chlorobenzoate respectively slowed down in the presence of added heavy metal ions Cd(II), Cu(II), Cr(VI), Hg(II), however different metals showed different toxicity patterns to the organisms. The consortia showed susceptibility towards some specific metals for example 3-chloro benzoate degradation was most sensitive to Cd(II) and Cr(VI), degradation of phenol and benzoate was sensitive to Cu(II) and Hg(II), however enhanced biodegradation of phenol and benzoate was observed on the addition of small amount of Cr(VI). Cu(II) was found to enhance the degradation rate of 2-chlorophenol and interestigly Hg(II) 1.0 to 2.0 ppm was found to enhance the degradation rate of 2-chlorophenol and 3-cholorobenzoate after an extended acclimation period indicating the adaptation of the consortia to mercury.

Heavy metals are increasingly found in microbial habitats due to natural and industrial processes. Thus, microorganisms have been exposed to metal polluted environments for long time, which has forced microorganisms to develop several mechanisms to tolerate, resist or detoxify these metal ions by efflux, complexation, or reduction of metal ions, to use them as terminal electron acceptors in anaerobic respiration (Dhanjal and Cameotra, 2010; Jaysankar et al., 2008; Sobolev and Begonia 2008: El-Deeb, B., 2009; Johncy et al., 2010; Chaalal et al., 2005; Patel et al., 2006).

Silver (1996) proposed four mechanisms for bacterial metal resistance as follows.

1. Reduced uptake to keep the toxic metal ion out of the cell.
2. Highly specific efflux pumping (i.e. removing toxic ions out of cell)

3. Intra and extracellular sequestration
4. Enzymatic detoxification such as converting a toxic metal ions into less toxic forms

Toxicity of metals depends upon the extent to which it penetrates the microorganism cells. There are several variables influencing the microbial resistance to heavy metals. These variables are briefly discussed in the subsequent paragraphs.

a. Effect of contact time - Contact time between microorganisms-pollutant influences the adaptation process of micro-organisms. Depending on the residence time of microorganisms in the metal environment, it has been found that 10 to 90% of the microorganisms become metal resistant. Doelman and Haanstra (1979) reported that a strain became lead tolerant after 2 years of contact with the metal, Kuo and Genthner (1996) also reported the adaptation of bacterial consortia to Cd(II), Cu(II), Cr(VI) and Hg(II) after a comparative extended adaptation period. Yeom and Yoo (2002) evaluated the enzymatic adaptation of *Alcaligenes Xylosoxidans Y234* to heavy metal during degradation studies of benzene and toluene in the presence of heavy metal ions. They established that co-culturing is an effective method to reduce the metal inhibition effect. The reasoning is that, if one inhibitory metal ion to enzyme A is present in a system, then the inoculation of a microorganism which already has enough enzyme A through microbial adaptation or the addition of a compound inducing enzyme A production will offset the inhibition effect.
b. Effect of cell cycle - Howlett and Avery (1999) investigated the relationship between the cell volume (as the volume of the cells vary during the cell cycle) and the susceptibility to copper toxicity in *Saccharomyces cerevisiae.* The correlation between the cell volume (by examining the forward angle light scatter –FSC) and the cell cycle stage (by examining the DNA content) indicated that the largest (the cells about to undergo mitosis) were most resistant to Cu and the smallest (newly divided cells) cells were also relatively resistant to copper. As the percent cell volume increased beyond 0-2% Cu resistance started decreasing and approach to minimal when the percent cell volume reached to 38-40% of the maximum cell volume. Then the resistance towards copper started increasing with further increase in cell volume and reached to maximum at the stage when cells having 98-100% of the maximum cell volume.
c. Effect of temperature - Tynecka et al. (1981) observed that the uptake of cadmium (Cd^{2+}) was highly reduced at low temperatures ($4^{0}C$), while the degradation of hydrocarbons is favored at higher temperatures. However, there are a number of cold adapted microorganisms reported that can efficiently degrade petroleum hydrocarbons at low temperatures (Braddock et al. 1997).
d. Effect of pH - pH affects metal toxicity to a great extent because metal ions could form complexes with the medium or buffered components or metal ions could precipitate as phosphates or sulphates, especially at pHs $\geq$ 7.0. Mergeay et al. (1985) replaced phosphate buffer by tris-HCl to avoid the precipitation of the metal ions. Tynecka et al. (1981) reported that the sensitive and resistant cells do not react in the same way to pH increase. Both types *of* cells accumulate more cadmium as the pH increases but the pattern of cadmium uptaking was different. Kelly et al. (1999) reported that the pH of the treatment influences the microbial population size, structure, activity, and that a decrease in pH exerts a stress on the soil microbial communities (Pennanen et al. 1996).
e. Metal tolerant strains - Microorganisms can naturally exhibit the multifunctional properties for hydrocarbon degradation and metal resistance properties or

microorganisms with hydrocarbon degradation capability can be genetically modified for metal resistance or vice-versa. Ueki et al. (2003) identified two vanadium binding proteins and expressed them in E. coli, which showed the capability of accumulating twenty times more copper than their control strain. Marine bacteria are also known to have metal resistance capabilities (De et al., 2003,). Keramati et al. (2011) isolated multi metal resistant bacteria highly resistant to mercury. El-Deeb (2009) reported a natural strain exhibiting the properties of organic pollutant degradation and tolerance toward heavy metals. Yoon and Pyo (2003) reported the development and characterization of a strain highly effective in degrading phenol and resistant to the effect of heavy metals ions. In order to restricted or minimized the effect of heavy metals on the hydrocarbon degradation capability of microorganisms the following actions are recommended (Riis et al., 2002; Malakul et al., 1998: Babich et al., 1977).

- Application of phosphate or sulfate compounds at the contaminated site to induce the precipitation of heavy metal ions.
- Addition of metal chelators.
- Increasing the pH of the soil. Addition of organic humus forming material

4. Conclusions and recommendations

Although extensive literature is available on petroleum hydrocarbon biodegradation, more research specifically oriented towards the effect of heavy metals on hydrocarbon biodegradation is required, particularly in the areas outline below:

1. Investigation of the nature of microbe interaction with metal ions in a variety of subsurface and sedimentary environments and under a range of terminal electron accepting metabolisms.
2. Development of technologies to evaluate the physical, chemical, and biological interactions of microorganisms with metal ions in the presence of petroleum hydrocarbons
3. Evaluation of the effect of metal ions on the hydrocarbon degrading microbial physiological activities and responses.
4. Integration of experimental and more advanced computational approaches to better predict and interpret biochemical processes during hydrocarbon degradation in the presence of heavy metal ions.
5. Establishment of the relevant parameters needed for kinetic and thermodynamic computational models focused on microbe/metal interaction during hydrocarbon bioremediation processes.
6. Development of technology to investigate the gene expressed by the microorganism in the presence of heavy metal ions. Development of laboratory bioreactor systems that more accurately reproduce field conditions.

5. References

Atlas R. M., Microbial Degradation of Petroleum Hydrocarbons: An Environmental Perspective, Microbilogical Reviews, p. 180-209 Vol. 45, No. 1, 1981

Angle J. S. and Chaney R. L, Cadmium Resistance Screening in Nitrilotriacetate-Buffered Minimal Media, Applied and environmental microbiology, Vol. 55, No. 8,1989, p. 2101-2104

Atlas R. M., Microbial degradation of petroleum hydrocarbons: An environmental perspective, Microbiological reviews, pp. 180-209, 1981

Babich H. and Stotzky G., Reductions in the Toxicity of Cadmium to Microorganisms by Clay Minerals,Applied and environmental microbiology, 1977, Vol. 33, No. 3, p. 696-705

Babich H.and Stotzky G., Toxicity of Nickel to Microbes: Environmental Aspects, Advances in applied microbiology, vol-29, 1983

Beveridge T. J. and Murray R. G. E., Uptake and Retention of Metals by cell Walls of Bacillus Subtilis, Journal Of Bacteriology, Sept. 1976, p. 1502-1518

Boening D. W., Ecological effects, transport, and fate of mercury: a general Review, Chemosphere 40 (2000) 1335-1351

Braddock J.F., Ruth M.L., Walworth J.L., McCarthy K.A. ,Enhancement and inhibition of microbial activity in hydrocarbon-contaminated arctic soils: implications for nutrientamended bioremediation. Environ Sci Technol, 1997, 31:2078–2084

Cameotra S. S., Singh H. D., Hazarika A. K. and Baruah J. N., Mode of uptake of insoluble solid substrates by microorganisms. II: Uptake of solid n-Alkanes by yeast and bacterial species, Biotechnology andBioengineering, Vol. XXV, 1983,pp 2945-2956

Cameotra S. S. and Makkar R. S. (1998) Synthesis of biosurfactants in extreme conditions. Appl Microbiol Biotechnol 50:520–529

Cameotra S. S., Makkar R.S., Kaur J. and Mehta S.K., Synthesis of Biosurfactants and Their Advantages to Microorganisms and Mankind, Biosurfactants, Edited by: Ramkrishna Sen, ISBN: 978-1-4419-5978-2, Publication date: February 16, 2010

Chaalal O., Zekri A. Y., Islam R., Uptake of Heavy Metals by Microorganisms: An Experimental Approach Energy Sources, 27:87–100, 2005

Cong L. T. N. , Mikolasch A., Awe S., Sheikhany H., Klenk H. P., Oxidation of aliphatic, branched chain, and aromatic hydrocarbons by *Nocardia cyriacigeorgica* isolated from oil-polluted sand samples collected in the Saudi Arabian Desert, Journal of Basic Microbiology 2010, *50*, 241–253

Costello J., Morbidity and Mortality Study of Shale Oil Workers in the United States, Environmental Health Perspectives,Vol. 30, pp. 205-208, 1979

Dhanjal S., Cameotra S. S., Aerobic biogenesis of selenium nanospheres by Bacillus cereus isolated from coalmine soil Microbial Cell Factories 2010, (9), 52

De J., Ramaiah N., Mesquita A. and Verlekar X. N., Tolerance to various toxicants by marine bacteria highly resistant to mercury, Marine Biotechnology. Vol. 5, pp. 185-193, 2003

Delin G.N., Essaid H.I., Cozzarelli I.M., Lahvis M.H., and Bekins B.A., <http://mn.water.usgs.gov/projects/bemidji/results/fact-sheet.pdf>

Dmitriev D. E. and Golovko A.K., modeling the molecular structure of petroleum resins and asphaltenes and their thermodynamic stability calculations, Chemistry for sustainable development,18, 171-180 (2010)

Doelman P. and Haanstra L., Effects of lead on soil bacterial microflora, soil Biol. Biochem., 11: 487-491, 1979

Ehrlich H. L., Microbes and metals, Appl Microbiol Biotechnol (1997) 48: 687-692

El-Deeb, B., Natural combination of genetic systems for degradation of phenol and resistance to heavy metals in phenol and cyanide assimilating bacteria Malaysian Journal of Microbiology, Vol 5(2) 2009, pp. 94-103

Forney F. W. and Markovetz A. J., Subterminal Oxidation of Aliphatic Hydrocarbons, Journal of Bacteriology, Vol. 102, No. 1, Apr. 1970, p. 281-282

Fritsche W. and Hofrichter M., http://www.wiley-vch.de/books/biotech/pdf/v11b_aero.pdf

Ganesh A., Lin J.,Diesel degradation and Biosurfactant production by gram positivr isolates, African journal of Biotechnology Vol. 8(21), pp. 5847-5854

Gautam K. K. and Tyagi, Microbial Surfactants :A Review, Journal of Oleo Science, vol.55, no.4, 155-166, 2006

Griffin L.F. and Calder J. A., Applied and environmental microbiology, May 1977, p. 1092-1096, Vol. 33, No. 5

Haferburg G. and Kothe E., Microbes and metals: interactions in the environment, Journal of Basic Microbiology 2007, 47, 453 - 467

Harrison J. J., Ceri H., and Turner R. J., Multimetal resistance and tolerance in microbial biofilms, nature reviews-microbiology, vol. 5, 2007

Hoffman D. R.,. Okon J. L., Sandrin T. R., Medium composition affects the degree and pattern of cadmium inhibition of naphthalene biodegradation, Chemosphere 59 (2005) 919–927

Howlett N. G., and Avery S. V., Flowcytometric investigation of heterogeneous copper sensitivity on asynchronously grown *Saccharomyces cerevisiae,* Fems microbiology Letter, 1999

Jaysankar D., Ramaiah N. and Vardanyan L., Detoxification of toxic heavy metals by marine bacteria highly resistant to mercury Marine Biotechnology. Vol.10, No. 4, pp. 471-477, 2008

Johncy R. M., Hemambika B., Hemapriya J., and Rajeshkannan V., Comparative Assessment of Heavy Metal Removal by Immobilized and Dead Bacterial Cells: A Biosorption Approach Global Journal of Environmental Research 4 (1): 23-30, 2010

Kelly J. J., Haggblom M. and Tate R. L., changes in soil microbial communities over time resulting from one time application of zinc: alaboratory microcosm study, soil biology and biochemistry, 31(10), pp. 1455-1465

Keramati P., Hoodaji M. and Tahmourespour A., Multi-metal resistance study of bacteria highly resistant to mercury isolated from dental clinic effluent, African Journal of Microbiology Research Vol. 5(7) pp. 831-837, 4 April, 2011

Kim S. J., Effect of heavy metals on natural populations of bacteria from surface microlayers and subsurface water, Marine Ecology - progress series, Vol. 26: 203-206, 1985

Kinawy A. A., Impact of gasoline inhalation on some neurobehavioural characteristics of male rats, BMC Physiology 2009, 9:21

Knox E. G. and Gilman E. A. Hazard proximities of childhood cancers in Great Britain from 1953-80, Journal of Epidemiology and Community Health 1997;51:151-159

Kuo C. W. and Genthner B. R. S., Effect of Added Heavy Metal Ions on Biotransformation and Biodegradation of 2-Chlorophenol and 3-Chlorobenzoate in Anaerobic Bacterial Consortia, Applied and Environmental Microbiology, 1996, p. 2317-2323, Vol. 62, No. 7

Mahatnirunkul V., Towprayoon S., Bashkin V., Application of the EPA hydrocarbon spill screening model to a hydrocarbon contaminated site in Thailand, Land Contamination & Reclamation, 10 (1), 2002

Makkar R.S., Cameotra S.S., An update on the use of unconventional substrates for biosurfactant production and their new applications, Appl Microbiol Biotechnology (2002) 58:428–434

Malakul P., Srivasan K. R., and Wang H. Y., Metal Toxicity Reduction in Naphthalene Biodegradation by Use of Metal-Chelating Adsorbents, Applied and environmental microbiology, Vol. 64, No. 11, Nov. 1998, p. 4610–4613

Menkes D. B. and Fawcett J. P., Too Easily Lead? Health Effects of Gasoline Additives, Environmental Health Perspectives, Volume 105, Number 3, March 1997

Mergeay M., Nies L. D., Schlegel H. G., Gerits J., Charles P., and Gijsegem F. V., Alcaligenes eutrophus CH34 Is a Facultative Chemolithotroph with Plasmid-Bound Resistance to Heavy Metals, Journal of Bacteriology, Apr. 1985, Vol. 162, No. 1, p. 328-334

Merroun M.L., Interactions between Metals and Bacteria: Fundamental and Applied ResearchCommunicating Current Research and Educational Topics and Trends in Applied Microbiology, A. Mendez-Vilas (Ed.),108 ©Formatex 2007

Mittal S. K., Ratra R. K., Toxic effect of metal ions on biochemical oxygen demand, Wat. Res. Vol. 34, No. 1, pp. 147-152, 2000

Mukherjee S., Das P. and Sen R., Towards commercial production of microbial surfactants, Trends in Biotechnology Vol.24 No.11, 1997

Muthusamy K., Gopalakrishnan S., Ravi T. K. and Sivachidambaram P., Biosurfactants: Properties, commercial production and application, Current Science, Vol. 94, NO. 6, 25 , 2008

Nieder M. and Shapiro J., Physiological Function of the Pseudomonas putida PpG6(Pseudomonas oleovorans) Alkane Hydroxylase: Monoterminal Oxidation of Alkanes and Fatty Acids, Journal of Bacteriology, Vol. 122, No. 1 ,Apr. 1975, p. 93-98

Nies D. H., Microbial heavy-metal resistance, Appl Microbiol Biotechnol (1999) 51: 730-750)

Obidike I.R., Maduabuchi I. U., Olumuyiwa S.S.V., Testicular morphology and cauda epididymal sperm reserves of male rats exposed to Nigerian Qua Iboe Brent crude oil, J. Vet. Sci. (2007), 8(1), 1–5

Orct T., Blanusa M., Lazarus M., Varnai V. M. and Kostial K., Comparison of organic and inorganic mercury distribution in suckling rat, J. Appl. Toxicol. 2006; 26: 536–539

Oudot J., Selective Migration of Low and Medium Molecular Weight Hydrocarbons in Petroleum-ContaminatedTerrestrial Environments, Oil & Chemical Pollution 6 (1990) 251-261

Patel J. S., Patel P. C., AND Kalia K., Isolation and Characterization of Nickel Uptake by Nickel Resistant Bacterial Isolate (NiRBI), Biomedical and Environmental Sciences, 19, 297-301 (2006)

Pennanen T., Frostegard, A., Fritze H., and Baath E., Phospholipid fatty acid composition and heavy metal tolerance of soil microbial communities along two heavy metal polluted gradients in coniferous forest, Appl. Environ. Microbial, 62(2), pp. 420-428

Qian K., Mennnito A.S., Edwards K.E., Ferrughelli D.T., Observation of vanadyl porphyrins and sulfurcontaining vanadyl porphyrins in a petroleum asphaltene by atmospheric pressure photonionization Fourier transform ion cyclotron resonance mass spectrometry, Rapid communications in mass spectrometry, 22, 2153-2160,(2008)

Rajapaksha R. M. C. P., Tobor-Kaplon M. A, and Baath E., Metal Toxicity Affects Fungal and Bacterial Activities in Soil Differently ,Applied and Environmental Microbiology, May 2004, p. 2966–2973

Riis V., Babel W., Pucci O. H., Influence of heavy metals on the microbial degradation of diesel fuel, Chemosphere 49 (2002) 559–568

Said W. A. and Lewis D. L., Quantitative Assessment of the Effects of Metals on Microbial Degradation of Organic Chemicals, Applied and Environmental Microbiology, May 1991, p. 1498-1503, Vol. 57, No. 5

Sandri T. R. and Maier R. M., Impact of Metals on the Biodegradation of Organic Pollutants, Environmental Health Perspectives, Vol. 111, No. 8, June 2003

Saurav K., and Kannabiran K., Biosorption of Cd (II) and Pb (II) Ions by Aqueous Solutions of Novel Alkalophillic Streptomyces VITSVK5 spp. Biomass, J. Ocean Univ. China, 2011 10 (1): 61-66

Silva A. A. L., Pereira M.P., Filho R. G. S.,Hofer E., Utilization of phenol in the presence of heavy metals by metal-tolerant nonfermentative gram-negative bacteria isolated from wastewater, Microbiologia, Vol. 49, Nos. 3-4, pp. 68 – 73, 2007

Silver S., plasmid determined metal resistance mechanism: range and overview, Plasmid., 27(1), 1-3, 1992

Singh P. , DeMarini D. M., Dick C. A. J., Tabor D. G., Ryan J. V., Linak W. P., Kobayashi T. and Gilmour M. I., Sample Characterization of Automobile and Forklift Diesel Exhaust Particles and Comparative Pulmonary Toxicity in Mice, Environmental Health Perspectives, Vol- 112, No. 8, June 2004

Sobolev D., and Begonia M. F.T., Effects of Heavy Metal Contamination upon Soil Microbes: Lead-induced Changes in General and Denitrifying Microbial Communities as Evidenced by Molecular Markers, Int. J. Environ. Res. Public Health 2008, 5(5) 450-456

Tynecka Z., Gos Z., Zajac J., Reduced cadmium transport determined by a resistance plasmid in Staphylococcus aureus, Journal of bacteriology, 147(2), 305-312, 1981

Ueki T., Sakamoto Y., Yamaguchi N. and Hitoshi Michibata, Bioaccumulation of Copper Ions by *Escherichia coli* Expressing Vanabin Genes from the Vanadium-Rich Ascidian *Ascidia sydneiensis samea,* Applied and environmental microbiology, Nov. 2003, Vol. 69, No. 11 p. 6442–6446

Utgikar V. P., Harmon S. M., Chaudhary N., abak H. H. T., Govind R., Haines J. R., Inhibition of Sulfate-Reducing Bacteria by Metal Sulfide Formation in Bioremediation of Acid Mine Drainage, Inc. Environ Toxicol 17: 40–48, 2002

Wataha J.C., Hanks C.T., and Craig R. G., *In vitro* effects of metal ions on cellular metabolism and the correlation between these effects and the uptake of the ions, Journal of Biomedical Materials Research, Vol. 28, 427-433 (1994)

Welch F., Murray V. S. G., Robins A. G., Derwent R. G., Ryall D. B., Williams M. L., Elliott A. J., Analysis of a petrol plume over England: 18–19 January 1997, Occup Environ Med 1999;56:649–656

Wenpo R., Chaoha Y., Honghong S., Characterization of average molecular structure of heavy oil fractions by ^{1}H nuclear magnetic resonance and X-ray diffraction, china petroleum processing and petrochemical technology,13(3), 1-7 (2011)

Yeom S. H. and Young J. Y. , Overcoming the inhibition effects of metal ions in the degradation of benzene and toluene by Alcaligenes xylosoxidans Y234, Korean J. of Chem. Eng., 14(3), 204-208 (1997)

Yoon, Pyo K., Construction and Characterization of Multiple Heavy Metal-Resistant Phenol-Degrading Pseudomonads Strains, J. Microbiol. Biotechnol. (2003), 13(6), 1001–1007

8

Bioremediation of Crude Oil Contaminated Soil by Petroleum-Degrading Active Bacteria

Jinlan Xu
School of Environmental and Municipal Engineering, Xi'an University of Architecture and Technology, Xi'an, Shaanxi, China

1. Introduction

Approximatelly 0.6 million tons of petroleum is poured into soil, groundwater, rivers and oceans every year in China. A large amount of the oil contaminated soil has not been remediated. The contaminated soil poses a severe threat to the environment and must be taken care of. Bioremediation has become one of the most popular and promising technologies with growing demand for the remediation of petroleum contaminated soils because pollutants can be completely removed at low cost.

Bioremediation uses microbes to degrade hydrocarbons in soil. In the absence of effective microbes, the effect of bioremediation on pollutants is limited because native microbes need long times to climate, and their low metabolic activities make a short term bioremediation difficult. Inoculation of petroleum-degrading active bacteria increases the number of effective microbes, and thereby accelerates the bioremediation process. Effective microbes can be found in oil contaminated soils.

This chapter summarizes the experimental outcome of the evaluation of the bioremediation of oil contaminated soil by several bacteria strains that were isolated from an oil contaminated soil from the north region of the Shaanxi province in China. This chapter includes the characterization of petroleum-degrading active bacteria and the effect of the oil environment on bacterial activities. Bioremediation depends on pH, temperature, oil concentration, nitrogen and phosphorous concentration, among others.

This chapter is structured in four sections as follows. The first and second sections present the separation, screening, and characterization of petroleum-degrading bacteria. The third and fourth sections presents the experimental observations obtained from the bioremediation of oil contaminated soil by the *Plesiomonas* SY_{23} bacteria.

2. Part I Separation and screening of petroleum-degrading bacteria

2.1 Summary

Seven bacteria strains were isolated from petroleum contaminated soil in the north of the Shaanxi province in China. These strains were studied and physiologically characterized. Preliminary results showed that the active strains were *Acinetobacter* SY_{21}, *Neisseria* SY_{22},

Plesiomonas SY_{23}, *Xanthomonas* SY_{24}, *Azotobacter* SY_{42}, *Flavobacterium* SY_{43}, and *Pseudomonas* SY_{44}. In addition, it was found that the biodegradation efficiency of total petroleum hydrocarbon (TPH) were higher than 80% after 8 days. The removal rates of TPH observed in this study were higher than the TPH removal rates previously reported (Truax et al, 1995; Ronald, 1996; Calabrese, et al, 1991). In addition, it was observed that more inoculums corresponding to a larger number of SY_{43} bacteria was present rendering a higher removal efficiency of TPH. Furthermore, the SY_{43} and SY_{23} strains adapted to the oil-contaminated soil setting establishing a local ecology, which rendered high removal efficiencies of TPH (88.4% and 73.4%, respectively), successfully remediating the contaminated soil.

2.2 Introduction

Areas of contaminated soil have increased rapidly in recent years due to the continuous growth and development of the oil industry. Meanwhile, the level of contamination becomes severe as time elapses.

Contaminated soils include many complex compounds such as alkanes, benzene, methylbenzene, and benzene, among others. These contaminants are toxic and usually categorized as carcinogenic substances. They can not be easily eliminated and eventually these contaminants will leach into the groundwater systems [1, 2]. Consequently, oil contamination is a serious environmental problem to our living ecosystem.

Bioremediation of oil contaminated soil requires low cost and does not lead to secondary pollution. In the last decades, relevant bioremediation techniques have been wildly studied. Most of these studies focused on enhancing bioremediation efficiency by increasing the activities of the native microorganisms by adding nutrients such as nitrogen (N) and phosphorous (P) [3]. These studies have shown that native microbes need a long time to domesticate due to slow growth rates. In addition, the low metabolic activities of these native microbes make a rapid decontamination difficult. Therefore, the application of bioremediation using indigenous microbes is restricted. Fortunately, petroleum-degrading active bacteria could be the solution to this problem. For instance, the Baltic General Investment (BGI) Corporation in America uses mixed microbes to improve bioremediation [4]. The research of Wilson (1993) presents bacteria strains which have shown high degradation rates of polycyclic aromatic hydrocarbons (PAHs) [5]. Bioremediation of petroleum contaminated soil is a complicated process, in which the pollutants characteristics, the ecological structure of microbes, and the environmental conditions must be considered. The adaptation characteristics of petroleum-degrading active bacteria can influence bioremediation. In this research, several strains were isolated from oil contaminated soils from the northern region of the Shaanxi province. The objectives of this work were to establish the characteristics of the strains and the kinetics of the bioremediation process.

2.3 Materials and methods

2.3.1 Crude oil and petroleum contaminated soil samples

Samples of the crude oil and the petroleum polluted soil used in this study were collected from oil wells located in the northern region of the Shaanxi province. The characteristics of the polluted soil samples are shown in Table 1.

No.	Total petroleum hydrocarbon (TPH) concentration (Conc.) (mg/kg dry matter)	pH	Water content (wt %)	Number of bacteria (CFU) (CFU/g dry matter)	
				Aerobic bacteria	Petroleum-degrading bacteria
Sample 1	209	8.89	8.3	9.2×10^3	1.3×10^3
Sample 2	148	8.66	15.5	1.3×10^5	5.6×10^3
Sample 3	28100	8.55	13.9	4.4×10^4	2.2×10^4
Sample 4	572	8.91	15.1	3.4×10^5	1.0×10^5

Table 1. Characteristics of Oil-Contaminated Soil Samples

2.3.2 Culture medium

The culture medium per unit used in this work was: 2 g of NH_4NO_3, 1.5 g of K_2HPO_4, 3 g of KH_2PO_4, 0.1 g of $MgSO_4 \cdot 7 H_2O$, 0.01 g of anhydrous $CaCl_2$, 0.01 g of $Na_2EDTA \cdot 2H_2O$, 1 g of crude oil, 1000 mL of distilled water, with a pH ranging from 7.2 to 7.4.

2.3.3 Methodology

The procedure for bacteria gathering, separation, and purification was as follows. Ten grams (10 g) of crude oil contaminated soil were quantitatively placed into a 100 mL of culture medium (250 mL flask). This mixture was shaken using a shaker model THZ-82 shaker, manufactured by Changzhou Guohua Electronic Appliance Ltd, China, at a speed of 180 r/min for 7 days at 30°C . Then, fifty milliliters (50 mL) of the medium were quantitatively added into a 100 mL of culture medium (250 mL flask). After which, the medium was shaken again at a speed of 180 r/min during 7 days at 30°C. This procedure was repeated 3 times.

Then, the culture medium solution was taken and streaked repeatedly 3 times on the plate, after which the strains were added and the plate was introduced in the refrigerator using a slant culture.

In this work the concentration of petroleum hydrocarbons was determined using a non dispersive infrared oil instrument [6]. Bacteria count was performed by the plate counting method [7]. The operational and performance parameters used for the evaluation of bioremediation included the concentration of TPH, removal efficiency (RE), and number of bacteria (CFU).

2.4 Results

2.4.1 The degradation ability of petroleum-degrading active bacteria

In order to test the effect of oil-degrading active bacteria on oil removal, oil was added to the soil samples. After which, active bacteria were added to the contaminated soil samples. Two blank samples were needed here, one of which was sterilized and the other was non-sterilized. The sterilized blank sample was prepared by adding 0.2 wt% of mercury chloride into one of the oil contaminated soil samples. A control test was conducted, which consisted of a contaminated soil sample in the absence of active bacteria but containing the indigenous or native microorganisms samples.

Fig 1 shows the relationship between oil removal efficiency and bioremediation time. The oil removal efficiency in the sample containing mercury (sterilized blank sample) chloride was approximately 9%. The oil material lost in the sterilized blank sample containing mercury chloride is attributed to oil volatilization and aerial oxidation. In the case of the blank sample containing only native microorganisms, the oil removal efficiency increased linearly during the first week and then it leveled off around 37% of oil removal. This indicates that the indigenous microbes contributed to the removal of 25% of petroleum hydrocarbon, compared to the removal efficiency of the soil sample containing mercury chloride.

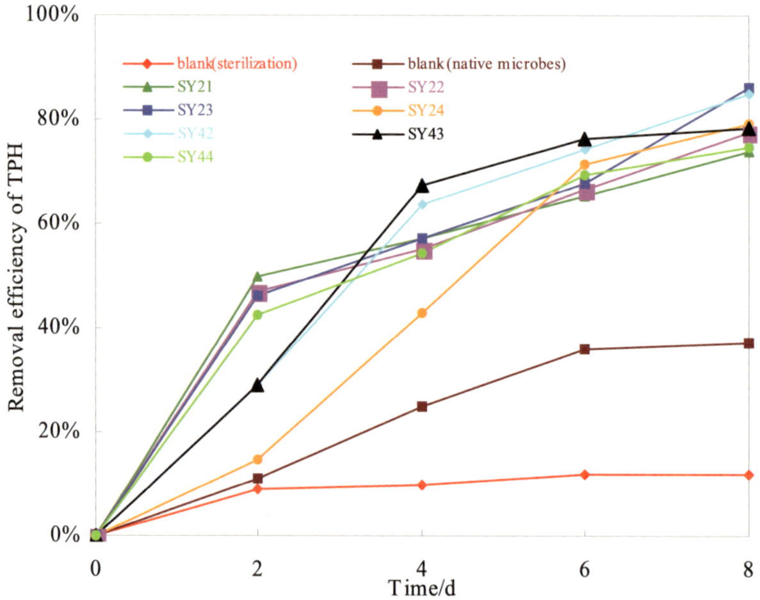

Fig. 1. The biodegradation efficiency of the strains evaluated

The soil samples inoculated with SY_{21}, SY_{22}, SY_{23}, and SY_{44} strains showed higher oil removal efficiencies. The removal efficiency of TPH reached up to 50% after 2 days of inoculation. The different strains inoculated in the contaminated soil showed the same removal rate up to 40%. The isolated strains rapidly adapted to the environment efficiently degrading hydrocarbons compounds. Fig. 1 shows that the biodegradation action of all the petroleum-degrading active bacteria tested follow the same trend. Thus the degradation efficiency of TPH increased gradually, which indicates that the contaminated soil provides an appropriated environment for bacterial growth. Eighty percent (80%) of hydrocarbons were degraded during the first 8 days of bioremediation while the degradation efficiency of the blank sample was only 37%. It is clear that the hydrocarbons were degraded mainly by the petroleum-degrading active bacteria. Fig.2 shows the variation of number of bacteria as a function of time. The number of bacteria increased gradually, as the bioremediation time progressed. The number of bacteria in the inoculated samples was higher that the number of bacteria in the blank sample. This indicates that the indigenous microbes were inefficient in degrading oil and that the inoculation of petroleum-degrading active bacteria shows potential in removing hydrocarbons contaminants.

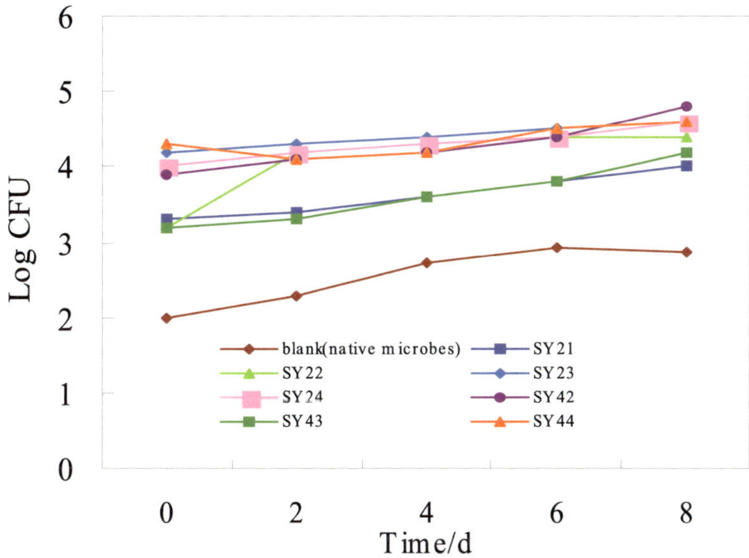

Fig. 2. The number change of 7 bacteria

The strains evaluated showed different oil degradation capacity and different patterns. Fig.3 presents the degradation rate of TPH by seven strains. It is observed that the rates of biodegradation of SY_{21}, SY_{22}, SY_{23}, and SY_{44} were higher in the second day, while the rates of SY_{42} and SY_{43} were higher in day 4, while in day 6, the highest degradation rate was shown by strain SY_{24}. The average degradation rates of TPH were within 0.01~0.1 $g \cdot kg^{-1} \cdot d^{-1}$, which are higher than degradation data previously reported [8-10], consequently these strains could degrade TPH more rapidly.

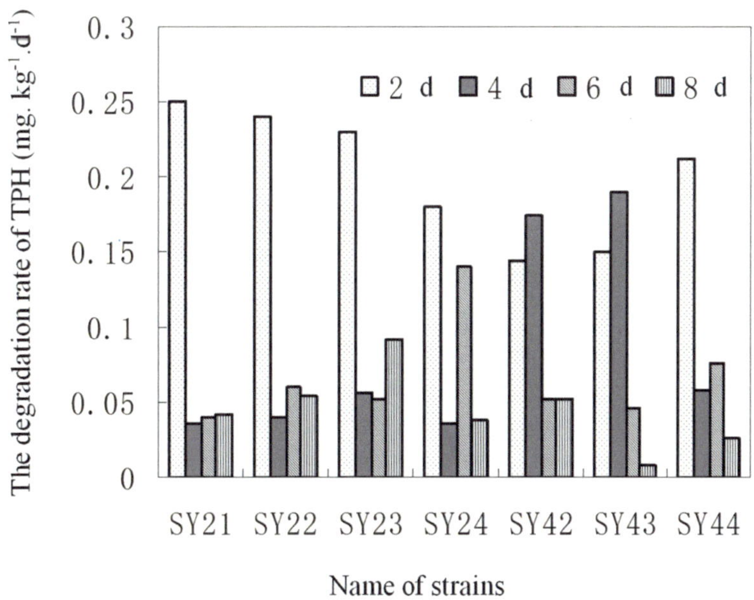

Fig. 3. The degradation rate of TPH by seven strains

Previous work [11] using marine filamentous bacteria reported rates of biodegradation in the range of 7.92×10^{-11} to 4.8×10^{-10} mg·cell^{-1}·d^{-1}. Similarly, other work[12] using pseudomonas showed that the highest oil degradation rate was from 1.44×10^{-10} to 3.77×10^{-9} mg·cell^{-1}·d^{-1}. Fig.4 presents the average degradation rate per cell of the seven strains used in this work. It shows that the highest degradation rate of 2.34×10^{-3} mg•cell^{-1}•d^{-1}was obtained for the strain SY_{43}, followed by strain SY_{23} with a degradation rate of 1.50×10^{-3} mg·cell^{-1}·d^{-1}. The degradation rate of the remaining strains ranged from 1.15×10^{-3} to 4.57×10^{-4} mg·cell^{-1}·d^{-1}. These results demonstrate that the rate of degradation shown by these strains is tens of thousands times higher than the degradation rate reported in previous work. Therefore, petroleum-degrading active bacteria show great potential for the bioremediation of TPH contaminated soils.

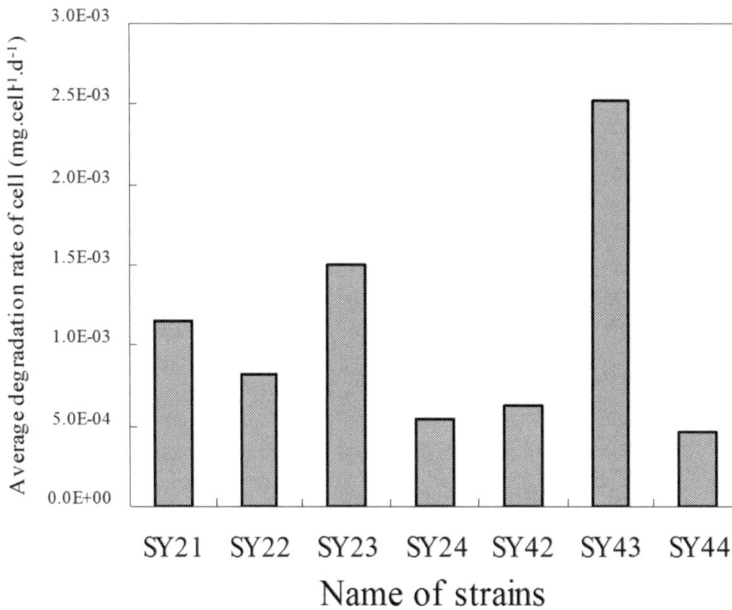

Fig. 4. Average degradation rate of cell by seven strains

2.4.2 Effect of concentration of TPH in contaminated soil on the biodegradation efficiency of petroleum-degrading active bacteria

Four oil samples containing different concentrations of TPH were prepared. The concentration of TPH in mud samples 1, 2, 3, and 4 were 125mg.L^{-1}, 56 mg L^{-1}, 3245 mg.L^{-1} and 209 mg.L^{-1}, respectively. In each sample the volume of inoculum was also varied as follows: 2 mL, 5 mL, and 10 mL. Fig.5 shows the number of bacteria and the removal efficiency (RE) of TPH observed in the mud samples during biodegradation. The number of bacteria increased as the volume of inoculum increased. Hence, the number of bacteria was enhanced by increasing the volume of inoculum. The removal efficiencies of TPH for sample 1 were 30%, 33%, and 70% for the corresponding inoculum volume of 2 mL, 5 mL, and 10 mL after 17 days of biodegradation. In addition, the RE of TPH of sample 2 was 20%, 42%, and 45% and the corresponding RE of TPH of sample 3 was 33%, 38%, and 45%.The petroleum removal efficiency in sample 4 was 43%, 80%, and 81%. These results indicate that the higher the inoculum volume the higher the RE of TPH. The number of bacteria and RE of TPH for the soil samples were low when the inoculum volume was 2 mL. However, when the inoculum volume was increased from 2 mL to 5 mL the removal efficiency of TPH increased. The RE of TPH was similar when the inoculum volumes were 5 mL and 10 mL, respectively. This performance might indicate that a high inoculum volume enhances the competition ability of petroleum-degrading bacteria in relation to native microorganism. It could also aid the quick bacterial adaptation to the environment, which also increases the TPH degradation efficiency.

After 17 days of testing, the RE of TPH when the inoculum volume was 10 mL for sample 1, sample 2, sample 3, and sample 4 were 70%, 44%, 45%, and 80% respectively. The average biodegradation rate from sample 1, sample 2, sample 3, and sample 4 were 3.86×10^{-4} mg. $cell^{-1} \cdot d^{-1}$, 3.43×10^{-4} mg. $cell^{-1} \cdot d^{-1}$, 8.11×10^{-4} mg. $cell^{-1} \cdot d^{-1}$, and 5.86×10^{-4} mg. $cell^{-1} \cdot d^{-1}$, respectively.

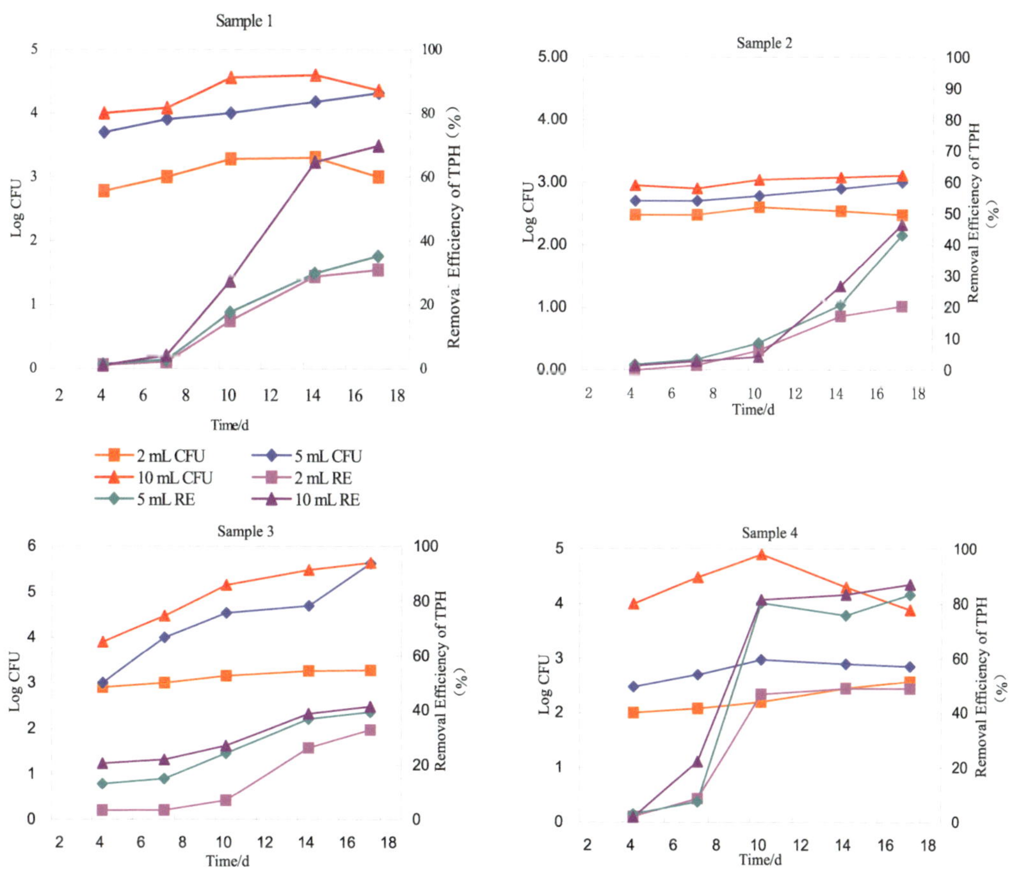

Fig. 5. The effect of inoculum volume and concentration of TPH in contaminated soil on biodegradation efficiency

2.4.3 Bioremediation of oil-contaminated soil by petroleum-degrading active bacteria

Figure 6 shows that significant removal of TPH took place in the soil samples inoculated with petroleum-degrading active bacteria compared with the removal of TPH in the blank sample. After 6 days of bioremediation, the percentage of TPH removal in the samples inoculated with bacteria was 17.2% and 19.2 % for SY_{43} and SY_{23} respectively; while the percentage of TPH removal in the blank sample was only 1.4%. At day 9, the removal efficiency of SY_{43} was 30.5%, which was higher than the removal efficiency observed for the strain SY_{23} (24.9%).

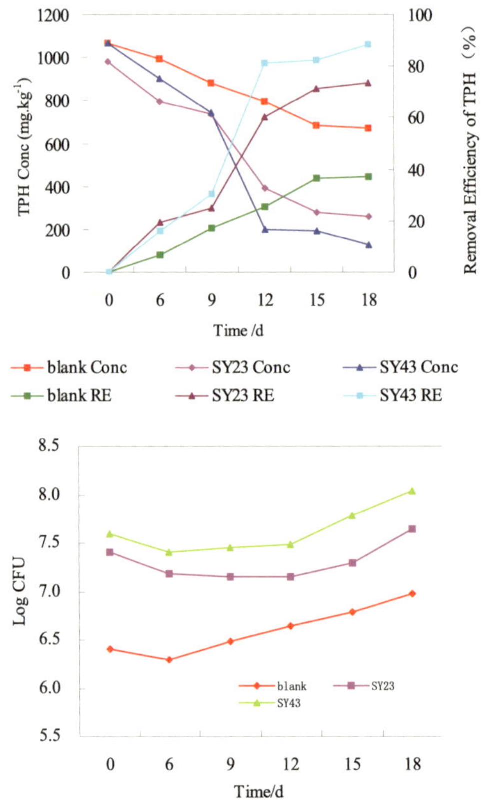

Fig. 6. TPH concentration and TPH removal efficiency (RE) as a function of bioremeditation time.

After 12 days of bioremediation, the TPH concentration of the sample having the strain SY_{43} declined to 213.3mg·kg^{-1} and the RE of TPH was 81.1%; while the TPH removal efficiency of the sample containing the strain SY_{23} and in the blank sample were 60.3% and 25.6%, respectively. After 18 days of inoculation, the removal efficiency in the three samples climbed up to 88.4%, 73.4%, and 37%, respectively. These observations indicate that the contaminated samples inoculated with petroleum-degrading active bacteria had higher oil removal efficiency than in the blank sample. The average degradation rates of TPH by SY_{43} and SY_{23} was 0.044 g·kg^{-1}·d^{-1}, 0.034 g·kg^{-1}·d^{-1} respectively; while the corresponding average degradation rate of the cells was 2.14x10^{-6} mg· cell^{-1}·d^{-1} and 1.64x10^{-6} mg· cell^{-1}·d^{-1} respectively. Figure 6 shows that the petroleum removal ability of SY_{43} is higher than that of SY_{23}. Fig.7 presents the number of bacteria during the bioremediation process for the strains

SY_{43}, SY_{23}, and the blank sample. The number of bacteria in the inoculated soil samples was 100 times bigger than the number of bacteria in the blank sample. The number of bacteria in the soil sample inoculated with SY_{43} was higher than the number of bacteria in the soil sample inoculated with SY_{23}. Thus, the number of petroleum-degrading active bacteria in the contaminated soil sample seems to be the main factor improving the bioremediation of the soil. The results show that both SY_{43} and SY_{23} could rapidly degrade TPH in the contaminated soil. The TPH degradation efficiency showed by the strain SY_{43} was better than the degradation efficiency achieved by the strain SY_{23}.

2.5 Conclusions

1. The bacteria strains isolated from petroleum contaminated soil in the north region of the Shaanxi province were *Acinetobacter* SY_{21}, *Neisseria* SY_{22}, *Plesiomonas* SY_{23}, *Xanthomonas* SY_{24}, *Azotobacter* SY_{42}, *Flavobacterium* SY_{43}, and *Pseudomonas* SY_{44}.
2. The TPH removal efficiency of these bacteria strains was 50% after 5 days of testing and 80% after 8 days of bioremediation. The average cell-degradation rate and the removal rate of TPH were thousand times higher compared to previously published results.
3. The number of active bacteria increases as the amount of inoculum increases, which enhances the competition ability of petroleum-degrading active bacteria against native bacteria. Simultaneously, this competition forces native bacteria to rapidly adapt to the environment improving their degradation ability, though, the bacteria strains show different petroleum removal trends even when the same inoculum amount was added, the average TPHs degradation rate was almost the same and the average degradation rate increases as the TPHs in contaminated soil increases.
4. Strains SY_{43} and SY_{23} show rapid degradation of TPH in contaminated soil. However, the degradation efficiency of SY_{43} was higher than the degradation efficiency of SY_{23}.

3. Part II Growth characteristics of highly petroleum-degrading bacteria

3.1 Summary

The growth characteristics of 7 strains isolated from oil contaminated soil, as well as their respective degradation efficiency for various hydrocarbons were investigated. Factors that can impact biological oil degradation efficiency were revealed in a series of experiments. The results indicate that isolated strains could rapidly degrade crude oil, showing high activity in the first 13 h of bioremediation. These strains could grow in paraffin wax, which indicates that these strains could degrade long chain hydrocarbons. Some of them (SY_{22}, SY_{23}, SY_{24}, SY_{42}, and SY_{43}) were able to use short chain hydrocarbons and aromatic hydrocarbons as substrate, so these five strains are the preferred ones for the bioremediation of oil contaminated soil. Suitable pH for the growth of these five strains was in the range from 7 to 9. NH_4NO_3 and oil concentrations should range from 1000 mg/L to 1500 mg/L in order to achieve optimum conditions for petroleum hydrocarbon degradation. Adding organic matter such as starch and glucose accelerated oil and PAH degradation capability of the strains SY_{22}, SY_{42}, and SY_{23} strains. The presence of metal ions, such as Ni^{2+} and Co^{2+}, in soil decreased the crude oil degradation efficiency of these strains, while metal ions, such as Fe^{2+} and Mn^{2+}, did not affect the oil degradation activities.

3.2 Introduction

Eight million tons (Mt) of petroleum is spilled into the environment every year worldwide. In China, 0.6 Mts of petroleum enters into soil, groundwater, rivers and ocean every year. Oil contamination is a severe threat for our environment and thereby attracts general concern. Consequently, the remediation of oil polluted sites has become an important issue worldwide.

Bioremediation has become one of the most promising technologies for soil remediation, because the cost of remediation of petroleum contaminated soils by biological techniques is low; in addition to the fact that through bioremediation the complete removal of oil can be achieved; while no secondary pollution is introduced. Microorganisms used for bioremediation are usually grouped as indigenous and exogenous microbes. The addition of nutrients increases the activity of native microorganisms, however bioremediation is boosted with the addition of exogenous bacteria. Native microbes need a long time to domesticate, and thereby show low growth rates and low metabolic activity, which make decontamination slow and ineffective. Therefore, the application of bioremediation using indigenous microbes is restricted. Thus, the screening of petroleum-degrading active bacteria to remediate oil polluted soil is a necessary task. In this experimental work, several strains (seven) were isolated from oil contaminated soil in the north region of the Shaanxi province. The effect of pH, nutrition (nitrogen and phosphorous), and pollution intensity on the oil degradation efficiency of the isolated strains were investigated. Furthermore, based on the complexity of the soil systems, the degradation capacity of the isolated bacteria of different kinds of petroleum hydrocarbons and the effect of adding organic co-substrate and metal ions on the bioremediation were studied.

3.3 Experimental section

3.3.1 Experimental material

3.3.1.1 The Source of samples

The tested crude oil and petroleum polluted soil were collected at oil wells in the northern region of the Shaanxi province. The strains separated from the oil contaminated soil were SY_{21}, SY_{22}, SY_{23}, SY_{24}, SY_{42}, SY_{43}, and SY_{44} [13].

3.3.1.2 Culture medium

The recipes of the different culture media used in this work are provided as follows.

Recipe of liquid or solid beef grease & peptone cultivation medium: 10 g of peptone, 3 g of beef grease, 5 g of NaCl, 1000 mL of distilled water, pH 7.0. The medium can be solidified using 20 g of agar.

Recipe of Liquid inorganic salts cultivation medium: 2 g of NH_4NO_3, 1.5 g of K_2HPO_4, 3 g of KH_2PO_4, 0.1 g of $MgSO_4 \cdot 7H_2O$, 0.01 g of anhydrous $CaCl_2$, 0.01 g of $Na_2EDTA \cdot 2H_2O$, 1000 mL of distilled water, and pH 7.0.

Crude oil culture medium: addition of crude oil into liquid inorganic salts culture medium.

The medium above were all sterilized for 30 min under 121°C.

3.3.2 Experimental methodoloy

3.3.2.1 Strains separation

Crude oil was added into the soil collected in the north region of the Shaanxi province intermittently, to progressively increase the oil concentration in the soil, so as to obtain bacteria with high capacity for oil degradation. The process was carried out at aerobic conditions under continuous shaking. Temperature was controlled at 30°C. Pure strains were isolated 21 days later by streaking them repeatedly on a plate.

The separated strains were identified to the genus level depending on their morphological and physiological-biochemical characteristics in general ways.

The preparation of the bacterium suspension was carried out by inoculating the strains into liquid beef grease & peptone medium, which was pre-sterilized under 121°C for 30min. The mixture of medium and bacteria was shaken for 36 h (180 r/min) under 30°C. Later, the mixture was centrifuged (180 r/min) and the resulting suspension was discharged while the residual sediment was washed 3 times using phosphate buffer. Finally, the washed sediments were diluted using phosphate buffer in order to adjust the number of the cells in bacterium suspension to be 1×10^8.

3.3.2.2 Study on the growth characteristics of the strains

Under sterile conditions, strains were inoculated in 200 mL liquid beef grease & peptone medium, which had been previously sterilized. Then the mixture of bacteria and medium was shaken at 30°C and 180 r/min. Afterwards, optical density D_{460} of the bacterium liquid using light (460 nm wavelength) was measured at regular intervals.

3.3.2.3 Determination of petroleum hydrocarbons

Oil degrading strains were inoculated into the crude oil medium, which was previously sterilized. This mixture was shaken and the pH adjusted below 3. Then, the medium was placed into funnels, shook and the total volume brought to 100 mL after adding 20 mL of CCL_4 in order to extract the hydrocarbons present in it. This mixture was kept static for segregation to take place (layered). The under layer was filtered and dried using anhydrous sodium sulfate and then it was placed into a 50 mL volumetric flask. The upper-layer was extracted using CCL_4 twice, then filtered, and placed into a 50 mL volumetric flask. The concentration of petroleum hydrocarbon was determined using a non dispersive infrared oil analyzer and the biodegradation of petroleum hydrocarbon η was determined using the following equation (1).

$$\eta = \frac{c_0 - c_x}{c_0} \times 100\% \tag{1}$$

In this equation (1), c_0 and c_x represent the residual concentration of petroleum hydrocarbon in blank samples and the test samples, in mg/L; respectively.

3.3.2.4 Strains' hydrocarbon degradation efficiency tests

Six hydrocarbon compounds which included normal octane, paraffin wax, benzene, methylbenzene, phenol, and naphthalene were added to the inorganic medium that was previously sterilized (121°C for30 min) using high pressure steam. Then, 5 mL of bacterium suspension were added to the medium. This mixture was shaken for 36 hours under 30°C and 180 r/min. After all the phenol was evaporated from the mixture, the bacterium suspension was added. Naphthalene mixed with acetone was also added into the bacterium suspension after all the acetone was evaporated from the mixture. Finally, the optical density D_{460} of the culture solution under 460nm was measured and the concentration of petroleum hydrocarbons was determined at regular intervals.

3.3.2.5 Evaluation of the factors influencing the activity of petroleum-degrading bacteria

Crude oil samples were mixed with petroleum ether to prepare a solution having a concentration of 60 g of crude oil/L of solution. Then, the mixture was filtrated using a 0.25 µm filter membrane. The filtrate was placed into a flask and the petroleum ether was completed evaporated from the flask. Then, pre-sterilized inorganic medium and 5 mL of bacterial suspension were added into the flask. Using this oil-contaminated soil samples as the starting point, several petroleum degradation experiments were performed at different conditions of pH, organic load, nitrogen source, carbon source and metal ions.

3.4 Results and discussions

3.4.1 Growth characteristic of petroleum-degrading active bacteria

3.4.1.1 Isolated strains cultivated in hydrocarbon medium

Table 2 shows bacterial growth, bacterial density, and the rate of oil biodegradation reached by each strain after 7 days of cultivation. The strains grew well in oil media and emulsified crude oil. The density of bacteria after 7 days of cultivation was observed to range between 1×10^7/mL and 1×10^9/mL. These results indicate that the strains used petroleum as the carbon source. The biodegradability (η) after 7 days of cultivation was between 43.8% and 58.9%, which exceeded the biodegradability of formerly reported petroleum-degrading bacteria B01(25.8%-32.8%) [11] and were close to that of O-8-3 *Pseudomonas*, marine bacteria SJ-06W, SJ-6, and SJ-16A-2 as previously reported [14].

Strain	Growth and emulsification	Bacteria quantity (CFU.mL^{-1})	η (%)	Identification
SY_{21}	Complete emulsification and dense liquid	5.3×10^7	43.8	*Acinetobacter*
SY_{22}	Forming oil film and flock	2.4×10^7	46.7	*Neisseria*
SY_{23}	Complete emulsification and dense liquid	3.6×10^9	58.9	*Plesiomonas*
SY_{24}	Complete emulsification and forming flock	1.2×10^7	45.0	*Xanthomonas*
SY_{42}	Complete emulsification and dense liquid	3.2×10^8	47.6	*Azotobacter*
SY_{43}	Forming oil film and flock	6.7×10^8	53.3	*Flavobacterium*
SY_{44}	Forming dispersed flocks and dense liquid	9.2×10^7	45.8	*Pseudomonas*

Table 2. Growth And Identification of the Isolated Strains Cultivated in Hydrocarbon Medium After 7 Days of Cultivation

These strains of petroleum-degrading active bacteria were all gram-negative bacteria. The strains SY_{21}, SY_{22}, SY_{23}, SY_{24}, SY_{42}, SY_{43}, and SY_{44} were identified as *acinetobacter, neisseria, plesiomonas, xanthomonas, zoogloea, flavobacterium,* and *pseudomonas*, respectively. Previous research has shown that gram-negative bacterium dominate in microbes that can degrade petroleum hydrocarbon [15]. The *Xanthomonas, zoogloea, flavobacterium,* and *pseudomonas* strains have been extensively studied and used.

3.4.1.2 The growth trend in liquid medium

Figure 8 shows bacteria growth as a function of time. The curves in Fig. 8 indicate that bacteria grow rates were low during the first 13 hours, after which the bacteria grow rates followed a logarithm growth period during the next 13~23 hours; and then turned into a slow down growth period during the following 23~40 hours. Finally, bacteria began to die after 40 hours of activity. Thus, the strains showed the highest activity during the 13 to 23 hours of life.

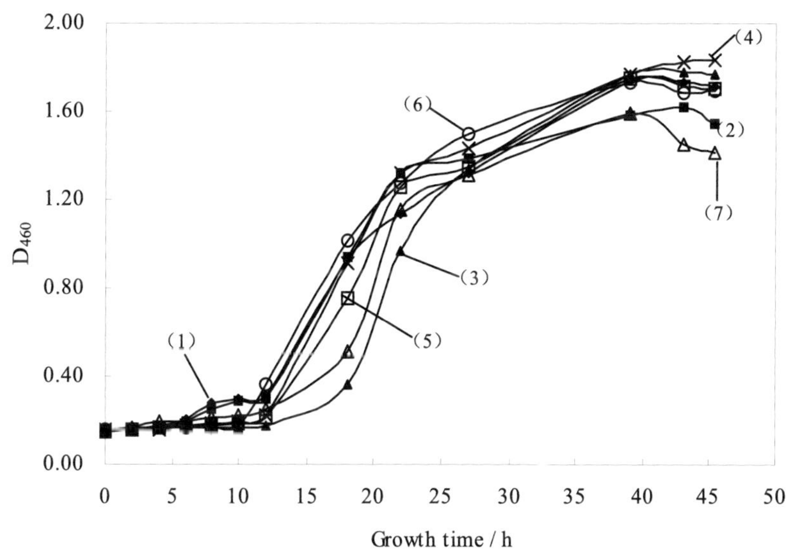

(1)SY_{21}; (2)SY_{22};(3)SY_{23};(4)SY_{24}; (5)SY_{42};(6)SY_{43}; (7)SY_{44}

Fig. 8. Growth trend of the seven isolated strains in liquid cultivation medium

3.4.1.3 Growth trend in agar medium

Figure 9 shows the growth trend of the same 7 strains in agar media.

Fig.9 shows the variation of the diameter of colony forming of the different strains as a function of time. The colonies of SY_{21} were formed after 4 hours of activity. After 9.2 hours the diameter reached 4 mm. This colony expanded continuously in the first 20 hours during which the average growth rate was 12.84 mm/d. In addition the colony growth was circular having an ivory-opaque color with an arid and disordered surface. The SY_{22} strain formed a circular ivory and semitransparent colony after 4 hours of inoculation. The surface of the colony was wet and orderly with a diameter of 3 mm after 9.2 hour. Similar to the previous case, the colony expanded continuously within the first 20 hours with an average growth rate of 8.41 mm/d. The colony formed by the strain SY_{23} was ivory and opaque with an arid and disordered surface. The SY_{23} colony was formed after 9.2 hours reaching a diameter of 2 mm after 15 hours. This colony also expanded continuously with an average growth rate at 2.49 mm/d. The SY_{24} formed an opaque and creamy yellow colony after 4 hours of inoculation. Its surface was flat and disordered, and the colony expanded continuously during the first 37 hours at a growth rate at 8.43 mm/d. After inoculation the strain SY_{42} formed a white transparent and circular colony. The surface of the colony was wet and orderly. The average growth rate of this colony was 5.30 mm/d. The colony made up by the SY_{43} strain was white-transparent and disordered with a wet and smooth surface. The average growth rate of this colony was 5.30 mm/d. The SY_{44} strain formed an ivory semitransparent circular colony. Its surface was wet, smooth, and orderly and reaching a diameter of 4.2 mm after 4 hours. The colony expanded continuously during the first 15 hours with an average growth rate of 24.32 mm/d. The average growth rate of the 7 isolated strains ranged from 2.49 to 32.4 mm/d.

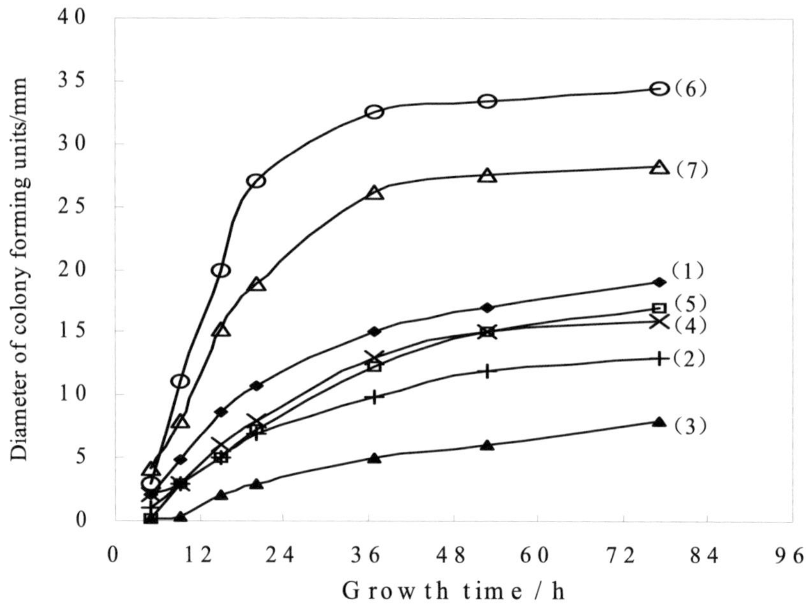

(1)SY_{21}; (2)SY_{22};(3)SY_{23};(4)SY_{24}; (5)SY_{42};(6)SY_{43}; (7)SY_{44}

Fig. 9. Growth tendency of seven isolated strains in agar cultivation medium

3.4.2 The degradation ability of petroleum-degrading active bacteria toward different types of hydrocarbons

The majority of petroleum-degrading bacteria can degrade only few kinds of hydrocarbons [15-16]. The middle-chain and long-chain normal alkane can be degraded by most petroleum-degrading bacteria. However, the short-chain hydrocarbons and aromatic hydrocarbons can only be degraded by few petroleum-degrading bacteria. For the majority of bacteria it is difficult to digest short-chain and aromatic hydrocarbons, which can even be toxic.

In this work, the hydrocarbon degradation capability of the 7 strains was evaluated using the following hydrocarbon compounds: octane, paraffin wax, benzene, methylbenzene, phenol, and naphthalene. The initial concentration of these hydrocarbon compounds were 125 mg/L, 64800 mg/L, 200 mg/L, 14.4 mg/L, 200 mg/L, and 330 mg/L, respectively. During these tests the temperature and pH were set at 30°C and 7, respectively.

Table 3 shows that the seven strains grew in the paraffin wax media (The optical density D_{460} measured range from 0.117 to 0.450). The degradation efficiency of paraffin wax shown by the SY_{43} strain was 81.3%, which was the highest degradation efficiency observed, while the strain SY_{21} showed the lowest efficiency at 43.7% degradation. The degradation efficiencies of the other strains were between SY_{43} and SY_{21}. All of these 7 strains showed a high degradation capability toward middle and long-chain alkane, as the 90% of paraffin wax consisted of C_{18}~C_{61} normal and isomeric alkanes [17]. The degradation efficiencies of naphthalene by the 7 strains were about 40%. The SY_{23} and SY_{24} strains showed high abilities to degrade benzene, methylbenzene, and phenol as the degradation efficiency reached from 80% to 90%. The majority of these strains showed low degrading efficiency toward normal octane, with the exception of strains SY_{24} and SY_{43}, which showed degradation rates of 54.4% and 56.8% respectively. These observations indicate that the following strains SY_{22}, SY_{23}, SY_{24}, SY_{42}, and SY_{43} are capable of degrading more than one hydrocarbon, which makes them potential candidate strains for the bioremediation of petroleum contaminated soil.

Hydrocarbon medium	D_{460}							η (%)						
	SY_{21}	SY_{22}	SY_{23}	SY_{24}	SY_{42}	SY_{43}	SY_{44}	SY_{21}	SY_{22}	SY_{23}	SY_{24}	SY_{42}	SY_{43}	SY_{44}
C_8H_{18}	0.103	0.013	0.011	0.116	0.017	0.249	0.015	35.2	12.8	12.0	54.4	21.6	56.8	20.8
Paraffin wax	0.300	0.322	0.132	0.117	0.320	0.409	0.450	43.7	60.1	47.3	47.3	66.6	81.3	62.8
Benzene	0.023	0.011	0.120	0.08	0.036	0.056	0.035	21.0	10.0	90.5	80.9	46.0	71.2	63.9
Naphthalene	0.073	0.034	0.032	0.048	0.030	0.030	0.040	44.7	42.6	35.0	34.4	40.8	43.5	42.6
Phenol	0.033	0.017	0.112	0.104	0.058	0.085	0.067	21.0	10.0	90.5	80.9	46.0	71.2	63.9
Xylene	0.014	0.052	0.075	0.090	0.023	0.007	0.007	8.3	11.1	84.7	93.8	9.7	6.9	4.2

Table 3. Growth Tendency And Degradation Efficiency of the Seven Isolated Strains in Different Hydrocarbon Medium

3.4.3 Factors influencing the hydrocarbon degradation efficiency of petroleum-degrading active bacteria

3.4.3.1 The effect of pH

In microorganisms, biochemical reactions are catalyzed by enzymes. It is well known that enzymatic reactions occur within a suitable pH range and microorganisms are sensitive to the alteration of pH. Thus, it is necessary to determine the optimum pH value suitable for petroleum degradation by bacteria. The pH value of soil normally ranges between 2.5 to 11.0. Thus, before the inoculation of strains into the crude oil media (petroleum concentration was fixed at 600 mg/L), the pH value was adjusted to 3, 5, 7, 9, and 11 for each medium. The experiments were carried out at a rotation speed of 180 r/min for 96 hours at 30°C. After which, the concentration of petroleum hydrocarbon was determined and the degradation efficiency was calculated. Figure 10 shows the degradation efficiency of 4 of the strains.

Fig. 10 indicates that SY_{22} and SY_{23} strains could degrade oil at a pH=9.0 with degradation efficiencies of 80% and 69.4%, respectively. Meanwhile, strains SY_{24} and SY_{42} had the ability of degrading oil at a pH value of 7.0 with degradation efficiencies of 73.1% and 74.9%, respectively.

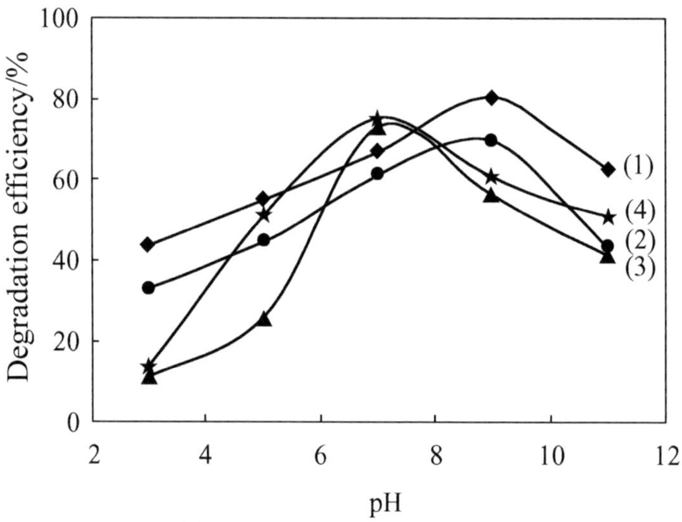

(1)SY_{22}; (2)SY_{23}; (3)SY_{24}; (4)SY_{42}

Fig. 10. Effect of pH value on the hydrocarbon degradation efficiency of some of the isolated strains

3.4.3.2 Effect of petroleum hydrocarbon concentration

Five (5) mL of bacterium suspension was inoculated in the crude oil media. The concentration of hydrocarbon was varied as follows: 200, 600, 1000, 1500, 3000 mg/L. The Bioremediation tests were conducted at 30°C and the pH was adjusted to a value of 8.0. The rotation speed was set at 180r/min for 96 hours. Table 4 summarizes the degradation efficiencies of the strains. The observations in Table 4 indicate that the hydrocarbon degradation efficiency shown by these strains exceeded 60% when the mass concentration of petroleum hydrocarbon was 1000 mg/L. The hydrocarbon degradation efficiency shown by the SY_{24} strain was reduced when the mass concentration of petroleum hydrocarbon increased to 1500 mg/L. The hydrocarbon degradation efficiency of all the strains was reduced when the mass concentration of petroleum hydrocarbon was increased to 3000

TPH concentration	η (%)			
(mg.L^{-1})	**SY_{22}**	**SY_{23}**	**SY_{24}**	**SY_{42}**
200	35.3	61.5	57.5	49.2
600	57.8	62.0	60.5	60.6
1000	61.6	64.8	63.5	63.5
1500	63.2	66.8	50.2	65.8
3000	46.0	56.2	43.0	37.4

Table 4. Effect of Petroleum Hydrocarbon Concentration on the Strains Hydrocarbon Degradation Efficiency

mg/L. These results indicate that excessive concentration of petroleum hydrocarbon restricted the growth of the strains and consequently reduced the hydrocarbon degradation efficiency. The SY_{23} strain showed the highest TPH degradation efficiency at all TPH concentrations, which indicates its endurance to the TPH toxicity.

3.4.3.3 Effect of different nitrogen sources

In order to determine the impact of the nitrogen source on the strain degradation efficiency, a series of different nitrogen sources were used in addition to NH_4NO_3 (200 mg/L) in the inorganic media. The concentration of the different nitrogen sources was set as 350 mg/L. The rotation speed was set at 180r/min for 96 hours at 30°C and at pH of 8.0. After 96 hours of inoculation the concentration of the residual petroleum was determined. Table 5 presents the hydrocarbon degradation efficiency as a function of nitrogen source. Table 5 indicates that all the strains showed the highest degradation efficiency when NH_4NO_3 was used as the nitrogen source, while the lowest degradation efficiencies were observed when $NaNO_3$ was used as the nitrogen source. This indicates that NH_4NO_3 is the best nitrogen source for the strains under evaluation, which is in agreement with the results presented by other researchers [18]. It is important to mention that the SY_{23} strain showed a high degradation efficiency when $(NH_4)_2SO_4$ and Urea were used as nitrogen sources, which points out that SY_{23} could be used for remediation situations were these nitrogen sources are readily available.

Nitrogen source	η (%)			
	SY_{22}	SY_{23}	SY_{24}	SY_{42}
NH_4NO_3	46.7	48.5	52.5	42.6
$(NH_4)_2SO_4$	20.2	47.1	30.0	30.9
$NaNO_3$	13.3	10.9	28.6	21.2
Urea	28.9	39.2	33.0	39.9

Table 5. Effect of Nitrogen Sources on the Petroleum Hydrocarbon Degradation Efficiency

3.4.3.4 Effect of different carbon sources

Glucose and starch were used as co substrate in the inorganic medium in which petroleum concentration was fixed at 1000 mg/L. These experiments were carried out at a rotation speed of 180 r/min for 96 hours at a temperature and pH of 30°C and 8.0, respectively.

Table 6 shows the effect of dosing co substrate on the removal efficiency of petroleum. Table 6 shows that the degradation efficiency of petroleum hydrocarbon by the SY_{22} and SY_{23} strains increased from 43.8% and 17.6% to 71.5% and 70.2%, respectively. These results show that the strains degradation efficiency was enhanced by the use of glucose and starch as carbon sources. The explanation is that glucose and starch can be used as co-metabolism medium during the petroleum degradation process [19]. The SY_{22} and SY_{42} strains had high degradation efficiency of naphthalene (42.6% and 40.8% as showed in Table 3), these efficiencies were improved after adding glucose and starch. The SY_{23} strain also showed high degradation efficiency of benzene, methylbenzene, and phenol with corresponding efficiencies of 90.5%, 84.7% and 90.5%.These observations indicate that the SY_{23} strain can degrade PAH in crude oil and that the degradation efficiency can be increased to a large extent after adding starch. Thus, bacteria activity can be improved by adding the appropriate carbon sources, in this case glucose and starch. It has been previously reported that using glucose as carbon source improves the degradation efficiency of PAH and if glucose is fed intermittently the abilities of bacteria to degrade petroleum hydrocarbon could be maximized [20].

Carbon source	η (%)			
	SY_{22}	SY_{23}	SY_{24}	SY_{42}
Oil	43.8	36.9	35.0	17.6
Oil+Starch	71.5	46.7	41.5	70.2
Oil+Glucose	58.2	60.8	30.7	35.6

Table 6. Effect of Carbon Sources on the Strains Petroleum Hydrocarbon Degradation Efficiency

3.4.3.5 Effect of different metal ions

The concentration of metal ions increased in oil fields due to the aging and mineralization of soil during the weathering process of petroleum contaminated soil. In order to find the impact of the presence metal ion on the biological removal of petroleum, metal ions of Fe^{2+}, Mn^{2+}, Ni^{2+}, and Co^{2+} were added into the inorganic salt liquid media in which the petroleum hydrocarbon concentration was 1000 mg/L.

The experiment was carried out at rotation speed of 180 r/min for 96 hours at 30°C and a pH of 8.0. Table 7 shows the calculated petroleum removal efficiencies and it clearly shows that the degradation efficiency of the strains declined significantly after adding Ni^{2+}. It seems that a high concentration of Ni^{2+} restricted the activities of the microorganisms. For instance, the degradation efficiency of SY_{43} strain was decreased from 53.3% to 10.9%. In contrast, adding Fe^{2+} improved the degradation efficiency of petroleum hydrocarbon (SY_{21} and SY_{23} strain). However, the activity of SY_{43} strain was restricted. After adding Mn^{2+} the degradation efficiency of oil by SY_{23} strain was enhanced by 12% while the degradation efficiency by SY_{21} and SY_{43} strain was not affected. The degradation efficiency of oil achieved by the SY_{21} and SY_{43} strains decreased 16% and 12% respectively after adding the metal ions of Co^{2+}. These results point out that the addition of metal ions such as Fe^{2+} and Mn^{2+} has a favorable influence on the oil degradation efficiency. On the contrary, the degradation efficiency of petroleum hydrocarbon was decreased after adding metal ions such Ni^{2+} and Co^{2+}.

Metal ion	η (%)		
	SY_{21}	SY_{23}	SY_{43}
-	58.2	38. 9	53.3
Fe^{2+}	64.1	46.7	40.2
Mn^{2+}	54.3	50.6	54.6
Ni^{2+}	33.3	20.0	10.9
Co^{2+}	42.5	43.0	45.1

Table 7. Effect of Metal Ions on the Strains Degradation Efficiency

3.5 Conclusions

1. Seven strains were isolated from petroleum contaminated soil. The degradation efficiency of oil after seven days of cultivation ranged from 43.8% to 58.9%, which

indicates high strains activity. These strains show a logarithm growth trend after 12 to 13 hours of inoculation with an average growth rate of the respective colonies between 2.49 to 2.4 mm/d.

2. The strains SY_{21}, SY_{22}, SY_{23}, SY_{24}, SY_{42}, SY_{43}, and SY_{44} were categorized as *acinetobacter, neisseria, plesiomonas, xanthomonas, zoogloea, flavobacterium* and *pseudomonas* respectively.
3. These strains are capable of using normal octane, paraffin wax, benzene, methylbenzene, phenol, and naphthalene as the solo carbon source. Five of these strains: SY_{22}, SY_{23}, SY_{24}, SY_{42} and SY_{43} show the ability of degrading more than one hydrocarbon, which make them potential candidates for the bioremediation of petroleum contaminated soil.
4. A pH value of 7.0 was optimum for the growth of strains SY_{21} and SY_{42} while a pH value of 9.0 was optimum for the development of strains SY_{22} and SY_{23}.
5. The strains show an optimum degradation of petroleum hydrocarbon when NH_4NO_3 was used as a nitrogen source in contaminated soil containing oil concentrations ranging from 1000 mg/L to 1500mg/L.
6. The degradation efficiency of strains SY_{22}, SY_{42}, and SY_{23} is significantly enhanced by the addition of starch and glucose.
7. The presence of metal ions such as Ni^{2+} and Co^{2+} in the oil contaminated soil decreases the strains degradation efficiency, while the presence of Fe^{2+} and Mn^{2+} does not affect the oil degradation by the strains, on the contrary might improve it.

4. Part III Bioremediation of oil contaminated soil by SY23 petroleum-degrading active bacteria - Impact of bulking agent and tillage on bioremediation of petroleum-contaminated soil

4.1 Summary

The impact of adding bulking agent and tillage on bioremediation of petroleum contaminated soil was studied. The results showed that the strain degradation efficiency of TPH was 76% in 48 days after adding bulking agents and tillage. This degradation efficiency was 15% higher than the degradation efficiency in the control experiment in which bulking agent was not added. Meanwhile, the degradation rate was 2.34 times higher in the sample containing bulking agent compared to the control sample. The addition of bulking agent increased water content in the soil as the bulking agent absorbed water, which improved the degradation of TPH. Gas chromatography-Mass Spectrometer (GC-MS) analysis indicated that the peak numbers of the GC profile decreased from 32 to 14 in 64 days after adding bulking agent. This result indicated that branched alkanes, alkene, carotane, and alkylnaphthalenes were thoroughly degraded. The peak numbers of the GC profile decreased 10 times. Furthermore residual hopanes and steranes were also thoroughly degraded after the combined treatment of the oil contaminated soil with bulking agents and tillage. The degradation efficiency of TPH decreased as tillage times decreased, the optimum tillage frequency was once per day with a shovel.

4.2 Introduction

Petroleum could lead to the contamination of ground water after petroleum sprays into soil. Bioremediation of petroleum hydrocarbon contaminated soil is considered a new technology with a broad prospect because of its low-cost [9]. During in situ bioremediation, sufficient dissolved oxygen (DO) is necessary to keep the degradation activity of bacteria. It has been reported that the mineralization of TPH in soil will be restrained when the DO concentration is lower than 15% (dry soil) or 5% (wet soil) [21]. Tillage is a mechanical way to improve the local environment in the soil. This method can change the physico-chemical characteristic of soil and stimulate the activity of bacteria by increasing the oxygen content in the soil and aiding the release of carbon dioxide [22, 23]. Bulking agents are cheap low density materials. The addition of bulking agents into the soil reduces the density, increases the porosity, and enhances the spread rate of oxygen in the soil [24]. This research evaluates the impact of adding a bulking agent and tillage on the bioremediation of petroleum-contaminated soil.

4.3 Materials and methods

4.3.1 Materials

4.3.1.1 Samples source

Oil contaminated soil samples were collected at contaminated sites surrounding oil wells from an oil field located in the northern region of the Shaanxi province, China. The samples were homogenized, sieved (pore size 2 mm), and stored in a dark and ventilated fridge at 4°C until use.

4.3.1.2 Microbial inoculations

Petroleum degrading active bacteria were collected from oil contaminated soil samples [25]. The solid SY_{23} strain was obtained as a concentrated bacteria suspension that was separated by centrifuge. The solid bacterial product was prepared after the solid SY_{23} strain was dried on sterilized condition at 37°C for 48 hours.

4.3.2 Experimental methods

4.3.2.1 Processing methods

In order to prepare a bacterial-rich suspension, certain amount of the solid bacterial product was added into a bacteria-free buffer. A substrate medium was added into the bulking agent to prepare a mixture. Later, the mixture was sterilized. Then, the bacterial-rich suspension was mixed with the bulking agent mixture which was cultured for 24 hours. The final product was ready for addition into the petroleum contaminated soil.

4.3.2.2 Analytical methods

The petroleum hydrocarbon (TPH) composition was determined using an OCMA - 350 non-dispersive infrared oil analyzer. Oil components were measured using a gas chromatograph (Model Trace 2000) and a Mass Spectrometer (Ms) model Voyager 5975B, manufactured by Agilent USA. The GC was equipped with a 30 m long capillary (DB-5) in which the stationary phase was 0.25 µm thick. The measurable range of relative molecular weights was 30 to 450. GC conditions for analysis were as follows: carrier gas: helium (0.8 $mL.min^{-1}$); Flame Ionization Detector (FID) temperature: 320°C; the initial temperature was 40°C and it was maintained for 5 minutes. Then, the temperature was increased to 300°C at a speed of 10 °C /min. Conductivity meter was used to measure conductivity and pH test paper was used to measure pH values. The conductivity and pH values must be analyzed under the conditions where the ratio of bulk factor of soil to water was 1:5(W/V)[8]. Finally, constant weight method was used to measure moisture content.

4.3.3 Design of the experiments

The objective of the experimental design was to investigate the bioremediation of petroleum hydrocarbon contaminated soil by bioaugmentation products through bioremediation simulation tests with the addition of bulking agent and tillage. The experiments were carried out in basin ports having diameters of 15 cm and depths of 15cm. A mass of 1000 g of soil was used in each group. Table.8 summarizes the experimental conditions. Each experimental group was conducted in triplicate. Experiments were carried out at room temperature, which varied in the range of 16 to 20°C. Samples were taken for analysis at time intervals to measure pH, conductivity, TPH concentration, and moisture content. The sample interval was increased when the rate of petroleum-degradation started to decrease.

Items	Sample number	Dose of bacteria agent(mg.kg^{-1})	Dose of Sawdust (g)	Treatment
Tillage and bulking agent	CK	0	0	Stagnant
	AJ	0.6	0	Stagnant
	BJ	0.6	80	Stagnant
	AF	0	0	Tilled once every day
	BF	0.6	0	Tilled once every day
	CF	0.6	80	Tilled once every day
Tillage frequency	F1	0.6	80	Tilled once every day
	F2	0.6	80	Tilled once every three days
	F3	0.6	80	Tilled once every five days
	F0	0.6	80	Stagnant

Table 8. Experimental Design

4.4 Results

4.4.1 The influence of bulking agent on the degradation of petroleum hydrocarbons

Figure 11 shows the experimental observations, and it indicates that the addition of bulking agent and tillage aid the reduction of petroluem concentration. Samples subjected to tillage had the lowest remaining TPH concentration. TPH concentration declined from 4.5 g/kg to 1.737 g/kg (57%) after tilling for 19 days with bulking agent. While, for sample BF without bulking agent only 23% of the TPH was degraded. The degradation rate in sample BF was only 0.968 g/kg. Thus, the removal rate of TPH was 2.34 times higher when bulking agent was added. The concentration of TPH in sample CF with bulking agent after 48 days of bioremediation had declined 3.044 g/kg (76%). Meanwhile the sample without bulking agent (BF) only declined 2.611 g/kg (61%). For the Stagnant samples with added bulking agent, the removal efficiency of TPH was 66.4% while the removal efficiency of the samples without bulking agent was only 50.9% after 48 days. These observations indicate that the addition of bulking agent accelerates the removal rate of TPH by bacteria. The bulking agent increases the permeability of the contaminated soil so that the oxygen transfer rate and the water-holding capacity are also enhanced. This provides an advantageous condition for bioremediation of oil contaminated soil. Research conducted by Xiaomei Ye [26] showed that the concentration of TPH declined 70% after 120 days of degradation which is in agreement with this study.

Figure 11 indicates that the removal efficiency of samples AF, BF, and CF were 22%, 33%, and 67% respectively after tilling for 26 days. Meanwhile, the removal efficiency in those samples without tilling was 2%, 22%, and 56% respectively. It is obvious that tillage has increased the removal efficiency of TPH. Tillage not only raises the content of oxygen in soil but also accelerates the speed of substrate transfer from outside into the biomembrane. The oxygen concentration in soil has a significant effect on the degradation of TPH. During pure oxygen aeration, the release rate of CO_2 is 0.013 mol/d. However, the production of CO_2 is reduced to 0.004 mol/d during air aeration [27]. In addition, tillage separate the soil block into small ones and mixes the soil. Therefore, tillage causes redistribution of C, N, and water in the contaminated soil. It has been reported [28] that tillage improves the removal efficiency of tetrachloroethylene and trichloroethylene in contaminated soil.

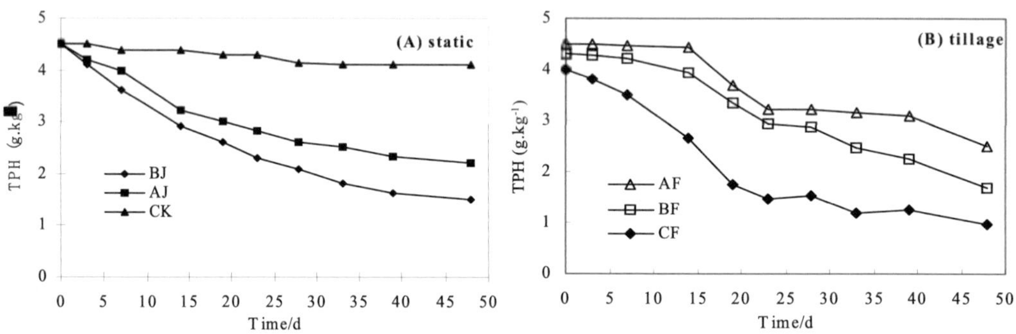

Fig.11. Influence of bulking agents on the degradation of TPH from oil-contaminated soil that was (A) left stagnant, (B) tilled once every day

4.4.2 Changing patterns of the oil component with bulking agent

In order to understand the way bulking agents work, extractions of soil samples after degradation for 64 days were analyzed by GC-MS. Table 9 and Figure 12 summarizes the results of the GC-MS. Fig.12 indicates that the removal efficiency of oil components in the samples with bulking agent were higher than in the blank sample. Comparing the GC-MS spectrum for sample AJ (bulking agent-free) with sample BJ (with bulking agent added), which were left stagnant for 64 days, 32 peaks were shown in the AJ soil sample (bulking agent-free) spectrum while the BJ showed only 14 peaks in the GC-MS. The branched alkanes, alkene, carotane and alkylnaphthalenes in sample BJ (containing bulking agent) were thoroughly degraded. But the removal rate of branched alkanes, alkene, carotene, and alkylnaphthalenes in sample AJ (bulking agent-free) were 57.2%, 67.8%, 68.8%, and 79.7% respectively. The addition of bulking agent promotes the growth rate of the active bacteria because bulking agent keeps water in the soil and aids oxygen transfer. So the removal rate of normal paraffin, hopanes, and steranes could also be promoted under these conditions.

After combined treatment of addition of bulking agent and tillage, 10 peaks in the GC-MS spectrum disappeared and residual branched alkanes, alkene, carotene, alkylnaphthalenes, hopanes, and steranes were also thoroughly degraded. As well as the removal of 99.7% of normal paraffin. Adding bulking agent promotes the interaction between bacteria and TPH. Likewise, the combined treatment with bulking agent and tillage demonstrate improved bioremediation.

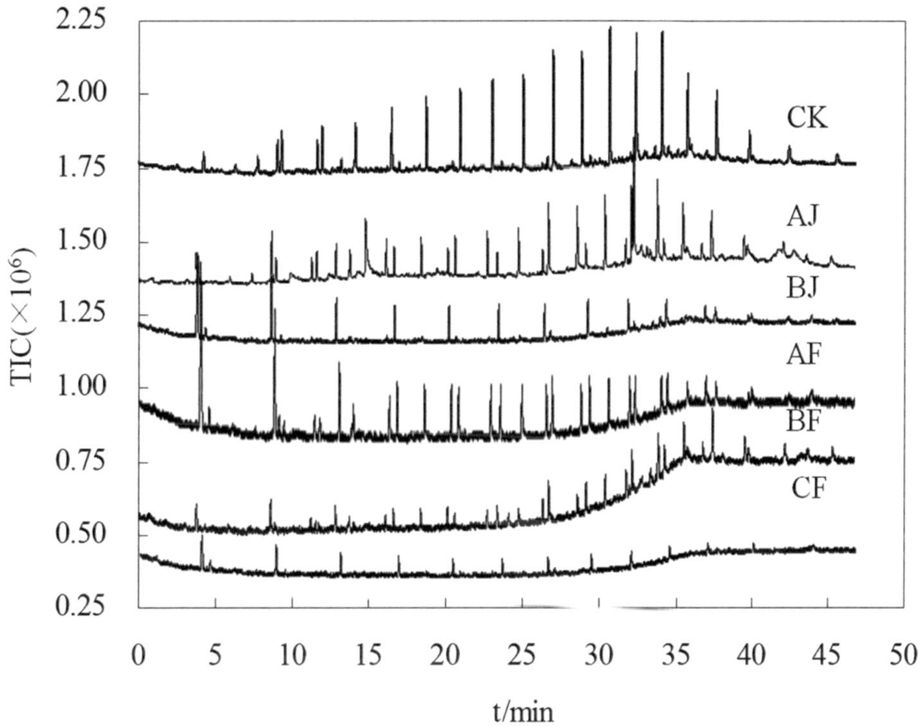

Fig. 12. TIC of remaining TPH concentration in soil on different treatment condition

Number	Peak area $A(\times10^3)$ and Degradation efficiency η (%)													
	Alkanes				Alkene		Crotane		Alkyl-naphthalenes		Hopanes		Steranes	
	n-Alkanes (m/z=57)		Branched alkanes (m/z=97)		(m/z=55)		(m/z=125)		(m/z=128)		(m/z=191)		(m/z=217)	
	A	η	A	η	A	η	A	η	A	η	A	η	A	η
Control	39502		3466		8074		845		251		542		156	
CK	13992	64.6	1684	51.4	2645	67.2	332	60.7	83	66.9	46	91.5	49	68.6
AJ	10875	72.5	1484	57.2	2601	67.8	264	68.8	51	79.7	261	51.8	27	82.7
BJ	627	98.4	9	99.7	33	99.6	0	100.0	0	100.0	78	85.6	6	96.2
AF	2139	94.6	143	95.9	341	95.8	29	96.6	0	100.0	20	96.3	0	100.0
BF	1409	96.4	75	97.8	137	98.3	9	98.9	0	100.0	127	76.6	13.7	91.2
CF	110	99.7	0	100.0	0	100.0	0	100.0	0	100.0	0	100.0	0	100.0

Table 9. GC-MS Analysis of Petroleum Hydrocarbon after Adding Bulking Agents

4.4.3 Influence tillage times on the degradation of TPH

Figure 13 shows the experimental matrix designed to explore the influence of tillage time and the addition of bulking agent on the degradation of TPH. In the samples evaluated the change of TPH followed the same trend. TPH concentration decreased as a function of time with the removal rate increasing gradually. TPH content in sample F1 (tilled once a day) was the lowest since day 3, followed by samples F2 and F3. The biodegradation level in sample F0 (without tillage) was the worst. The removal efficiency of TPH in F1 was up to 92.64% while the removal efficiency in samples F2, F3, and F0 was 80.46%, 69.14%, and 60.26% after 36 days of biodegradation. Likewise, the removal efficiency of TPH decreased as tillage times decreased. The optimum tillage frequency was once a day every day using a shovel. If economic factors are considered, then a minimum of once tillage each 3 days would be acceptable.

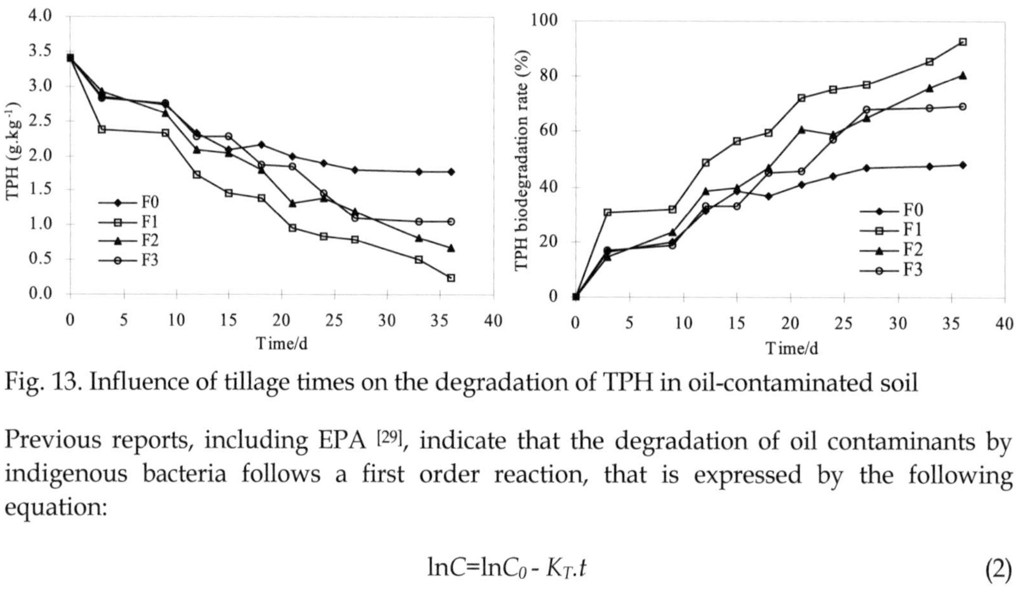

Fig. 13. Influence of tillage times on the degradation of TPH in oil-contaminated soil

Previous reports, including EPA [29], indicate that the degradation of oil contaminants by indigenous bacteria follows a first order reaction, that is expressed by the following equation:

$$\ln C = \ln C_0 - K_T . t \quad (2)$$

In equation (2), C represents concentration of TPH in the soil in mg • kg^{-1}; K_T represents substrate removal constant in d^{-1}; and t represents degrading time in days.

A plot of lnC as a function of time results in a linear regression with $\ln C_0$ as the intercept and K_T as the slope. Then, the half-life time of the petroleum contaminant in the soil can be calculated as $t_{1/2}=\ln 2/K_T$, as shown in Table 10. These data indicates that the removal rate increases with tillage times.

Sample Number	Tillage frequency	Equation	R^2	K_T (d^{-1})	Half-life (d)
F1	Tilled once every day	$c=3402e^{-0.0636t}$	0.95	0.0636	11
F2	Tilled once every three days	$c=3402e^{-0.0441t}$	0.97	0.0441	16
F3	Tilled once every five days	$c=3402^{-0.0352\,t}$	0.95	0.0352	20
F0	Stagnant	$c=3402e^{-0.0177t}$	0.90	0.0177	40

Table 10. Half-life of the Petroleum Contaminant in Soil at Different Tillage Times

4.5 Discussion

To explain how the removal efficiency of TPH is increased by the addition of bulking agents, pH, electric conductivity, and soil moisture were analyzed. The samples under evaluation had almost the same pH ranging from 6.0 to 6.7 during the evaluation period. Similarly, the electric conductivity in the samples ranged from 200 to 300 μs/cm during testing, with the exception of water content, which was different for the samples tested. Figure 14 summarizes the results.

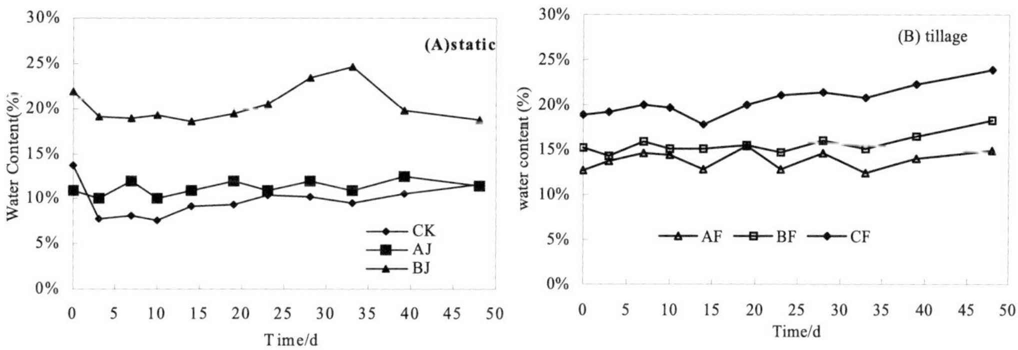

Fig. 14. Influence of bulking agents on water content of oil-contaminated soil that was (A) left static, (B) tilled once every day

Fig.14 displays the curves of water content of the oil-contaminated soil sample that was left stagnant and for the samples that were tilled once every day. The experimental observations revealed that although the dose of water was the same for all the samples, the water content in the samples was quite different. The water content of the samples containing bulking agent was kept in 20% while the content of water in the blank sample was only about 15%. The bulking agent or saw dust shows high capacity for water absorption. The water saturation rate in these samples is very fast, with saturation times of just 3 min. The water absorption capacity of saw dust is 4.45 times bigger than its own weight [30]. In the samples containing saw dust, it was observed that the saw dust absorbs the water as soon as it is dosed into the sample. However in the case of the samples free of saw dust, water added would transfuse along the holes in soil. So that the water content in the blank sample (saw dust-free) is lower than in the samples containing saw dust. Furthermore, Figure 14(B) indicates that during the evaluation period, in the tilled samples without saw dust, the water content is unstable, which can be explained by the fact that in this sample, tillage accelerates the speed of water evaporation. Saw dust retains the water in the soil, in addition to enhancing the air permeability in the contaminated soil sample which benefits the degradation of TPH by bacteria.

4.6 Conclusions

1. Adding bulking agent enhances the retention of water and the air transfer in the soil. Experimental results show that the degradation efficiency of TPH was 76% in 48 days after adding bulking agents and tillage, which was 15% higher than the degradation efficiency in the samples without bulking agent. Likewise, the degradation rate is 2.34 faster in the samples containing bulking agent that in the samples without bulking

agent. GC-MS analysis indicates that the peaks (concentration) of the pollutants in the GC spectrum decreased from 32 to 14 in 64 days after bioremediation of adding the bulking agent.

2. Tillage not only raises the content of oxygen in the soil but also accelerates the speed of transfer of substrates from outside into the biomembrane. The removal rate of TPH in samples that were tilled was 1.2 times higher than in the samples without tillage. Especially the concentration peaks of the pollutant components in the GC spectrum decreased 10 times after 64 days of degradation. Similarly, residual branched alkanes, alkene, carotene, alkylnaphthalenes, hopanes, and steranes were also thoroughly degraded after the application of the combined treatment with bulking agent and tillage. The degradation efficiency of TPH decreased as tillage times was decreased; the optimum tillage frequency is once every day with a shovel.

5. Part IV Impact of bioaugmentation dosage and temperature on bioremediation of petroleum-contaminated soil

5.1 Summary

The influence of bioaugmentation products dosage and temperature on bioremediation was studied. The results showed that the degradation rate was related positively to the amount of inoculation. When the inoculums dose was increased to 0.6 $mg.kg^{-1}$ the degradation efficiency of petroleum hydrocarbon (TPH) increased to 87% in 48 days. The results of GC-MS indicated that the petroleum constituents in the oil-contaminated soil were 82.7% *n*-alkane, 16% alkene and the balance corresponded to others hydrocarbons, such as carotane, alkylnaphthalenes, hopanes, and steranes. The concentration peaks in the GC spectrum decreased from 32 to 14 within 40 days of bioremediation. This result indicated that branched alkanes, alkene, and alkylnaphthalenes were thoroughly degraded. However, linear alkanes chains, hopanes, and steranes were still remaining in the soil. In addition, the longer *n*-alkane chains were degraded with relatively higher rate, leaving the shorter part of the n-alkane as the residual fraction at the end of the bioremediation test. Moreover, it was observed that the concentration of residual *n*-alkane (short chains) decreased as the inoculation volume increased. This indicates that an increase in the amount of bioaugmentation products into the contaminated soil clearly improves the biodegradation efficiency. It was also determined that temperature has an important effect on degradation efficiency. For instance it was observed that at 30°C, the concentration of TPH in the soil sample was reduced by 80%, whereas at a temperature of 20°C, the removal efficiency of TPH was only 60%, which indicated that higher temperatures favors TPH degradation and accelerates bioremediation.

5.2 Introduction

As the development of the petroleum industry in China increases, soil contamination by petroleum products also increases in both: areal extension and concentration. Common pollutants in the contaminated soil are hydrocarbon compounds such as: benzene, methylbenzene, and dimethylbenzene, which are highly toxic and some of them may be carcinogenic. These hydrocarbons could be difficult to degrade after entering the soil [1]. Furthermore, once these pollutants are established in the soil, they could enter into the surrounded underground water and river basins. Bioremediation of petroleum hydrocarbon contaminated soil is considered a new technology with a broad prospect because of its low-cost and pollution-free characteristics. Previous research has focused on the stimulation of the activity of the indigenous microorganisms by adding nutrient substances such as nitrogen (N) and phosphorous (P) [3]. However, bioremediation would be restricted by the slow-growth rate and limited amount of the indigenous microorganisms, which can be overcome through the addition of active microorganisms that can easily adapt to the contaminated soil environment. Several studies on this issue have been published. For example, a patented technology by the BGI (Baltic General Investment Corporation, USA) promotes the degradation rate of petroleum hydrocarbons by using mixed microorganisms [4]. Wilson (1993) [31] also reported some strains which quickly degrade PAHs. Xiao-Fang (2007) [32] found that exogenous microbes increase the degradation of petroleum contaminated soil. Another research [33] showed that by splashing bacteria into the petroleum contaminated soil, with oil content ranging from 5% to 45%, degradation rates reached up to 90% after 150 days of testing. The addition of petroleum-degrading active bacteria and microbial inoculation into petroleum contaminated soil is an in-situ bioremediation technology which has been wildly investigated. Efficient bioremediation have been obtained by the application of microbial inoculation-formulations, which have been recently developed, into actual bioremediation projects [34]. This section of the chapter summarizes the influence of bioaugmentation dosage, dosage sequence, and temperature on the bioremediation of petroleum hydrocarbon contaminated soil and provides a theoretical foundation for practical applications.

5.3 Materials and methods

5.3.1 Materials

5.3.1.1 Soil samples

The oil contaminated soil samples used in these experiments were collected from areas surrounding producing wells from an oil field located in the northern part of the Shaanxi province In China. The soil samples were free from additional dirt, pulverized, and well mixed before the experiments.

5.3.1.2 Microbial inoculations

Oil contaminated soil around oil wells was used to isolate bacteria which utilized crude oil as the only carbon source. Bacteria at hand in this oil contaminated soil samples were screened and separated and ranked as highly active petroleum-degrading bacteria [14]. The strain SY_{23}, which is commonly reported in the literature [14] was separated under aseptic conditions that was then dried and sterilized at 37°C for 48 hours. At this stage, the bioaugmentation product was ready for use.

5.3.2 Experimental methods

5.3.2.1 Bioremediation methods

In order to prepare a bacterial-rich suspension, certain amount of the solid bacterial product was added into a bacteria-free buffer. A substrate medium was added into the bulking agent to prepare a mixture. Later, the mixture was sterilized and the bacterial-rich suspension was mixed with the bulking agent mixture, which was cultured for 24 hours. The final product was ready for addition into the petroleum contaminated soil.

5.3.2.2 Analytical methods

The composition of the petroleum hydrocarbon (TPH) was determined by OCMA-350 non-dispersive infrared oil analyzer. The concentration of the petroleum components was determined by GC-MS. A CG 7890, Model Trace 2000, was manufactured by Agilent was used and the mass spectrometer 5975, model Voyager, manufactured by Agilent was employed for the analyses. The GC was equipped with a 30m long DB-5 capillary, with a stationary phase of 0.25μm. The range of relative molecular mass was from 30 to 450, and the first 6.10 min corresponded to the solvent peak. The conditions of the GC analysis were: temperature was increased at a rate of 10°C/min from 100°C to 200°C, and then the temperature was increased at rate of 5°C/min up to 280°C. The contact temperature between the Ms and GC was 250oC. Conductivity was determined using a conductivity meter and pH was determined using pH test paper. Finally, moisture content was determined by the constant weight method.

5.3.3 Design of experiments

An experimental matrix was designed to evaluate the effect of different inoculation schemes and dosage of bioaugmentation products on the bioremediation of petroleum hydrocarbon contaminated soil. The experiments were carried out in basin ports having a diameter of 15cm and depth of 15cm. In every group, 1000g of soil was used. Table.11 summarizes the experimental observations. Each group was conducted by duplicate. Every group was turned over once a day and 20mL of water were added every day [35]. During this set of experiments, the moisture content in the environment was maintained at 18% using a humidifier. Samples were taken at constant time intervals for pH, conductivity, petroleum concentration, and moisture content analyzes. The sampling intervals were increased when the petroleum-degradation rate started to decrease.

Number	Bioaugmentation products dose ($mg.kg^{-1}$)	Temperature (°C)	Moisture (%)	Bioaugmentation products inoculations	Treatment measures
Experiments of bioaugmentation products dosage	0(CK)	20	18	No addition	Turn over
	0.01	20	18	Addition one time	Turn over
	0.2	20	18		Turn over
	0.4	20	18		Turn over
	0.6	20	18		Turn over
Experiments of bioaugmentation products inoculations	0.2	20	18	Stepwise addition (two times)	Turn over
	0.4	20	18		Turn over
	0.6	20	18		Turn over
Experiments of temperature	0.2	20	18	Addition one time	Turn over
	0.2	37	18		Turn over

Table 11. Experimental Observations

5.4 Results and discussions

5.4.1 Influence of bioaugmentation products dosage on the degradation of petroleum hydrocarbons in soil

Figure 15 and Figure 16 show the residual TPH concentration and the TPH degradation rate in soil as a function of bacteria dosage and time. The initial TPH concentration in soil was 4.3 g•kg^{-1}. The dosage of bioaugmentation products was 0.01, 0.2, 0.4, and 0.6mg•kg^{-1} respectively. Figure 16 indicates that the TPH degradation rate after 8 days of biodegradation for the corresponding bacteria dosage conditions were 2%, 10%, 25%, and 40% respectively. While, the degradation rate of the blank sample without bacteria (CK) was only 3%. The TPH degradation efficiency during 48 days of testing was 47%, 68%, 79%, and 87%; while the degradation efficiency in the blank sample (CK) did not achieve even 20%.

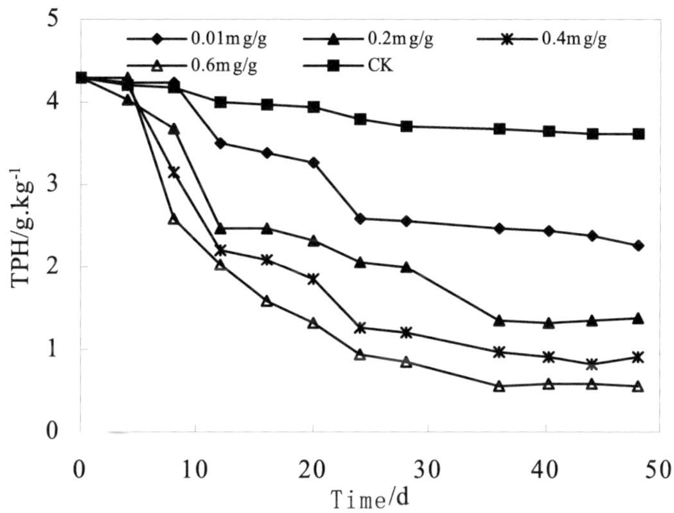

Fig. 15. Residual TPH concentration in soil as a function of bacteria dosage and time.

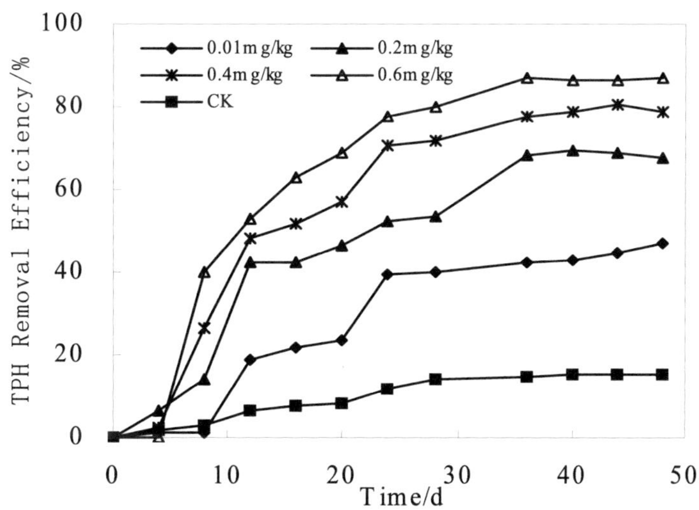

Fig. 16. TPH degradation rate as a function of bacteria dosage and time.

Previous research [36] showed that the key problem of bioremediation by adding exogenous bacteria is how to solve the competition between exogenous and indigenous bacteria. Experimental observations show that exogenous bacteria survive well with the indigenous flora. As demonstrated through this experimental study, the TPH degradation rate in soil is positively related to the dosage of bioaugmentation products. The highest degradation rate was achieved when the bioaugmentation products dosage was 0.6 mg•kg^{-1}. Furthermore, at low dosage (0.01 mg•kg^{-1}) of bioaugmentation products did rise significantly the degradation rate after 20 days of testing; due to exogenous bacterial growth.

Previous reports [37] indicate that the degradation of petroleum contaminated soil by bacteria can be described by a first order reaction: $\ln C = \ln C_0 - K_T.t$

Where : C is petroleum content in soil mg•kg^{-1}; K_T is substrate removal constant d^{-1}

A plot of lnC as a function of time renders a straight line with intercept $\ln C_0$ and slope K_T. The dynamic equations and the half-life of petroleum contaminants in soil were calculated using the following expression: $t_{1/2}=\ln 2/K_T$. Table 12 shows the half-life of the petroleum contaminants in soil as a function of bacteria dosage.Table 12 also shows that the TPH remediation cycle in soil is 300 days shorter than in the blank or control sample (CK). The substrate removal constant K_T indicates that biodegradation increases as the dosage of bacteria increases while the half-life of the petroleum contaminants is reduced. The optimum dosage was 0.6 mg•kg^{-1}.

Number	Dose (mg•kg^{-1})	Dynamic equations	R^2	K_T	Half-life (d)
1	CK	$c=4246e^{-0.00225t}$	0.96	0.00225	308
2	0.01	$c=4023e^{-0.0302t}$	0.96	0.0302	23
3	0.2	$c=4318e^{-0.0373t}$	0.96	0.0373	19
4	0.4	$c=4203^{-0.0454\,t}$	0.96	0.0454	15
5	0.6	$c=4299e^{-0.0604t}$	0.97	0.0604	11

Table 12. Half-life of the Petroleum Contaminants in Soil at Different Dosage Conditions

To investigate the effect of microbial inoculums on the degradation of hydrocarbons, samples were subjected to extraction after 48 days of degradation. The composition of the extract liquors were analyzed through GC-MS. Fig.17 shows the GC-MS spectra. The GC-MS spectra confirm that most of the petroleum constituents in soil were degraded by the microbial inoculums. When the dose was increased to 0.6 mg•kg^{-1}, the peak lines were almost parallel with the base line. Fig.17 clearly shows that increasing bacteria dose improves TPH degradation and accelerates bioremediation.

GC-MS indicated that the dominant petroleum constituents in the oil-contaminated soil were 82.7% n-alkane, 16% alkene, and the balance corresponded to other hydrocarbons, such as carotane, alkylnaphthalenes, hopanes, and steranes. Monocyclic benzene series were not found in the hydrocarbon analyzed, which is in agreement with Xiao-Fang's study (2007)[32]. Table 13 presents the (TPH) degradation rate as a function of bacteria dosage for a period of 48 days. Specific charge (m/z) measurement was used to determine chemical structure. The specific charge (m/z) for $[CH_3\text{-}CH_2\text{-}CH_2\text{-}CH_2]^-$ is 57, for $CH_3{-}CH_2{-}C(CH_3)_2{-}CH_2{-}CH_2{-}$ is 97, and for $[CH_3\text{-}CH{=}CH\text{-}CH_2]^-$ is 55. So that, Table 13 indicates that branched alkanes and alkenes are easier to degrade than linear alkanes. The content of linear alkanes decreases as bacteria dosage increases. The degradation rate of linear alkanes was increased to 97.4% when the dose of bacteria was increased to 0.6mg•kg^{-1}. The residual components shown in Table 13 were mainly linear alkanes.

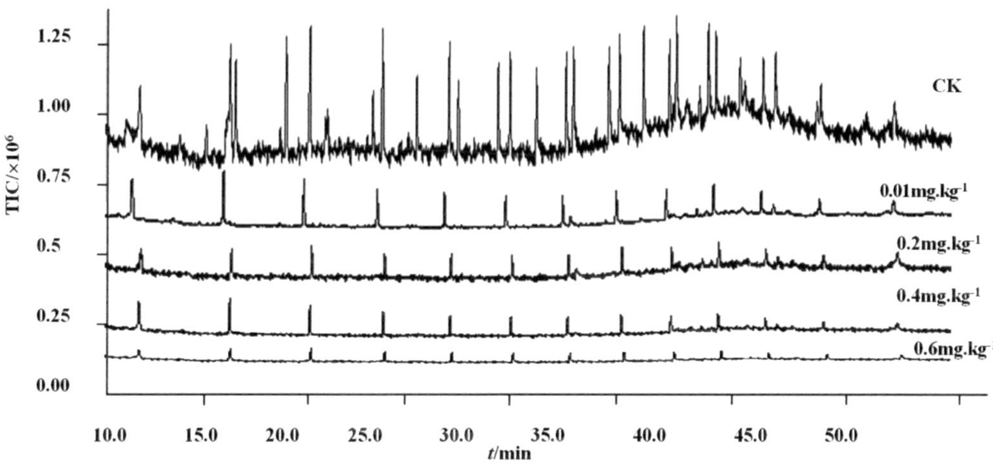

Fig. 17. TIC of remaining TPH concentration in soil at different dose condition

Bacteria dose (mg•kg^{-1})	Degradation rate of different HC in 48 days (%)					
	Alkanes				Alkenes (m/z =55)	
	Linear alkanes (m/z =57)		Branched alkanes (m/z =97)			
	A×10^3	η(%)	A/×10^3	η(%)	A(×10^3)	η(%)
CK	2303.0		142.0		410.0	
0.01	278.0	87.9	0.0	100.0	6.0	98.5
0.2	261.0	88.7	0.0	100.0	5.0	98.7
0.4	81.0	96.5	0.0	100.0	0.0	100.0
0.6	61.0	97.4	0.0	100.0	0.0	100.0

Table 13. Peak Area and Removal Rate of Alkane & Alkene in Soil at Different Dosage of Bacteria

The hydrocarbon degradation rate was monitored when bacteria dosage was 0.2mg•kg^{-1}. Samples were taken for GC-MS analysis at the following degradation times: 0d, 4d, 8d, 40d, and 48d respectively. Fig.18 and Table 14 show the experimental observations. Table 14 indicates that the peak values on the GS-MS spectrum are reduced after 4 days of bioremediation, which indicates that hydrocarbon biodegradation is taken place GC-MS analysis shows that 98.5% of alkenes were degraded, followed by 87.9% degradation of linear alkanes and 41 % degradation of branched alkanes. The degradation rate of branched alkanes, carotene, and alkylnaphthalenes was increased up to 74.9%, 91.0%, and 100% respectively in 8 days of testing. At the same time the peak of the phytan (Ph, 2, 6, 10, 14-tetramethylhexadecane) in the GC-MS spectrum, (shown by the arrows in Figure 18) almost disappeared. The Ph and Pristane (Pr, 2, 6, 10, 14-Tetramethylpentadecane) were not detected after 48 days of bioremediation. Thus, the presence of these components in the contaminated soil sample is taken as an indication of the biodegradation of the petroleum hydrocarbon in the contaminated soil. In Fig. 18, the peak of Ph could be seen in the CK line after 48 days of biodegradation. However, the Ph peak disappeared after inoculating bioaugmentation products. This is strong evidence that bioaugmentation products could degrade petroleum. The concentration peaks from the GC-MS spectra decreased from 32 to 14 after 40 days of bioremediation. This result points out that branched alkanes, alkene, and alkylnaphthalenes were thoroughly degraded. However, linear alkanes, hopanes, and steranes were left in the soil as pollutants. The residual part of the n-alkane corresponds to the short chains because bacteria degraded the longer n-alkane chains, which is in agreement with previous studies[37], which have reported that the degradation rate of n-alkanes is reduced as the chain length increases. Previous work [37, 38] claims that marine filamentous fungi have higher degradation capacity for eicosane, tetracosane, and hexacosane than bacteria. So, finding bacteria capable of degrading long chains of n-alkane is a key issue in bioremediation of petroleum hydrocarbon pollutants. The strain SY_{23} in this experiment is a plesiomonas. Hydrocarbon biodegradation tests have shown that SY_{23} bacteria have high capacity for degrading benzene, methylbenzene, phenol, and long chains of n-alkanes. During these experiments it was observed that the color of the raw contaminated soil changes from brown to yellow after the addition of bacteria; which could also indicate that biodegradation of petroleum hydrocarbon is taken place. Bioremediation rate could be increased if inoculation volume is increased. In addition, it was observed in

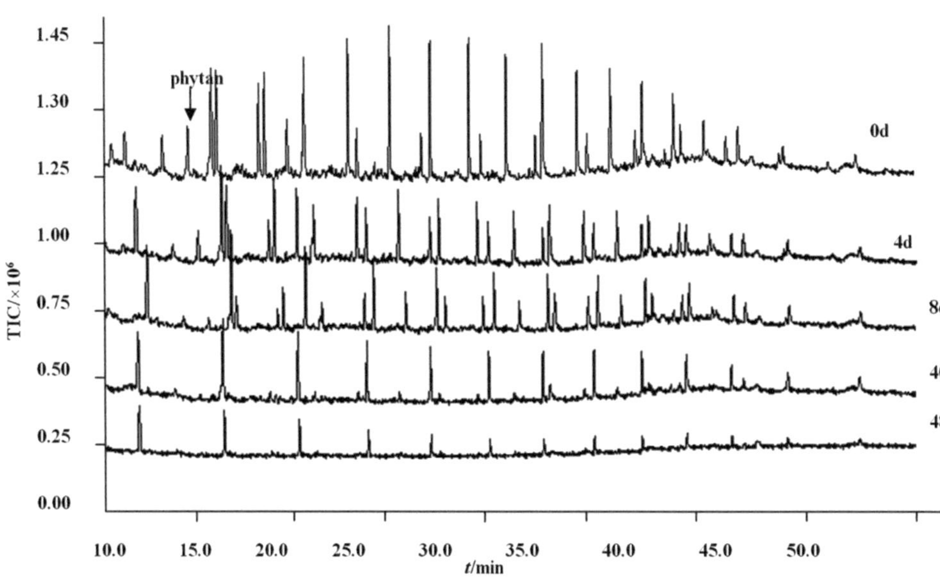

Fig. 18. TIC of TPH concentration in soil during bioremediation

this study that hopanes were not degraded during the first 8 days of testing, however it reaches 53% degradation after 40 days. This shows that hopanes are stabilized in the contaminated soil. In general, hopanes are hardly degraded if their concentration in the contaminated soil is too high. In this work, the concentration of hopanes was low (58mg/kg) and 60% of this concentration in the soil was removed.

Testing Time	Peak area $A\times10$ and removal rate η(%)of different kind of alkenes													
	Alkanes													
	Linear alkanes (m/z =57)		Branched alkanes (m/z =97)		Aalkenes (m/z =55)		Carotane (m/z =125)		Alkylnaph thalenes (m/z =128)		Hopanes (m/z =191)		Steranes (m/z=217)	
	A	η	A	η	A	η	A	η	A	η	A	η	A	η
0d	9255.0		981.0		1998.0		119		26.0		84.0		11.2	
4d	6524.0	87.9	577.0	41.0	6.0	98.5	113	5.0	4.0	86.4	82.0	2.3	9.53	14.9
8d	3489.0	88.7	246	74.9	5.0	98.7	10.7	91.0	0.0	100.0	82.0	2.3	5.7	49.1
40d	981.0	96.5	0.0	100.0	0.0	100.0	0.0	100.0	0.0	100.0	39.0	53.0	3.0	73.2
44d	242.0	97.4	0.0	100.0	0.0	100.0	0.0	100.0	0.0	100.0	0.0	100.0	0.0	100.0

Table 14. Analysis of Petroleum Hydrocarbon after Inoculation of Microorganisms

5.4.2 The Influence of bioaugmentation product injection scheme on the degradation of petroleum hydrocarbon

Samples were divided into two groups A and B. In group A, the bioaugmentation products were injected all at once in one application. In group B, half of the bioaugmentation products were injected at the beginning of the test, and the other half was injected after 20 days of testing. The total bioaugmentation products in the two groups were the same. Three series of experiments were conducted for each group having a different concentration of the bioaugmentation products in each series as follows: 0.2 mg•kg^{-1},0.4 mg•kg^{-1}, and 0.6 mg•kg^{-1} respectively. Figure 19 summarizes the GC-MS results after 48 days of degradation. Figure 19 indicates that the value of concentration peak in the GC-MS spectra for group A was lower than for group B. The CG-MS results show that the main residual components were linear alkanes and hopanes. The peak area of the linear alkanes in group A was 170000, while the peak area of the linear alkanes in group B was 261000 when the bioaugmentation products dose was 0.2 mg•kg^{-1}. The peak area of the linear alkanes in group A was 81000 while the peak area of the linear alkanes in group B was 175000 when the bioaugmentation products dose was 0.4 mg•kg^{-1}, and finally the peak area of the linear alkanes in group A was 61000 while the peak area of the linear alkanes in group B was 142000 for a dose of 0.6mg•kg^{-1}. These observations indicate that the best sequence of bioaugmentation products application is to inject all products at once, which is in agreement with previous studies [39]. Thus, the addition of the total amount of bioaugmentation products aids the quick adaptation of the bacteria to the environment, that results in petroleum components removal efficiency. Injecting the bioaugmentation products in only one application degraded about 50% of TPH in the soil after 20 days of testing; while only 30% of TPH were degraded in group B. Degradation rates were 68%, 79%, and 87% when the concentration of the bioaugmentation products were 0.2 mg•kg^{-1}, 0.4 mg•kg^{-1}, and 0.6 mg•kg^{-1} respectively. In Group B, the late injection of the remaining half of the bioaugmentation products after 20 days of testing did not increase the degradation of TPH. Finally, in group A the degradation rates after 48 days of testing were 54%, 63% and 74%.

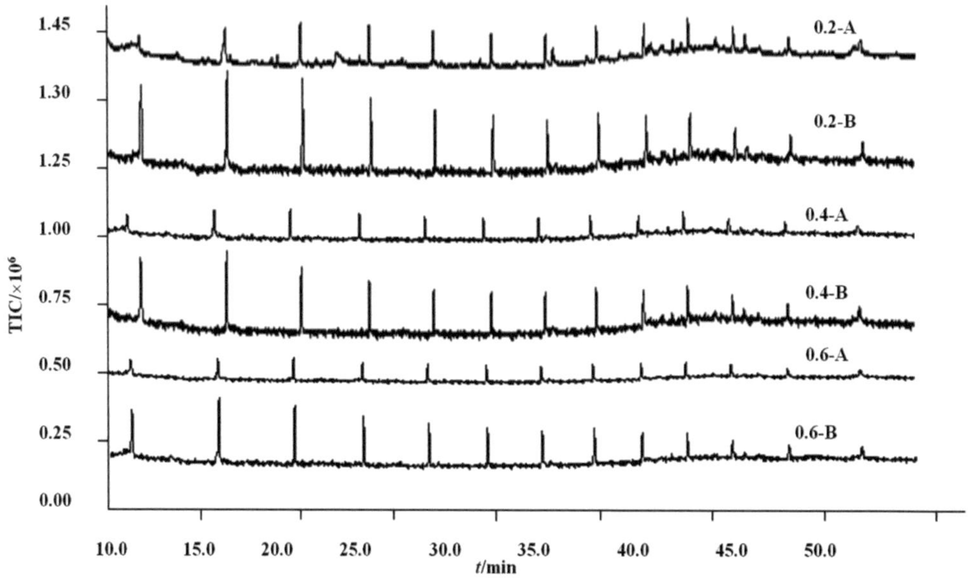

Fig. 19. TIC TPH concentration in the soil at different injection schemes of the bioaugmentation products

5.4.3 Influence of temperature on the biodegradation of petroleum hydrocarbon

Several studies [30] indicate that temperature is an important variable that determines the efficient degradation of petroleum hydrocarbon in contaminated soil. Two groups of experiments were carried out at two different temperature conditions. In the first group, the

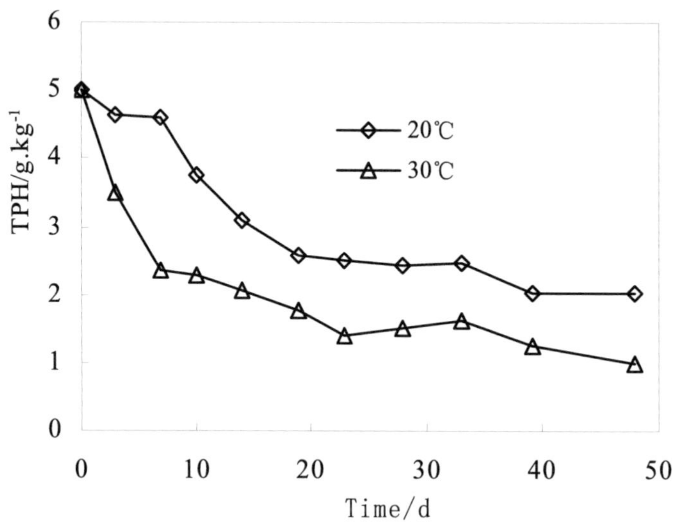

Fig. 20. TPH concentration in the soil as a function of temperature and degradation time

average temperature was 20°C while in the second group of experiments the average temperature was 30°C. The experiments were conducted between May 1st to June 27th in 2006. Figure 20 presents the residual TPH concentration in the soil as a function of temperature and testing time. These experiments indicatd clearly the effect of temperature on TPH degradation. At 30°C, the degradation of TPH reached 29.8% after 3days of treatment, whereas at 20°C, the degradation of TPH reached only 7%. This indicates, that at higher temperatures, bacteria grow faster, which accelerates bioremediation. For the experiments conducted at 20°C, the degradation rate of TPH increased rapidly after 10 days, which points out that the adaptation of bacteria in the low temperature environment took longer, which resulted in a total TPH degradation of only 60% at the end of the testing period. This overall degradation percentage (at 20°C) was 20% lower than the total degradation of TPH reached in the experimental group conducted at 30°C.

5.5 Conclusions

1. Exogenous bacteria survive quite well in the presence of indigenous flora. The degradation rate of TPH contained in the soil was related to the dosage of bioaugmentation products. In this work, the highest degradation rate was achieved when the dose of bioaugmentation products was 0.6 mg•kg^{-1}.
2. GC-MS analysis indicates that the dominant petroleum constituents in the oil-contaminated soil were 82.7% n-alkane, 16% alkene, and the balance corresponded to of others hydrocarbons, such as carotane, alkylnaphthalenes, hopanes, and steranes. The height of the GC spectrum peaks decreased from 32 to 14 after 40 days of bioremediation; this indicates that branched alkanes, alkene, and alkylnaphthalenes were thoroughly degraded. However, linear alkanes, hopanes and steranes were left in the soil. Furthermore, long chains of n-alkane were degraded, while short chains of n-alkane were left as residual hydrocarbon in the soil. It was determined that increasing inoculation increases the capacity of the bacteria to attack short chains n-alkanes. The application of bioaugmentation products into soil in one dosage significantly improved the biodegradation efficiency rather than the application of bioaugmentation products in several dosages.
3. Temperature is the dominant variable that determines the efficiency of bioremediation of petroleum hydrocarbon contaminated soil. At higher temperatures, bacteria grow and metabolize faster. At 20°C, the degradation of TPH increased rapidly after 10 days of testing, which indicates that the bacteria slowly adapted to the low temperature environment. The total degradation percentage of TPH rate was 60% at 20°C, which was 20% lower than the degradation percentage of the experimental group tested at a higher temperature (30°C).

6. References

[1] Huang T L, Ren L. Simulating and modeling of the runoff pollution of petroleum pollutants in loess plateau [J] China Environmental Science, 2000, 20(4): 345-348 (in Chinese).

[2] Ren L, Huang T L. Study on petroleum runoff pollution in loess plateau [J].China Water & Wastewater, 2000, 16(11): 1-5(in Chinese).

[3] He L J, Wei D Zh, Zhang W Q. Research of microbial treatment of petroleum contaminated soil [J]. Advance in Environmental Science, 1999, 7(3): 110-115 (in Chinese).

[4] Baltic General Investment Corporation, 1996. Method of decontamination of a hydrocarbon-polluted environment by the use of bacterial composition [P]. US patent 5494580.

[5] Wilson S C, Jones K C. Bioremediation of soil contaminated with polycyclic aromatic hydrocarbons: a review [J]. Environmental pollution, 1993, 81(3): 229-249.

[6] National Standard of GB/T 16488-1996, 1996. Method of non-disperse infrared luminosity [S]. State Environmental Protection Ministry (in Chinese).

[7] Yu Sh X, Wu G Q, Meng X T. Manual of microbial test of environmental engineering [M]. Bei jing: China environmental science press, 1990: 216-218.

[8] Truax, D. D., R. Britto, et al. Bench-scale studies of reactor-based treatment of fuel-contaminated soils [J].Waste Management, 1995, 15(5): 351-357.

[9] Ronald M. Atlas. Petroleum biodegradation and oil spill bioremediation [J].Oceanographic Literature Review, 1996, 43(6): 628.

[10] Calabrese J, Kosteck P T. Hydrocarbon contaminated soil [M]. New York: Lewis Publishers, 1991: 411-421.

[11] Lin F G, Yu Zh G, Li H, et al. Study on degradation of crude oil by marine filamentous fungi [J]. ACTA oceanolocica sinica. 1997, 19(6): 68-76 (in Chinese).

[12] Muraue. Technology and application of extracting oil: state of status quo and perspectives[J].Engineering, 1999, 97:128-131(in Japanese).

[13] Leahy J G, Colwell R R. Microbiol degradation of hydrocarbons in the environment[J]. Microbiol Rev, 1990, 54(3): 305-315.

[14] Xu Jinlan, Huang Tinglin, Tang Zhixin, et al. Isolation of petroleum degradation bacteria and its application to bioremediation of petropleum-contaminated soil [J]. Actz Scientiae Circumstantiae, 2007, 27(4): 622-628 (in Chinese).

[15] Chen Bie, Liu Zutong. Biostransformation of petroleum hydrocarbons by marine filementous fungi [J].Acta Petrolei Sinica (Petroleum Process Section), 2002, 18(3): 13-17 (in Chinese).

[16] Wang Liansheng, Zhong Huijuan. Study advance of degradation of crude oil by marine microorganisim [J]. Acta Oceanolocica Scientiae Circumstantiae, 1990, 9(1): 52-59 (in Chinese).

[17] Zhao Tianbo, Tang Ruikun, Ji Dekun. The determination of n, i-paraffins and their carbon number distribution in paraffin wax and microcrystal wax with on column gas chromatography[J].Petrochemical Technology, 1996, 25(9): 646-650 (in Chinese).

[18] Wen Hongyu, Liao Yinzhang, Li Xudong. Degradation characteristics of naphthalene by strain N-1 [J].Chin J Appl Environ Biol, 2006, 12(1): 96-98 (in Chinese).

[19] Shen Yiyong, Zhang Yinan, Liu Zufa, et al. Preliminary study on biodegradation of MGP-wastewater influenced by adding glucose[J]. Acta Scientiarum Naturalium Universitatis Sunyatseni, 2005, 44(6): 114-117 (in Chinese).

[20] Kishore D, Ashisk M. Crude petroleum-oil biodegradation efficiency of Bacillus subtilis and Pseudomonas aeruginosa strains isolated from a petroleum-oil contaminated soil from North-East India [J].Bioresource Technology, 2007, 98(7): 1339-1345.

[21] Rhykerd R L, Crews B, MeInnes K J, et al.Impact of bulking agents, forced aeration, and tillage on remediation of oil-contaminated soil[J].Bioresoure Technology, 1999, 67: 279-283.

[22] Hillel D. Soil structure and aggregation: Introduction to soil physics[M]. London: Academic press, 1980, 40-52.

[23] Melope M B, Grieve I C, Page E R.Contributions by fungi and bacteria to aggregate stability of cultivated soils[J].Soil Sci, 1987, 38: 71-77.

[24] Morgan P, Lee S A, et al. Growth and biodegradation by white-rot fungi in soil[J]. Soil Biol Biochem, 1993, 25: 279-287.

[25] YE Xiao-mei, CHANG Zhi-zhou, ZHU Wan-bao, et al. Bulking agents on the rate of crude oil degradation in sludge[J]. Environment Herald, 1999, 2: 21-22.

[26] Freijer J L. Mineralization of hydrocarbons in soil under decreaseing oxygen availability[J]. Environ Qual, 1996, 25: 296-304.

[27] Nocentini M, Pinelli D, Fava F. Bioremediation of a soil contaminated by hydrocarbon mixtures:the residual concentration problem[J]. Chemosphere, 2000, 41: 1115-1123.

[28] YANG Li-xia. Analysis on the water containing capacity of the water contention agent and the addition agent[J]. SCI/TECH Information Development & Economy, 2006, 16(1): 162-163.

[29] Liang Shengkang, Wang Xiulin, Wang Weidong, et al. Screening highly efficient petroleum-degrading bacteria and their application in advanced treatment of oil field wastewater [J]. Environmental protection of chemical industry, 2004, 24(1): 41-45 (in Chinese).

[30] Leahy J G, R R Colwell. Microbiol degradation of hydrocarbons in the environment[J]. Microbiol Rev, 1990, 54(3). 305-315.

[31] Wilson SC, Jones KC. Bioremediation of soil contaminated with polycyclic aromatic hydrocarbons: a review [J]. Environmental pollution, 1993, 81(3): 229-249.

[32] Xiao-fang Wei, Zhong-zhi Zhang, Shao-hui Guo. Ex Site-Bioremediation of Petroleum-Contaminated Soil With Exogenous Microbe [J]. Journal of Petrochemical Universities, 2007, 20(2): 1-4.

[33] Xiao-fang Hu, Fu-jun Xia, Nan-wen Zhu. Study on bioremediation of Oil-Contaminated Soil [J]. Environmental Chmistry, 2006, 25(6): 593-597.

[34] Mark A.S, James S.B, Chery A.P, et al. Evaluation of two commercial bioaugmentation products for enhanced removal of petroleum from a wetland [J].Ecological Engineering, 2004, 22: 263-27.

[35] Zhixin Tang. The Study of Bioremediation of Petroleum Polluted Soil in Northwest Loess Area [D]. Xi'an: Xi an University of Architecture and Technology, 2007, 19-33.

[36] Bouchez T, Patureau D, Dabert P, et al. Ecological study of a bioaugmentation failure [J]. Environmental Microbiology, 2000, 2:179-190.

[37] Pai A, Willumense, Erik A. Kinetics of degradation of surfactant-solubilized fluoranthene by a Sphingomonas paucimobilis [J]. Environmental Science & Technology, 1999, 33: 2571-2578.

[38] Fengxiang Lin, Zhanguo Yu, Hong Li. The Research of Marine Filamentous Fungi Degrading Oil [J]. Acta Oceanologica Sinica, 1997, 10(5): 68-76.
[39] Yi Zhong, Guang-he Li, Xu Zhang. A study of the bioremediation effects on the petroleum contaminated soil [J]. Earth Science Frontiers, 2006, 13(1): 128-133.

9

Microbial Outlook for the Bioremediation of Crude Oil Contaminated Environments

Rachel Passos Rezende, Bianca Mendes Maciel,
João Carlos Teixeira Dias and Felipe Oliveira Souza
Universidade Estadual de Santa Cruz
Brazil

1. Introduction

The oil and petrochemical industry in the course of the exploration, production, refining and transportation operations required for the commercialization of oil and its derivatives have contaminated with toxic wastes the environment, which requires urgent remediation. The biological treatment of these wastes aims the rational exploitation of microorganisms that use hydrocarbons as the source of carbon and energy. This biotechnological strategy includes one of the processes most commonly used by industries today: *bioremediation*. Bioremediation makes possible the recovery of oil contaminated sites through biostimulation, which increases the hydrocarbon-degrading ability of native microorganisms, thereby transforming or mineralizing the pollutants.

The study of microbes in bioremediation systems makes possible the selection of microorganisms with potential for the degradation and production of compounds with biotechnological applications in the oil and petrochemical industry. These applications include the use of biosurfactants for advanced oil recovery, cleaning of tanks, and the dispersion of oil spills on water environments. The current concern for the environment has stimulated research on biosurfactants, which are considered to be more acceptable because of their higher biodegradability and lack of toxicity compared with chemical synthesized surfactants.

The study of native microbial diversity is fundamental to understand the roles played by microorganisms in contaminated sites and for determining their interactions with the environment. Such information can generate economic and strategic benefits in several areas.

2. Oil bioremediation

Petroleum is a complex mixture of aliphatic, alicyclic, aromatic hydrocarbons, and smaller proportions of heteroatom compounds, such as sulfur, nitrogen, and oxygen. Crude oil also contains organometallic complexes containing nickel and vanadium in much smaller proportions compared to the other constituents; however these organometallic compounds are problematic during crude oil refining (Head et al., 2003). The susceptibility of these compounds to biodegradation varies with the molecular concentration (molecular weight), composition, and electronic stability (Chosson et al., 1991). For instance, it is difficult for microorganisms to use long-chain alkanes with several branches as a carbon source because of the increased hydrophobicity of the molecule, which makes the molecule more resistant to oxidization during microbial assimilation. Thus, in these situations the biodegradation rate is reduced. Another relevant factor is the bioavailability of nutrients, particularly of nitrogen, phosphorus, and trace elements. The lack of such nutrients in ancient oil reservoirs explains their continued existence, without any significant degradation.

Different microorganism species have different enzymatic abilities and preferences for the degradation of oil compounds. Some microorganisms degrade linear, branched, or cyclic alkanes. Others prefer mono- or polynuclear aromatics, and others jointly degrade both alkanes and aromatics.

Although the constituent atoms in oil molecules generally provide nutrients for microorganisms, these same molecules can be bioinhibitors because of their intrinsic toxic properties. The lighter alkanes, for example, those in the C_5 to C_9 range, are toxic to many organisms because of their solvent effects, which tend to break the structure of microbial lipid membranes. Aromatic hydrocarbons, particularly molecules with more than five condensed aromatic rings called polycyclic aromatic hydrocarbons (PAHs), are especially harmful to microorganisms. These molecules are lipophilic and interact with the cell membrane.

Several methods can be employed to remove oil wastes and derivatives from soil and water. These include physical (spray, vapor extraction, stabilization, solidification), chemical (photo-oxidation, dissolution, detergent use), and biological methods (bioremediation). All these methods can be used in the treatment of contaminated sites depending on the priorities and circumstances of each case. Physical treatments have the advantage of separating contaminants without destroying or chemically modifying them but also have limitations. Most of these techniques are too costly to be implemented on a large scale and require continuous monitoring and control to achieve optimal performance. In addition, they do not usually result in complete contaminant destruction, removal, or degradation. Biological processes, in contrast, have the promise of complete contaminant removal and are usually simpler and less expensive than other alternatives.

Bioremediation is an alternative process that involves accelerated natural biodegradation. Bioremediation techniques can be performed either *in-situ* or *ex-situ*. In *in-situ* processes, the biological remediation is conducted at the contaminated site, whereas in *ex-situ* processes, the contaminated medium is extracted and processed at off-site purification facilities. Usually, bioremediation techniques involve biostimulation, i.e., the addition of nutrients (mainly nitrogen, phosphorus, and potassium), and the adjustment of environmental conditions (pH, temperature, moisture, aeration) to facilitate native microbial growth. Microorganisms with the ability to quickly degrade specific contaminants can also be added in a technique known as bioaugmentation. Thus, bioremediation can be defined as a process that exploits microbial diversity and metabolic versatility to transform chemical contaminants into less-toxic products. This alternative offers significant advantages over other processes because the costs are reduced and local pollution is minimal.

Among the various treatments methods for the removal of oily wastes, *landfarming* is the most widely used, as this method makes possible to transform pollutants into products that are less toxic to humans and the environment. Landfarming consists on the *ex-situ* treatment of the oily sludge waste by means of biodegradation using native microorganisms. Initially, the soil is distributed over a large flat surface area above an impervious layer. The oily sludge landfarm is divided into cells containing soil, usually at different treatment stages, in which the oily wastes are placed for biodegradation (Figure 1).

In oily sludge landfarm systems, the soil is periodically aerated by plowing and enriched with nutrients to stimulate native aerobic microbial metabolism, thereby favoring biodegradation (Ausma et al. 2003). This soil is treated until the contaminant concentrations are below or at acceptable limits as established by environmental control agencies.

Fig. 1. Schematic representation of a *landfarm.* **(A)** Landfarm cell during the plowing process that promotes aeration and biostimulation by nutrient addition. (B) Representation of a landfarm cell at equilibrium. **(C)** Representation of a landfarm cell newly inoculated with oily sludge during the aerobic metabolism promoting plowing process.

Biopiling is another technique that is increasingly applied to biological waste treatment by the petrochemical industry (Figure 2). Biopiles involve the transfer of contaminated soil into cells suspended above the ground, where organic contaminants are degraded by aerobic microbial activity. This technique is different from landfarming because it allows control over the levels of moisture, pH, temperature, oxygen, nitrogen, and phosphorus. Thus, the nutritional and physical-chemical soil conditions ideal for bioremediation are maintained. In biopiles, the mass transfer efficiency, i.e., the transfer efficiency of air, water, and nutrients, makes it possible to obtain better results in contaminant reduction. Structural materials that can be used to increase mass transfer in biopiles include sand, straw, wood chips, sawdust, and dry manure (Seabra 2005). To improve aeration in biopiles, air pipes are used for artificial ventilation or soil can be raked periodically. Aeration determines treatment success because aerobic degradation processes are more efficient than anaerobic processes. Biopiles are very efficient when applied to sandy soils if compared to clay soils because clay soils can form agglomerates that reduce soil permeability, thereby making the mass transfer difficult (Seabra, 2005).

As in landfarms, the microbial metabolism in biopiles is stimulated by the addition of nutrients whose concentrations must be monitored, as nutritional excess can inhibit microbial activity (Khan et al., 2004). Metabolic activity is constantly monitored by measuring the reduction of contaminants via chromatography. Biopiles are simple to implement and advantageous because they require smaller areas and relatively short treatment compared with other techniques such as landfarms.

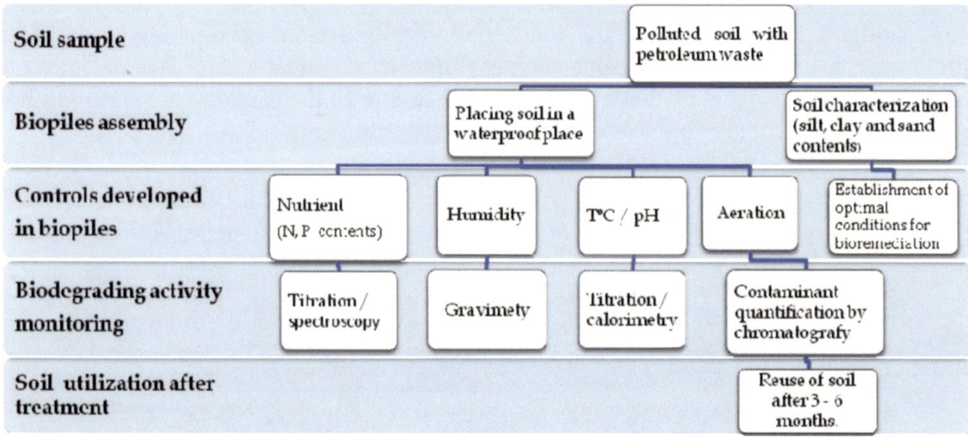

Fig. 2. Schematic representation of the structure and application of the biopile technique (Silva, 2004, adapted).

3. Biosurfactants

Compounds with surface-active properties synthesized by microorganisms are called biosurfactants and consist of metabolic byproducts from a wide variety of bacteria, fungi, and yeasts. Biosurfactants are amphipathic molecules composed of hydrophobic and hydrophilic groups that are able to spontaneously associate between liquid interfaces (oil/water and water/oil) with different degrees of polarity. As a result, the surface and interfacial tensions are reduced, and large molecular aggregates, called micelles, are formed. These properties give biosurfactants the following attributes: detergency, emulsification, lubrication, foaming ability, solubilization, and phase dispersion.

Micelle formation is associated with the critical micellar concentration (CMC) of the surfactant. Below its CMC, the surfactant is predominantly in monomer form, but when the concentration approaches the CMC, a dynamic equilibrium leads to micelle formation, with an individual micelle typically representing an association of 30 to 200 monomers (Maniasso, 2001). The CMC depends on the surfactant structure and is commonly measured to assess the surface-active efficiency in oil solubilization between the aqueous phases (Desai & Banat, 1997).

The activity of biosurfactants is commonly quantified by measuring the change in surface and interfacial tensions and by calculating the CMC. The surface tension of distilled water is 72 mN/m and the addition of a surfactant reduces its surface tension to 30 mN/m. Most biosurfactants render a surface tension in aqueous systems of approximately 30 mN/m and an interfacial tension in oil-water systems of approximately 1 mN/m. The CMC of biosurfactants is estimated by continuously increasing the concentration of biosurfactants in aqueous solutions and measuring the surface tension of the aqueous system after each addition of biosurfactant using a tensiometer. The CMC is reached when the surface tension value levels off. Thus, the CMC for most biosurfactants ranges from 1-2,000 mg/L (Desai & Banat, 1997; Nitschke & Pastore, 2002).

Because biosurfactants possess significant advantages over chemically synthetized surfactants, they have been widely studied and used in various industries, such as the petrochemical industry, in the food industry, as additives in the manufacture of cosmetics, for biological control, in medicinal therapy, in bioremediation, and several other applications (Banat et al., 2000; Desai & Banat, 1997; Fiechter, 1992; Karanth et al., 1996; Makkar & Cameotra, 2002; Nitschke & Pastore, 2002; Ron & Rosenberg, 2001). The advantages of biosurfactants include low toxicity, high biodegradability, tolerance to temperature, pH, and salinity; high foaming activity, excellent surface and interfacial activity (they decrease surface tension at low concentrations), production from renewable substrates; and their potential to be structurally modified by genetic engineering or biochemical techniques. Furthermore, there is a large and chemically diverse group of biosurfactants with distinct physical properties. The broad diversity enables biosurfactants to be better selected for specific applications.

Hydrocarbon-degrading microorganisms play a key role in bioremediation, and biosurfactant production from these processes is important. There are two mechanisms by which biosurfactants enhance the hydrocarbon degradation rate. First, biosurfactants can solubilize hydrophobic compounds between the micelle structures, effectively increasing hydrocarbon solubility and their availability to the cell. Second, biosurfactants can cause the cell surface to become more hydrophobic because of an increased cell surface association with the hydrophobic substrate (Al-Tahhan et al., 2000).

3.1 Selection of biosurfactant-producing microorganisms

The increasing industrial interest in microbial surfactant production has motivated many researchers to study and develop fast and safe methods for the selection of biosurfactant-producing microorganisms. Among these methods, the drop-collapse test (Jain et al., 1991) deserves special attention because it is fast, practical, and reproducible. Youssef et al. (2004) compared different methods for detecting potential biosurfactant-producing microorganisms, and the drop-collapse test proved to be a reliable method for the selection of microorganisms. In this method, a small amount of water is applied to a hydrophobic compound. In the absence of surfactants, a drop forms because the polar water molecule is repelled by the hydrophobic surface. However, if a surfactant is added to the water, the accompanying interfacial tension reduction results in dispersion (collapse) of the water drop on the hydrophobic surface. Based on this principle, the drop-collapse test can be used to select potential biosurfactant-producing microorganisms.

Bodour and Miller-Maier (1998) developed a modified drop-collapse test method for the quantitative analysis of biosurfactants. In this technique, a standard concentration curve for each surfactant is calculated. Drops of a known surfactant solution at various concentrations are placed on a solid and flat surface. After one minute, the diameter of each drop is measured, and a standard curve of drop diameter (Y axis) as a function of surfactant concentration (X axis) is plotted. As the surfactant concentration increases, the drop diameter also increases until the CMC is reached, at which point the diameter remains constant. This curve is used to determine the surfactant concentration from unknown samples by comparing the diameters found with the standard drop diameter. The authors demonstrated that this method had greater sensitivity and reproducibility than surface tension measurements.

Biosurfactant-producing microbial strains can be selected by isolating native microorganisms from environments naturally contaminated with oil. In this case, the drop-collapse technique can be used for screening purposes. Figure 3 illustrates a simplified method for the isolation and selection of potential biosurfactant- producing microorganisms.

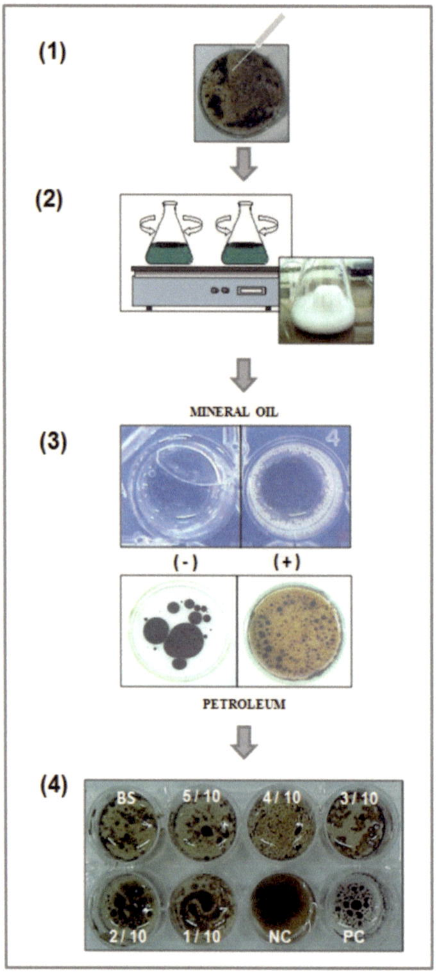

*DM composition: 2% (w/v) agar, 0.1% (w/v) KH_2PO_4, 0.1% (w/v) K_2HPO_4, 0.1% (w/v) NH_4NO_3, 0.05% (w/v) $MgSO_4$, saturated $FeSO_4$ solution and saturated $CaCl_2$ solution added to the medium at 0.001% (v/v), pH 7.2. **EM composition: 0.5% w/v beef extract, 1% w/v peptone, 0.5% w/v NaCl, and pH 7.0 (Li et al., 2000)

Fig. 3. Schematic representation of an experiment for the selection of biosurfactant-producing microorganisms. **(1)** Isolation of microorganisms able to grow in degradation medium (DM*) containing 2% petroleum as the sole carbon source. **(2)** Microbial colonies isolated from the DM medium are inoculated into 100 mL flasks containing 5 mL of enrichment medium (EM**) and incubated at room temperature (28°C ± 2°C) under constant agitation at 130 rpm for five days. At this stage, it is possible to observe the formation of foam produced by the biosurfactant. **(3)** To select microorganisms that exhibit surfactant activity, the drop-collapse method is performed in mineral oil and petroleum after the fifth day of growth. Then 5-mL aliquots of supernatant from each microbial suspension grown in EM are placed in the center well of an ELISA plate containing 1.8 mL of each type of oil previously deposited. The results are visually determined after 1 minute. In positive samples, there is a collapse of the oil drop. The photographs on the left depict the negative control (water and oil), and the pictures on the right indicate the test samples that were positive for biosurfactant production. **(4)** A quantitative analysis of biosurfactant production in petroleum with the drop-collapse test after increasing dilutions (5/10, 4/10, 3/10, 2/10, and 1/10) of supernatants from microbial cultures in sterile distilled water. BS. - Pure biosurfactant (undiluted) with added petroleum; NC - Negative control (water and oil); PC - Positive control with 1% SDS and petroleum.

3.2 Microbial growth kinetics related to biosurfactant production

Because of the amphipathic structure of the surfactant molecules, two biochemical pathways are necessary for biosurfactant production. One pathway is related to the synthesis of the hydrophobic portion, and the other is related to the synthesis of the hydrophilic portion of the molecule. These pathways produce a hydrocarbon and a carbohydrate, respectively. Several pathways, involving specific enzymes, are used for the synthesis of these two precursor groups. In many cases, the regulatory genes for specific enzymes are the first to be expressed, and despite the diversity among biosurfactants, there are some common characteristics in their synthesis and gene expression regulation. The study of molecular genetics in biosurfactant production is still new and many aspects remain unclear. However, the genetic mechanisms for the synthesis and regulation of rhamnolipid and surfactin in *Pseudomonas aeruginosa* and *Bacillus subtilis*, respectively, have been extensively studied (see review by Sullivan, 1998).

Generally, the genes required for biosurfactant production are controlled by *quorum sensing* systems. By definition, a *quorum sensing* system is a mechanism by which bacterial cells regulate specific genes in response to critical concentrations of signal molecules produced with increasing cell density (Diggle et al., 2002). Therefore, a high cell density is necessary for biosurfactant production to occur. Thus, signal molecule production becomes sufficient for binding to specific autoinducers and consequent gene activation. However, this mechanism does not explain biosurfactant production in environments contaminated with hydrocarbons, where the cell density is presumably low but biosurfactant production is high. In these environments, each cell creates its own microenvironment within each micelle, stimulating biosurfactant production independent of increased cell density (Sullivan, 1998).

Genetic, nutritional, and environmental factors are all related to biosurfactant production. Each microorganism requires an appropriate medium for growth. Thus, the choice of the ideal medium for biosurfactant production depends on the microbial strain used. Therefore, the best conditions for an individual strain's development must be assessed.

Among the nutritional factors that influence biosurfactant production, the available sources of carbon and nitrogen in the medium are crucial. Production can also be increased with the addition of hydrophobic compounds, such as aliphatic and even aromatic hydrocarbons, to the culture medium during the stationary growth phase (Desai & Banat, 1997; Déziel et al., 1996). However, some microorganisms produce biosurfactants only when consuming hydrocarbons, i.e., only after all of the soluble carbon in the medium has been consumed. In such cases, the use of hydrocarbons is essential for the acceleration of biosurfactant production. Nitrogen metabolism is also directly related to biosurfactant production, especially among denitrifying microorganisms. It should be noted that for some bacteria, the culture medium composition may interfere with the biosurfactant chemical composition (Desai & Banat, 1997; Zhang & Miller, 1995).

The kinetics of biosurfactant production depends on the substrate used for microbial growth. Biosurfactant production may occur during the microbial active growth period or it may take place during the stage of limited growth. In the first case, biosurfactants production begins simultaneously with microbial growth when the substrate is used. In contrast, production under limited growth conditions occurs only during substrate scarcity (Desai & Banat, 1997; Maciel et al., 2007;). Figure 4 shows different stages of bacterial growth and the corresponding kinetics for biosurfactant production.

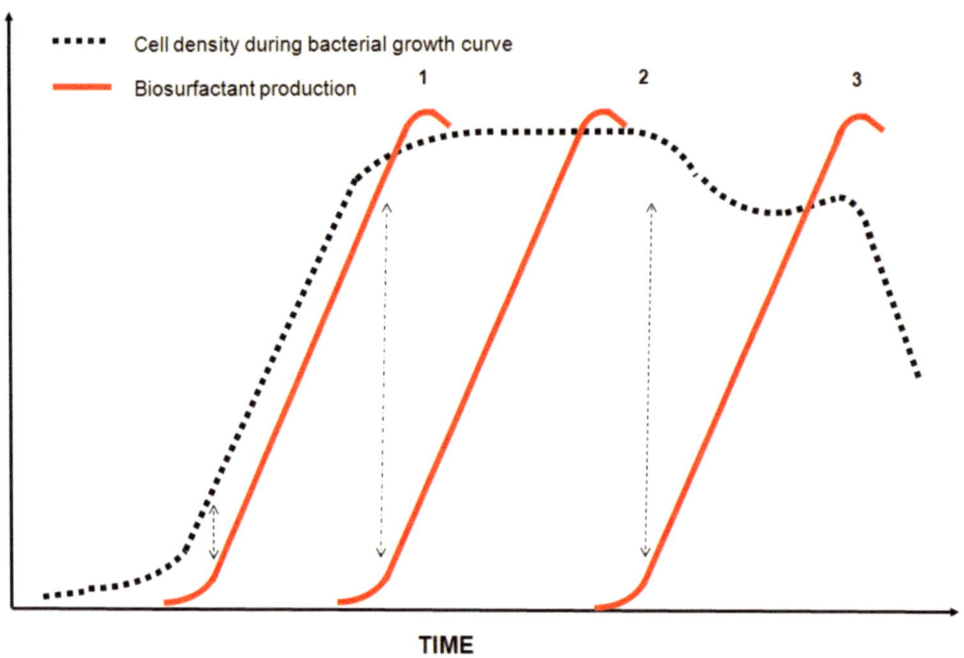

Fig. 4. Schematic illustration of the three different stages of bacterial growth and the corresponding kinetics for biosurfactant production. **(1)** Biosurfactant production associated with microbial growth in *Acinetobacter calcoaceticus*. The arrow indicates the start of biosurfactant production associated with the onset of the logarithmic phase in the bacterial growth curve. Biosurfactant production peak occurs in the early stationary phase in the bacterial growth curve. **(2)** Biosurfactant production in *Pseudomonas aeruginosa* under limited

growth conditions. The arrow indicates the onset of biosurfactant production associated with the late logarithmic phase and early stationary phase of the bacterial growth curve. The peak of biosurfactant production occurs at the end of the stationary phase in the bacterial growth curve. **(3)** This biosurfactant production curve corresponds to different bacterial isolates under nutritional stress. These bacteria include the *Achromobacter xylosoxidans, Cellulosimicrobium cellulans, Alcaligenes* sp., and the *Achromatium oxaliferum*. The arrow indicates the start of biosurfactant production associated with the end of the stationary phase and the beginning of the decline phase for the bacterial growth curve. Biosurfactant production under these conditions causes a prolongation of the bacterial culture survival, which reflects a second stationary phase during the bacterial growth curve.

Environmental factors, such as pH, temperature, agitation, and oxygen availability, should be also carefully adjusted in biosurfactant production to provide the best conditions for microbial growth. In general, the optimum pH range for biosurfactant production is between 6.5 and 8.0, although biosurfactant activity remains stable over a wide pH range. Temperature and aeration should also be carefully adjusted because most biosurfactant-producing bacteria grow well at room temperature (28°C ± 2°C), and heat treatment can alter the stability of some biosurfactants. However, some bacteria, such as thermophilic bacilli, are capable of producing biosurfactants above 40°C without any alterations in stability (Banat, 1993) and therefore can be used in deep-water oil recovery. The aeration rate should also be adjusted depending on the microorganism.

3.3 Economic viability of biosurfactant applications

Economic viability is often a major constraint for biotechnological processes, especially in biosurfactant production. Biosurfactants have to compete with surfactants of petrochemical origin in three aspects: cost, functionality, and production capacity to meet the needs of the application. A high production cost is incompatible with microbial-enhanced oil recovery (MEOR), which requires large amounts of biosurfactants. Some factors may decrease the production cost in these cases: (i) selection of microorganisms adapted to the polluted area, (ii) decreased biosurfactant recovery cost, and (iii) microbial growth on more affordable substrates (Makkar & Cameotra, 2002).

To minimize production costs, agro-industrial waste products have been used as substrates for biosurfactant production. These waste products include post-harvest waste from cassava, soybeans, beets, sweet potatoes, potatoes, sorghum, wheat, and rice; husks from soybeans, corn, and rice; sugar cane and cassava bagasse; coffee industry waste products (pulp, husk, and coffee grounds); industrial wastes from the juicing of apples, grapes, pineapples, bananas, and carrots; candy manufacturing waste products; and other substrates such as sawdust, corn grain, tea manufacturing waste, and chicory roots (Makkar & Cameotra, 2002; Pandey et al., 2000). Other substrates have been suggested for biosurfactant production, such as syrup, whey, and distillery waste products (Makkar & Cameotra, 2002).

4. Application of microbiological culturing and metagenomic methods focused on oil degradation

To allow adaptation to various environments, microorganisms have a rich taxonomic, metabolic, physiological, and molecular diversity in nature. Therefore, studies based on the isolation of microorganisms that biodegrade oil and its fractions are certainly of great importance. In contaminated environments, such as oily sludge landfarming or biopile soil, different microorganisms with different metabolic pathways and desirable biodegradation or bioconversion attributes can be found. In these studies, the test compound, in this case a hydrocarbon should be the energy source in the culture medium to exert a selection pressure on microorganisms with the potential to metabolize it. Bioprospecting studies can be then conducted with the aim of exploiting potential bioremediation or bioconversion biotechnology.

Studies on the impact of environmental changes on microbial populations and their activities have often been restricted to evaluating basic parameters including the total number of microorganisms, microbial biomass, respiration rate, and total enzyme activities involved in organic carbon and nitrogen mineralization. Currently, these studies can be conducted using methods based on DNA sequence analysis, especially analysis of the 16S rDNA gene in bacteria and archaea and the 18S gene in fungi. These genes can be amplified by PCR (Polymerase Chain Reaction) and subsequently characterized by electrophoresis using different molecular techniques such as ARDRA (Amplified Ribosomal DNA Restriction Analysis), T-RFLP (Terminal Fragment Length Polymorphism), RAPD (Random Amplification of Polymorphic DNA), RISA (Ribosomal Intergenic Space Analysis), DGGE/TGGE (Denaturing Gradient Gel Electrophoresis / Temperature Gradient Gel Electrophoresis, and SSCP (Single-Strand Conformation Polymorphism to obtain microbial community profiles (Ranjard et al., 2000; Kozdrój & Van Elsas, 2001, Konstantinidis & Tiedje, 2007, Maciel et al. 2009).

In DGGE (denaturing gradient gel electrophoresis), for example, DNA fragments with the same size but with different sequences can be separated. Thus, each band formed during gel electrophoresis represents a single operational taxonomic unit (OTU), i.e., a single microorganism. Thus, changes in environmental microbial diversity can be observed with the DGGE banding profiles. Separation is based on the decreased electrophoretic mobility of a partially denatured double-stranded DNA molecule in a polyacrylamide gel containing an increasing linear denaturing gradient (a mixture of urea and formamide). One of the primers used in the PCR reactions must contain a G-C clamp in its sequence to prevent complete denaturation of the double-strands. Fragments with GC-rich sequences need higher denaturing gradients so that AT-rich sequences can separate the double-strands completely (Figure 5).

However, high-impact molecular studies of microbial communities and the development of metagenomic libraries are dependent on DNA extraction. This critical step of the procedure should be the first and most important step for such studies. The success of the evaluation is directly related to the type and quality of DNA extraction (Maciel et al., 2009). There are several techniques to extract the total DNA from environmental samples; however, no method is universally applicable, as each type of sample requires an individually optimized extraction method. Four important variables must be considered during DNA extraction: the amount of extracted DNA, purity, DNA integrity, and representativeness. The extraction of total DNA from a sample is necessarily a balance between rigorous extraction, which is required for representation of all microbial genomes in the sample (direct lysis); minimal DNA fragmentation; and avoidance of co-extracted contaminants.

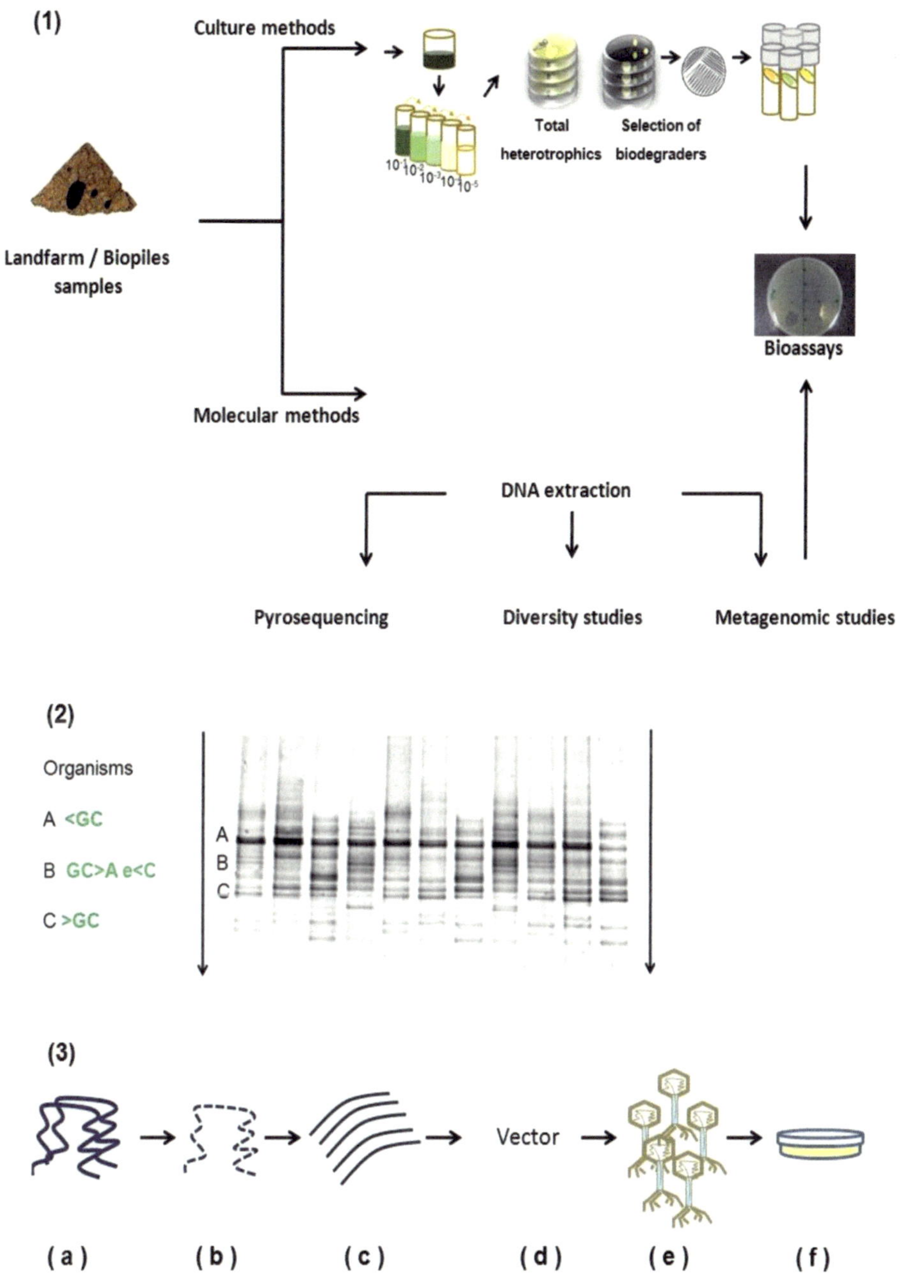

Fig. 5. Application of molecular and culture techniques for the study of microorganisms from *landfarming* and biopiles soils with biotechnological potential. **(1)** Culturing methods - After soil homogenization, a small aliquot is serially diluted and cultured in media for counting of the total number of heterotrophic microbes and in medium containing the hydrocarbon test compound as sole carbon source for biodegrader selection. The

microorganisms are then isolated in pure culture and stored for bioprospecting studies (bioassays). Molecular methods - Different molecular methods can be applied after DNA extraction, depending on the purpose: sequencing methods are used to study genome structure, microbial diversity methods are used to assess environmental impact on microbial communities, and metagenomic studies are used to prospect genes of biotechnological interest (bioassays). **(2)** An example of a microbial diversity study by the DGGE technique (denaturing gradient gel electrophoresis). Photograph of a DGGE were the OTU (operational taxonomic unit) in A has a standard denaturing gel pattern different from that of B and C because B and C have a higher GC content (guanine and cytosine). **(3)** Schematic representation of steps for metagenomic library construction (modified from http://www.epibio.com): (a) extraction of total DNA from contaminated soil sample, (b) DNA fragmentation by restriction enzymes, (c) tip repairing to obtain DNA fragments with appropriate sizes to be used in the vector, (d) inserting the DNA fragments into a vector system (fosmid), (e) packaging of the recombinant vector in envelope proteins from *Lambda* bacteriophage, and (f) transformation into competent *E. coli* cells.

The rational exploitation of the microbial potential is now possible not only through culturable microorganisms but also with molecular strategies, where it is possible to clone microbial DNA extracted from different environments in *Escherichia coli* hosts. These strategies, combined with general metagenomic naming, use BAC (*bacterial artificial chromosome*), cosmid, or fosmid vectors for cloning of different-sized inserts and require the use of different methods for host cell transformation (Yung et al., 2009; Singh et al., 2009). Most libraries constructed in vectors that contain large DNA fragments (40-100 Kb) obtained directly from environmental microbial communities, such as soil, sediment, or sludge, are used to investigate the expression of genes involved in the synthesis or metabolism of certain compounds (Rondon et al., 2000; Singh et al., 2009).

After the DNA extraction, the library construction involves DNA fragmentation by restriction enzymes or mechanical breaking, DNA fragment insertion into a suitable vector system, and transformation of recombinant vectors into a suitable host (Figure 5). The vector choice depends on the DNA quality, average insert size, number of host vector copies, and testing strategy. Libraries with large DNA inserts (~ 20 Kb) are constructed with cosmid, fosmid, and BAC vectors, among others. Libraries with smaller inserts (less than 10 Kb) are constructed with plasmid vectors (Handelsman, 2004).

Fosmid vectors fit fragments of up to 45 Kb, offering excellent gene coverage. This vector is particularly suitable when the genes to be cloned are organized in large operons of up to 25 Kb, as is the case for microbial genes that encode the biosurfactant synthesis. After vector ligation, the vector and insert assembly are encapsulated using an envelope protein from the *Lambda* bacteriophage. The resulting particles are adsorbed to the host cells and are responsible for introducing the DNA. Cloning efficiency must be verified by plating serial dilutions generated with the phage capsids and *E. coli* cells. Generally, this plating is performed in LB medium with chloramphenicol added as a selective antibiotic. After collection and storage, the clones can be tested for degradation of innumerous aromatic compounds.

It is important to note that most metagenomic studies aim to isolate the genes responsible for biosurfactant synthesis and to use culture media that favor the production of this secondary metabolite (Franzetti et al. 2008; Pornsunthorntawee et al., 2008). However, as discussed earlier, most of the genes for surfactant molecules are regulated by *quorum sensing* systems that trigger the synthesis with increasing cell density, thus delaying the process. To circumvent this limitation, cloning and heterologous expression can be performed using vectors with constitutive promoters, such as the T7 promoter from the pCC2FOS vector, which eliminate gene regulation by the *quorum sensing* system and therefore may be a promising alternative for biosurfactant production on an industrial scale.

5. Conclusion

The search for culturable microorganisms with desirable attributes for the effective bioremediation of crude oil contaminated environments and the successful biodegradation of hydrocarbons remains the main focus for biotechnology research, because of the natural metabolic ability found in impacted environments. However, no single microbial species is capable of degrading all oil components; thus, a consortium between strains is required. In addition, most microorganisms are not culturable, and those that may exhibit a diverse catabolic potential are still unknown. The use of culture-independent methods has shown a vast microbial diversity in these environments. Metagenomics offers an alternative to the exploitation of heterologous gene expression important in degradation processes. It is worth noting that there are two sides to biotechnology the culture- and the culture-independent methods and that each complements the other.

6. References

Al-Tahhan, RA., Sandrin, TR., Bodour, AA. & Maier, RM. (2000). Rhamnolipid-induced removal of lipopolysaccharide from *Pseudomonas aeruginosa*: effect on cell surface properties and interaction with hydrophobic substrates. *Applied and Environmental Microbiology*, Vol. 66, No. 8, (August 2000), pp. 3262-3268, ISSN 0099-2240.

Ausma, S., Edwards, GC. & Gillespie, TJ. (2003). Laboratory-scale measurement of trace gas fluxes from landfarm soils. *Journal of Environmental Quality*, Vol. 32, No. 1, (January-February 2003), pp. 8-22, ISSN 0047-2425.

Banat, IM. (1993). The isolation of a thermophilic biosurfactant producing *Bacillus* sp. *Biothechnology Letters*, Vol. 15, No. 6, (June 1993), pp. 591-594, ISSN 0141-5492.

Banat, IM., Makkar, RS. & Cameotra, SS. (2000). Potential commercial applications of microbial surfactants. *Applied Microbiology and Biotechnology*, Vol. 53, No. 5, (May 2000), pp. 495-508, ISSN 0175-7598.

Bodour, AA. & Miller-Maier, RM. (1998). Application of a modified drop-collapse technique for surfactant quantitation and screening of biosurfactant-producing microorganisms. *Journal of Microbiological Methods*, Vol. 32, No. 3, (May 1998), pp. 273-280, ISSN 0167-7012.

Chosson, P., Lanau, C., Connan, J. & Dessort, D. (1991). Biodegradation of refractory hydrocarbon biomarkers from petroleum under laboratory conditions. *Nature*, Vol. 351, (June 1991), pp. 640-642, ISSN 0028-0836.

Desai, JD. & Banat, IM. (1997). Microbial production of surfactants and their commercial potential. *Microbiology and Molecular Biology Reviews*, Vol. 61, No. 1, (March 1997), pp. 47-64, ISSN 1092-2172.

Diggle, S.P., Winzer, K., Lazdunski, A., Williams, P. & Cámara, M. (2002). Advancing the quorum in *Pseudomonas aeruginosa*: MvaT and the regulation of *N*-acylhomeserine lactone production and virulence gene expression. *The Journal of Bacteriology*, Vol. 184, No. 10, (May 2002), pp. 2576-2586, ISSN 0021-9193.

Déziel, E., Paquette, G., Villemeur, R., Lépine, F. & Bisaillon, J.G. (1996). Biosurfactant production by a soil *Pseudomonas* strain growing on polycyclic aromatic hydrocarbons. *Applied and Environmental Microbiology*, Vol. 62, No. 6, (June 1996), pp. 1908-1912, ISSN 0099-2240.

Fiechter, A. (1992). Biosurfactants moving towards industrial application. *Trends in Biotechnology*, Vol. 10, p. 208-217, ISSN 0167-7799.

Franzetti, A., Bestetti, G., Caredda, P., La Colla, P. & Tamburini, E. (2008). Surface-active compounds and their role in the access to hydrocarbons in *Gordonia* strains. *FEMS Microbiology Ecology*, Vol. 63, No 2, (February 2008), pp. 238–248, ISSN 0168-6496.

Handelsman, J. (2004). Metagenomics: application of genomics to uncultured microorganisms. *Microbiology and Molecular Biology Reviews*, Vol. 68, No. 4, (December 2004), pp. 669-685, ISSN 1092-2172.

Head, IM., Jones, DM. & Larter, SR. (2003). Biological activity in the deep subsurface and the origin of heavy oil. *Nature*, Vol. 426, (November 2003), pp. 344-352, ISSN 0028-0836.

Jain, DK., Thompson, DLC., Lee, H. & Trevors, JT. (1991). A drop-collapsing test for screening surfactant producing microorganisms. *Journal of Microbiological Methods*, Vol. 13, No. 4, (August 1991), pp. 271-279, ISSN 0167-7012.

Khan, FI., Husain, T. & Hejazi, R. (2004). An overview and analysis of site remediation technologies. *Journal of Environmental Management*, Vol. 71, pp. 95–122, ISSN 0301-4797.

Karanth, NGK., Deo, PG. & Veenanadig, NK. (1996). Microbial production of biosurfactants and their importance. *Microbiology*, Vol. 18, No. 1, pp. 1-18, ISSN 1350-0872.

Konstantinidis, KT. & Tiedje, JM. (2007). Prokaryotic taxonomy and phylogeny in the genomic era: advancements and challenges ahead. *Current Opinion in Microbiology*, Vol. 10, No.5, (October 2007), pp. 504-509, ISSN 1369-5274.

Kozdrój, J. & Van Elsas, JD. (2001). Structural diversity of microorganisms in chemically perturbed soil assessed by molecular and cytochemical approaches. *Journal of Microbiological Methods*, Vol. 43, No. 3, (January 2001), pp. 197-212, ISSN 0167-7012.

Li, G., Huang, W., Lerner, DN. & Zhang, X. (2000). Enrichment of degrading microbes and bioremediation of petrochemical contaminants in polluted soil. *Water Research*, Vol. 34, No. 15, (October 2000), pp. 3845-3853, ISSN 0043-1354.

Maciel, BM., Dias, JCT., Santos, ACF., Argôlo Filho, RC., Fontana, R. Loguercio, LL. & Rezende, RP. (2007). Microbial surfactant activities from a petrochemical landfarm in a humid tropical region of Brazil. *Canadian Journal of Microbiology*, Vol. 53, No. 8, (August 2007), pp. 937-943, ISSN 0008-4166.

Maciel, BM., Santos, ACF., Dias, JCT., Vidal, RO., Dias, RJC., Gross, E., Cascardo, JCM. & Rezende, RP. (2009). Simple DNA extraction protocol for a 16S rDNA study of bacterial diversity in tropical landfarm soil used for bioremediation of oil waste. *Genetics and Molecular Research*. Vol. 31, No. 8(1), (March 2009), pp. 375-388. ISSN 1676-5680 (Electronic).

Makkar, R.S. & Cameotra, SS. (2002). An uptake on the use of unconventional substrates for biosurfactant prouction and their new applications. *Applied Microbiology and Biotechnology,* Vol.58, No. 4, (March 2002), pp. 428-434, ISSN 0175-7598.

Maniasso, N. (2001). Micellar media in analytical chemistry. *Química Nova,* Vol. 24, No. 1, (January / February 2001), pp. 87-93, ISSN 0100-4042.

Nitschke, M. & Pastore, GM. (2002). Biosurfactants: Properties and applicationss. *Química Nova,* Vol. 25, No. 5, (September / October 2002), pp. 772-776, ISSN 0100-4042.

Pandey, A., Soccol, CR. & Mitchell, DA. (2000). New developments in solid-state fermentation: I - bioprocesses and products. *Process Biochemistry,* Vol. 35, No. 10, (July 2000), pp. 1153-1169, ISSN 1359-5113.

Pornsunthorntawee, O., Wongpanit, P., Chavadej, S., Abe, M. & Rujiravanit, R. (2008). Structural and physicochemical characterization of crude biosurfactant produced by *Pseudomonas aeruginosa* SP4 isolated from petroleum-contaminated soil. *Bioresource Technology,* Vol. 99, No.6, (April 2008), pp. 1589–1595, ISSN 0960-8524.

Ranjard, L., Poly, L. & Nazaret, S. (2000). Monitoring complex bacterial communities using culture-independent molecular techniques: application to soil environment. *Research in Microbiology,* Vol. 51, No. 3, (April 2000), pp. 167–177, ISSN 0923-2508.

Ron, EZ. & Rosenberg, E. (2001). Natural roles of biosurfactants. *Environmental Microbiology,* Vol. 3, No. 4, (April 2001), pp. 229-236, ISSN 1462-2920.

Rondon, MR., August, P., Bettermann, AD., Brady, S., Grossman, TH., Liles, M., Loiacono, KA., Lynch, BA., MacNeil, IA., Minor, C., Tiong, CL., Gilman, M., Osburne, MS., Clardy, J., Handelsman, J. & Goodman, R (2000). Cloning the Soil Metagenome: a Strategy for Accessing the Genetic and Functional Diversity of Uncultured Microorganisms. *Applied and Environmental Microbiology,* Vol. 66, No. 6, (June 2000), pp. 2541-2547, ISSN 0099-2240.

Seabra, PN. (2005). *Aplicação de biopilhas na biorremediação de solos argilosos contaminados com petróleo.* Doctoral Thesis. Instituto de Química - Universidade Federal do Rio de Janeiro, Rio de Janeiro – RJ, Brazil *(in portuguese).*

Silva, EP. (2004). *Preliminary assessment of the potential application of technology for bioremediation of soil biopilhas of Guamaré – RN.* Master's Thesis. Federal University of Rio Grande do Norte, Natal, Brazil *(in portuguese).*

Sullivan, ER.(1998). Molecular genetics of biosurfactant production. *Current Opinion in Biotechnology,* Vol. 9, No. 3, (June 1998), pp. 263-269, ISSN 0958-1669.

Singh, J., Behal, A., Singla, N., Joshi, A., Birbian, N., Singh, S., Bali, V. & Batra N. (2009). Metagenomics: Concept, methodology, ecological inference and recent advances. *Biotechnology Journal,* Vol. 4, No. 4 (April 2009), pp.:480-494, ISSN 1860-7314.

Youssef, NH., Duncan, KE., Nagle, DP., Savage, KN., Knapp, RM. and Mc Inerney, MJ. (2004). Comparision of methods to detect biosurfactant production by diverse microorganisms. *Journal of Microbiological Methods,* Vol. 56, No. 3, (March 2004), pp. 339-347, ISSN 0167-7012.

Young, PY., Burke, C., Lewis, M., Egan, S., Kjelleberg, S. & Thomas, T. (2009). Phylogenetic screening of a bacterial, metagenomic library using homing endonuclease restriction and marker insertion. *Nucleic Acids Research,* Vol, 37, No. 21 (November 2009), pp. 2-8, ISSN 0305-1048.

Zhang, Y. & Miller, RM. (1995). Effect rhamnolipid (biosurfactant) structure on solubilization and biodegradation of *n*-alkanes. *Applied and Environmental Microbiology*, Vol. 61, No. 6, (June 1995), pp. 2247-2251, ISSN 0099-2240.

10

Crude Oil Metagenomics for Better Bioremediation of Contaminated Environments

Ines Zrafi-Nouira[1], Dalila Saidane-Mosbahi[1], Sghir Abdelghani[2], Amina Bakhrouf[1] and Mahmoud Rouabhia[3,*]

[1]*Laboratoire d'Analyse, Traitement et Valorisation des Polluants de l'Environnement et des Produits, Faculté de Pharmacie de Monastir,*

[2]*CEA Genoscope, Evry,*

[3]*Groupe de Recherche en Écologie Buccale, Faculté de Médecine Dentaire, Université Laval,*

[1]*Tunisie*

[2]*France*

[3]*Canada*

1. Introduction

Our planet suffers more and more from various pollution problems. The marine environments are always regarded as sewers without end and are then subjected to different types of toxic rich rejects leading to oceans and seas degradation. The marine environment becomes at the same time the witness and the actor of the history of the planet, and its chemical composition witnesses all the complexity of its evolution. Coastal regions are often where we find various pollutants. Seawater (Jaffrennou et al. 2007; Pérez-Carrera et al. 2007), marine sediments (Wakeham, 1996), and interstitial water (Pérez-Carrera et al., 2007) have shown this pollution. Petroleum hydrocarbons are among the most toxic compounds poured at sea. Known as the most significant pollutant, crude oil can persist for years (Burns & Teal 1979), with dangerous effects on coastal environments and negative effects on both the ecosystem and the marine biodiversity (Clark 1992; Rice et al., 1996). In marine environment, crude oil is subjected to physico-chemical and biological modifications, which enhance hydrocarbons solubility in the water and consequentially cause extensive damage to marine life, natural resources, and human health.

It is estimated that most petroleum compounds have carcinogenic properties (Ericson et al. 1998; Aas et al., 2000; Shaw & Connell 2001). Because of their toxicity, several hydrocarbons are classified by the Agency of Environmental Protection, the World Health Organization, and the European Union as top priority pollutants. Based on their distribution and their high toxicity, hydrocarbons are considered as the principal organic markers of the anthropogenic activity in the ecosystems (Laflamme & Hites, 1978; Bouloubassi & Saliot, 1991; Budzinski et al., 1997; Yunker et al., 1996; Fernandes et al., 1997).

* Corresponding Author

Crude oil is composed of hundreds of compounds (Bertrand & Mille 1989). The major fractions are the non-aromatic hydrocarbons (NAH) and the aromatic hydrocarbons (AH). The NAH fraction includes aliphatic hydrocarbons, branched isoalkanes such as pristane and phytane, and cycloalkanes which are represented by an unresolved complex mixture (UCM). Of great interest to several research groups are non-aromatic and aromatic oil hydrocarbon fractions, such as the contents of total aliphatic (TA), total resolved or n-alkanes, total aromatics, and polyaromatic hydrocarbons (PAHs). n-Alkanes represent a group of non-polar and photocatalytically stable organic compounds. Their characteristics provide useful information on the origin of biological and/or petrogenic sources of pollution. At the same time, aromatic hydrocarbons, and particularly PAHs, represent a group whose characteristics and individual proportions are useful in comparative studies of petrogenic and pyrolytic sources (Clark 1992; González et al., 2006; Rice et al., 1996). Characterization of hydrocarbons and their specific sources are important for understanding the crude oil fate and behaviour in marine environment and for predicting their potential long-term impact, which allow taking effective clean-up measures in the specific marine compartment. Therefore, there is an increasing need to develop easy and accurate methods for removing hydrocarbon from seawater in the case of an oil spill. The removal of petroleum hydrocarbon or its transformation into less toxic products by bioremediation is a less invasive and less expensive process if compared to classical decontamination. However, the use and optimization of bioremediation treatments in the water compartment require knowledge of the seawater microbial communities directly and indirectly involved in the degradation of hydrocarbons.

Multiple initiatives have been developed to resolve the problem of petroleum pollution. An array of procedures has been developed including physical, chemical, and biological techniques. Among these procedures, bioremediation is currently used alone or associated to physicochemical procedures. Biological methods of rehabilitation of polluted sites represent an interesting alternative. These techniques are based on the microorganism's capacities to degrade petroleum compounds (Harayama et al., 1999). Indeed, within the last two decades, the use of molecular techniques has led to a significant improvement in our knowledge of microbial diversity in different complex environments. Hence, it becomes necessary to characterize microbial communities in polluted environments; especially when the pollutant is as complex as crude oil. Thus, characterization of the microbial diversity and the identification of microbial key players in the degradation of crude oil could be useful in defining new strategies for bioremediation (Prince et al., 1993).

Prokaryotes have been extensively documented in hydrocarbon-contaminated environments (sediment, seawater, soil, activated sludge, ice, estuary and river) as well as during oil degradation. Comparisons of the microorganism's phylogenetic diversity from multiple petroleum contaminated environments are of great importance. This information allows improving our knowledge on the active core members of this group and to understand whether the composition of this bacterial group changes in response to specific petroleum substrate and environmental parameters. The ability to understand such changes and to correlate them to microbial activities through population and phylogenetic inventories is especially important to understand naturally dynamic environmental parameters. This will help in setting up strategies for optimizing environmental conditions that could improve the hydrocarbon biodegrader's effectiveness towards bioremediation.

In fact, to combat petroleum pollution and for an effective intervention in case of accidental or chronic discharge, the study of the bioremediation process is necessary. Bioremediation rests on the use of the natural capacities of certain organisms to metabolize pollutants (Serrano et al., 2007). It is necessary to improve our knowledge on the microbial communities involved in the metabolisation of hydrocarbons and to follow their process dynamics to optimize the bioremediation techniques. However, in the majority of the natural ecosystems, the study of the microbial communities, their densities, and their diversity is difficult (El Fantroussi et al., 2003; Curtis & Sloan, 2004, Elloumi et al., 2008). Because microorganisms are not easy to isolate due their diversity, their organization in consortia, their dynamic and specific cultivation characters, molecular techniques must be used to identify the effective players (microorganisms) in bioremediation. These techniques promote the identification, phylogenetic diversity analyses, and the study of the metabolic diversity of the microbial communities present in petroleum contaminated sites. Currently, several researches aim to establishing correlations between the microbial diversity, the nature, and levels of hydrocarbon in the polluted environment (Head et al., 2006). Thus, the molecular approach could allow the establishment of inventories of the microorganism's composition in polluted environments and to elucidate the "black box" of this composition in terms of abundance and thereafter in terms of adaptation and functionalities.

2. Crude oil as an important source of energy in the world

2.1 World production of oil and its derivatives

One hundred years ago oil exploitation began, first as a source of energy and later to include oil as a source of raw material. As a source of energy, crude oil went through an uninterrupted progression of its extraction during more than one century, driven by development of transport and industry. The annual report of the OPEC in 2008 showed an increase in the oil production worldwidwe. Indeed, the annual production of oil products was about 11 million tons per day (11 Mt/d) in year 2000 with an increase in the production of 1.9%/year in the current decade. Estimates of the worldwide oil consumption, suggest an increase of about 44.7 million barrels per day (that is to say 6.4 Mt/day) between 1999 and 2020, which corresponds to an increase in oil consumption from 74.9 Mb/d (either 10.7 Mt/day) in 1999 to 119.6 Mb/d (or 17 Mt/day) in 2020. The annual growth is thus estimated at 2.3 % /year, whereas it was only of 1.6 %/year between 1970 and 1999. Fig.1 shows the distribution of the worldwide total oil (Gasoline/Naphtha, Fuel oil, Middle distillates and others product) production. It is estimated that one trillion barrels have already been harvested, and about 3 trillion barrels of oil remain to be recovered worldwide. Oil production is expected to peak sometime between 2010 and 2020, and then fall inexorably until the end of this century.

2.2 Sectors of use

Fig. 2 presents the worldwide consumption of petroleum during the 20th and 21st centuries. The increase in the world production of oil is in direct relationship to the world requirement of oils and its derivatives. This need is imposed by various sectors (Fig. 2), for instance, the transportation sector is the greatest consumer of petroleum products, and its avidity for petroleum will increase until year 2030. The other sectors are mainly: industry, electricity generation, agriculture, and marine activities that include marine transportation, off shore oil production and fishing, among others.

Our society faces the challenge of increasing the production of goods and services for a growing population using new process technology that should be energetically efficient and environmentaly friendly. This also will be the case for the petroleum industry. The primary target of the petroleum industry is to enhance and maintain a continuous oil production.

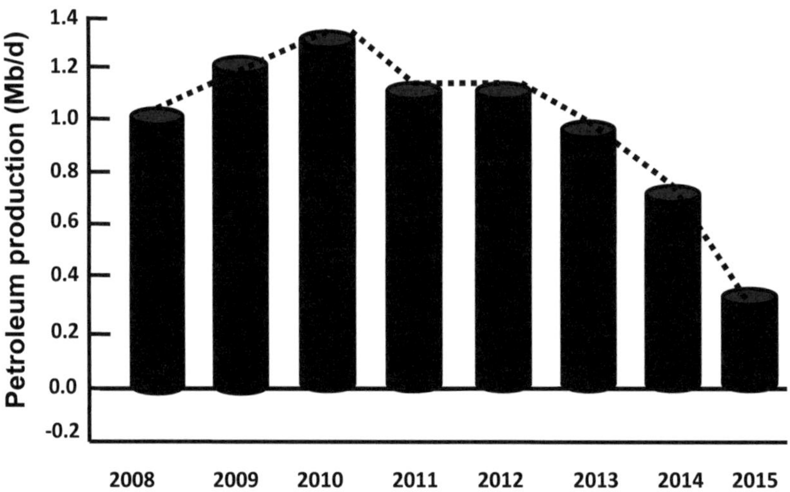

Fig. 1.Estimation of petroleum production worldwide. Adapted from OPEC (2008) with modifications.

3. Crude oil as a principal source of pollution in the world

The problems involved in the excessive production and consumption of oil are increasingly numerous and at various levels, which is directly related to the impact of industrialized factories on the environment (Killops & Killops, 2005).

3.1 Various aspects of marine oil pollution

Oil is a great source of pollution to the marine environments in the world, which can largely influence the ecological balances and the economic activities in the areas of interest. The National Research Council estimated in the review "Oil in the Sea: inputs, fates and effects, 2002 ", that the quantity of oil introduced annually into the oceans through different ways is 1.3 million tonnes/year. Take into consideration that one ton of oil can cover approximately a surface of 12 Km^2.

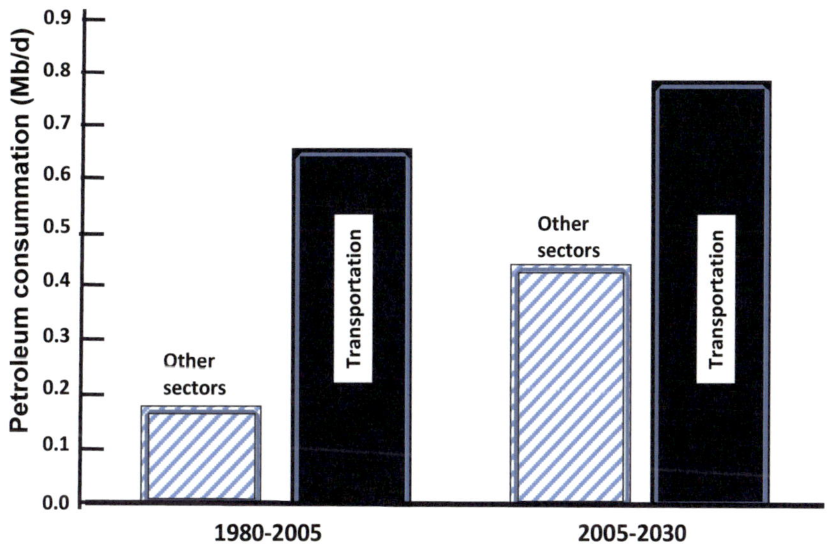

Fig. 2. Estimation of the worldwide consumption of petroleum products. Adapted from OPEC (2008), with modifications.

Oils are generally rejected close to the refineries or in the ports (telluric pollution), but also in open sea by the discharge of water from tankers ballast (approximately two million tons of hydrocarbons can be rejected by deballasting each year) or accidentally by spillage (pelagic pollution) (Marteil, 1974). Oil spills in the sea due to accidents, represent a fewer proportion of the total hydrocarbons poured in seawater each year. An imperceptible but daily source of oil discharge in the sea is represented by wastewater discharges (industrial and domestic wastes). The total hydrocarbon content in wastewater ranges from 200 to 1800 mg/day/habitant (Louati et al., 2001). Multiple examples in the world confirmed that the most significant catastrophes are related to oil pollution such as those recorded in 1967 (Spain, France, USA) and in 2010 (Gulf of Mexico, USABP platform). This catastrophe refers to the April 20, 2010 explosion and subsequent fire on the Deepwater Horizon semi-submersible Mobile Offshore Drilling Unit (MODU), which was owned and operated by Transocean and drilling for BP in the Macondo Prospect oil field about 40 miles (60 km) southeast of the Louisiana coast, in this catastrophe a total of 4.9 million barrels of crude oil were released into the Deepwater Horizon, making this the biggest oil spill to have occurred in US-controlled waters (Jarvis, 2010) (Source: Alice-Azania Jarvis http://www.independent.co.uk/environment/bp-oil-spill-disaster-by-numbers-2078396.html). War is also a significant source of pollution. Indeed, the Gulf war in the Middle East, in 1991 the destruction of crude oil field's and spills from oil tankers that were bombarded caused the pouring of significant quantities of crude oil into the marine environment (Bingham, 1992). The total assessment was estimated at 800 000 tons poured, 40 million tons of water-logged soils with oil, and 700 km of polluted coasts. It was estimated that since 1960 500 million of gallons of oil were poured by oil tankers in the European seas and the Pacific. In China, pollution by hydrocarbons is primarily in lakes and rivers as a result of the industrialization increase.

3.2 Fate of crude oil in the marine environment

The distribution of hydrocarbon pollutants in the environment comes from different sources. Once the oil spill takes place and in spite of their relative chemical stability and their persistent character, hydrocarbon pollutants are carried from an environmental compartment to another by undergoing in each one of them specific transformations such as physical, chemical, and biological modifications (Fig. 3).

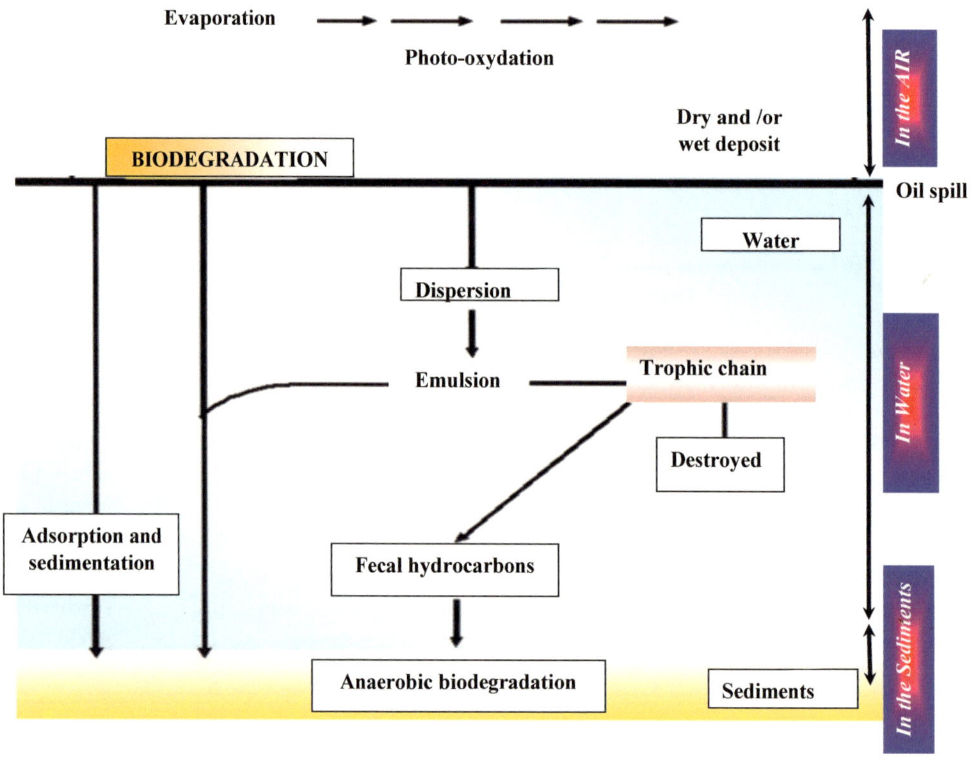

Fig. 3. Physico-chemical and biological parameters involved in marine environment after a crude oil spill (Bertrand & Mille, 1989, with modifications).

3.2.1 In the air

Atmospheric rejections of hydrocarbons have as the principal source the anthropic activity and specifically the heating by fuel, coal, and wood combustion. Indeed road transport (black fume from diesel engines); oil industries (refining, catalytic cracking of oil), production of aluminum or iron; treatment of wood by creosote; the asphalt of roads, and the cigarette smoke represent important sources of this anthropic activity. The contribution of the biogenic sources in the atmospheric contamination is supported by volcanic activities, decomposition of organic matter and the control of fires (drill, agriculture). In the air, hydrocarbons are distributed between the gas and particulate phases (volatile, semi-volatile or particulate) according to their vapor pressures, their molecular weights, and the ambient temperature (Atkinson & Arey, 1994; Oda et al., 1998). The transportation of hydrocarbons by air masses can be done locally or on a large scale. In fact, the presence of hydrocarbons has been detected at several hundreds of kilometres or even several thousands of kilometres from their emission source (Ramade 1992; Bouchez et al., 1996). Hydrocarbon carrying into the air depends mainly on the intensity of the emission, its altitude, the size, and the chemical stability of the molecules concerned (Ramade, 1992). The presence of hydrocarbons in the air represents a significant source of contamination for the marine environments due to the phenomenon of deposit.

3.2.2 In water

Significant discharges of petroleum products occur in continental and littoral water (oceans). According to Palayer (1997) the origin of water pollution by organic compounds derives from seven sources, which include:

- Runoff waters, which mainly contaminate the surface water
- Domestic and industrial effluents rejected after passage through wastewater stations
- Industrial effluents (washings of produced gases, coals cooling, etc.)
- Spent oils such as oil draining following rejection
- Dry and wet deposits coming from the atmosphere
- Water pipelines with an interior coat of tar that could salt out hydrocarbons in distributed water as the pipelines age.
- Accidents related to oil activities during oil extraction, refining, and oil spills.
- Natural oil escapes

To these sources cited by Palayer, is convenient to add water drainage and the scrubbing of the agricultural grounds contaminated with hydrocarbon products (McKiney et al., 1999). It is also possible to include as contamination sources other discharges related to the oil activity specifically the degasification of oil tankers. The multiplicity of the sources of water pollution by hydrocarbons makes very difficult the identification of the main sources of contamination.

The distribution and concentration of hydrocarbons in contaminated waters are influenced by several parameters such as the water pH, the salinity, and content of organic matter in suspension (Fernandes et al., 1997). In general, hydrocarbon concentration in bulk water is relatively low compared to hydrocarbon concentrations in sediments (Zrafi-Nouira et al., 2008; Zrafi-Nouira et al., 2009; Zrafi-Nouira et al., 2010) owing to the fact that hydrocarbon contaminants are absorbed easily on the organic matter in particulate and colloidal form (Baumard et al., 1998a). In the aqueous compartment, hydrocarbons are subject to degradation similar to the hydrocarbon degradation observed in the atmosphere. For some hydrocarbon compounds, volatilization ensures the transfer of hydrocarbons from the water surface towards the atmosphere as is the case of low-weight molecular compounds. While high-molecular weight compounds tends to precipitate and settle on the sediments where they accumulate. Finally in water, the oil accumulation by the living organisms also contributes to the reduction in the concentration of hydrocarbons (Baumard et al., 1998b).

3.2.3 In the sediments

Hydrocarbon concentration in sediments is in general relatively higher than hydrocarbon concentration in air and water, however the concentration of hydrocarbon in sediments are in direct relation with its concentration in the watery and atmospheric compartments (Gao et al., 1998). Simpson et al (1998) stated that hydrocarbon contamination in sediments and its compositions are related to its origin. However, many environmental processes contribute to skew this "signature" while acting selectively on each compound as in the case of degradation by microorganisms. In addition, the distribution of hydrocarbons, in the sediments, between the aqueous phase and the solid phase depends on the coefficient of division n-octanol/water (log Kow) of each compound as well as the nature of the sediments (content of clay, silts, and carbon) (Palayer, 1997). After adsorption on the sedimentary particles, hydrocarbon is accumulated, and reach concentrations higher than the concentrations of hydrocarbon found in the water column, even though contaminant concentration decreases as the distance from the source increases following a logarithmic trend (Tuvikene, 1995). The adsorption of hydrocarbons in the sediments makes them more stable and less accessible to biological degradation. Thus, these pollutants can persist in the sediments several months even several years. That explains why sediments are considered tanks of pollutants (Ollivon et al., 1993; Djomo et al., 1995; Baumard et al., 1998). There are two phases of hydrocarbon accumulation in the sediments:

- A reversible phase (unstable) in which there is attachment of hydrocarbons on the sediments surface. This phase reflects the biodisponibility of hydrocarbons for delivery in the aqueous circulation compartment by bioturbation (agitation of the sediments by invertebrates) and suspension or diffusion. Thus, in this phase the biogeochemical cycle of these compounds could be initiated (Ramade, 1992).
- An irreversible phase (stable) which is a slow process that corresponds to the diffusion of the contaminant towards the sedimentary organic fraction.

Thus, the accumulation of hydrocarbons in the sediments is not definitive. The delivery of sedimentary hydrocarbons in solution or suspension can have harmful effects on the watery organizations (Beckles et al., 1998). However, only a small fraction of sedimentary hydrocarbons is subject to dissolution. Hydrocarbons are less persistent in the aerobic surface zone of the sediments than in the anaerobic deeper layers (Wilcok et al., 1996). The examination of the spatial distribution of hydrocarbons in the sediments is a determining factor in the evaluation of the biodisponibility of these compounds for the watery organizations and the identification of the possible sources of contamination (Gao et al., 1998). However, it has been shown that variation of hydrocarbons concentration in the sedimentary compartment does not reflect necessarily a pollution variation, but rather a change of the properties in the hydrocarbon/sediment interaction (Schilderman et al., 1999).

4. Bioremediation of hydrocarbon pollution in marine environment

It is clear from the description of the hydrocarbons fate in the different marine compartments that petroleum compounds are subjected to natural degradation by the native microorganisms. This degradation limits the dispersion of petroleum hydrocarbon in water and minimizes the pollution levels in some degree. However, natural degradation of hydrocarbons is not sufficient and can not eliminate the bulk of the pollution fraction in the environment. Thus, the application of biotechnology in the elimination of crude oil pollution is necessary. In fact, environmentally-related biotechnological processes were pioneered in the petroleum industry. Oil spill bioremediation technologies use modern environmental techniques that are based on natural processes to remove spilled oil from the environment without undesirable environmental impacts.

4.1 Bioremediation

Bioremediation regroup strategies that allow microorganisms the elimination of oil pollution. In fact, few pollutants resist indefinitely the attack of natural microbiota. However, the time required for the natural degradation of pollutants is extremely variable and can reach several decades. This explains the necessity of human intervention to accelerate this process. Biological methods are often cheaper than other types of treatments (Van Hamme et al., 2003). Microorganisms, which are present in natural environments, have a significant potential for the biological degradation of hydrocarbons. Bioremediation is a process that can be adopted *in situ* or *ex situ*. It's a process that uses indigenous microorganisms in recycling specific pollutants through a series of complex chemical reactions. Metabolized organic waste is transformed into water, carbon dioxide, and other sub-products that provide bacteria the energy needed for their development (Prince, 1993; Shwannell et al., 1996). The result of a total biological degradation is the elimination of pollutants without producing toxic or dangerous residue. The de-pollution by biological means attempts to accelerate or stimulate the processes of biological degradation in order to reduce the contents of contaminants. Though this approach remains a subject of controversy, several bioremediation processes have been carried out worldwide following accidental oil discharges. Romantschuk et al. (2000) showed the effectiveness and the reproducibility of the *in situ* bioremediation of a ground contaminated by crude oil. Margesin & Schinner (2001) confirmed the effectiveness of the bioremediation on site. In fact, they applied an in-situ bioremediation for the decontamination of a ground polluted by crude oil in the Alpine Glacier area.

Bioremediation represents an interesting alternative to the physico-chemical treatments of remediation; consequently, the development of natural techniques for the elimination of pollutants has been encoraged. Bioremediation can be carried out in various forms: natural attenuation, biostimulation, and bioaugmentation.

4.2 Different forms of bioremediation

4.2.1 Natural attenuation

The natural attenuation of pollution refers to the processes contributing to the reduction of pollution in a specific site. It consists on the involvement of the indigenous bacteria present in the environment without any intervention. This approach is passive, thus the process of biological degradation depends only on the natural conditions of the site. A significant reduction of pollution can be attained by this approach but it takes long periods of time. This approach was evaluated by Margesin & Schinner (2001) in a crude oil polluted site. The evaluation of the natural attenuation of polluted sites is an essential task. It aims to predict the fate of the pollution source in the site that allows planning the actions to be undertaken. Even if this technique of decontamination does not require direct human intervention, it is however necessary to intervene in order to eliminate or neutralize the source of pollution and to supervise permanently the site until the end of treatment (Mulligan et al., 2004). This type of bioremediation is not expensive because it does not require additional technical intervention but it requires long time for the treatment to be effective. For instance, Serrano et al. (2007) demonstrated that this process, carried out as a pilot of bioremediation, allowed the restoration of a ground polluted with 2700 hydrocarbon ppm in 200 days. Biodegradation by natural attenuation, cannot however be allotted only to the microorganism action. In fact, abiotic processes, such as evaporation, dissolution, dispersion, emulsification, adsorption, and photo-oxidation, can also contribute to the natural attenuation of the pollution.

4.2.2 Biostimulation

Biostimulation consists in modifying the natural conditions of a site to increase the rates of biological degradation. Indeed, the intrinsic capacities of an environment to be able to carry out natural attenuation are correlated to several parameters. The first parameter is the biodisponibility of the pollutant which can be affected by the pollutant affinity to the mineral and organic fractions of the environmental matrices. Biodisponibility of the pollutant can be influenced by the organic matter concentratin, the pH, the mineralogy, the temperature, and the type of pollutant.

The second parameter is the oxidation-reduction environment of the ecosystem. Thus, the availability as acceptors or donors of electrons can influence the microbial activities. The biodisponibility of the nutriments (nitrogen and phosphorus) and the microbial associations in the ecosystem can also control the natural attenuation (Röling & Van Verseveld, 2002). Biostimulation improves biodegradation by modifying the environmental conditions through the addition of surfactants, nutrients, water, and chemical species acting as donors or acceptors of electrons. Perfumo et al. (2007) studied the effect of temperature and the addition of nitrogen, phosphate, inorganic potassium, and surfactant on the bioremediation of a ground polluted by hexadecanes. After 40 days of treatment, the grounds treated by the addition of nitrogen, phosphorus and inorganic potassium showed 10% of increase in the degradation compared with the grounds subjected to only natural attenuation. In the same way, Margesin & Schinner (1997) showed the effectiveness of biostimulation in the biodegradation of oil (diesel). In 1991, Hinchee et al. carried out the biostimulation of natural flora in a polluted ground by pumping oxygen. During *in situ* bioremediation, the control and intervention on pH, oxygen addition, and agitation is difficul to apply which makes *in situ* bioremediation expensive. The addition of nutriments is generally adopted. Bragg et al. (1994) carried out a biostimulation by fertilization of the marine environment to stimulate the natural biomass in areas of Alaska following the accident of the Exxon Valdez. Biostimulation of marine environment can be carried out by:

i. Bioventilation: introduction of air within the contaminated zone,
ii. Biosparging: injection of air under pressure,
iii. Pumping and treatment: pumping and purification of the water contaminated by the oil spill,
iv. Addition of small quantity of mineral nutrients to the indigenous microflora. However, this method presents the disadvantage of being no selective and stimulates all types of microorganism present in the site.

4.2.3 Bioaugmentation

This method is based on the addition of *in vitro* cultivated inoculum in the polluted site. The inoculum contains one or more adapted microorganisms, with tested capacities for the degradation of pollutants. Indeed, sometimes, the endogenous microbial populations of polluted ecosystem do not present the metabolic tools to carry on the complete degradation of the pollutants. Thus, bio-augmentation makes possible to overcome this deficiency by incorporating microorganisms with adapted metabolisms on the level of the contaminated ecosystems. (El Fantroussi & Agathos, 2005). Bioaugmentation can be done by the addition of a stock alone or a group of stocks "consortia". Da Silva & Alvarez (2004) showed an improvement of anaerobic degradation of a mixture composed of Benzene/Toluene/Ethylbenzene/Xylene (BTEX) and ethanol, contained in a polluted aquifer, following the contribution of methanogene consortia. However, multiple factors, especially biotic (predation, competition), can lessen the effectiveness of such techniques while leading to the decline of the introduced populations (Van Veen et al., 1997). In order to mitigate these ecological constraints, physical protection systems of the sown microorganisms, such encapsulation in gellan gum, can be used (Moslemy et al., 2002). Genetic engineering can participate to solve the stability problems of the exogenous stocks. In this context, endogenous microorganisms can be genetically modified in their degradation genes necessary to the improvement of their purifying capacities. Thus, Watanabe et al. (2002) demosntrated an improvement of the decomposition of phenol after the addition of an endogenous microflora genetically modified by the introduction of a gene coding for a phenol-hydroxylase. Certain studies, moreover, established the possibility of improving the effectiveness of the bioremediation by the directly addition of the genes of interest in the polluted ecosystems. In this case, gene transfers between donor bacteria of catabolic plasmid and the endogenous populations allow the biological degradation of pollutants (Bathe et al., 2005). The addition of microorganisms in the environment is not frequently practiced because it is always difficult to control (Van Veen et al., 1997). It is then significant to improve our knowledge on the fate, the persistence, the activity, and the dispersion of the microorganisms injected in polluted sites. In addition, the study of biotreatability must be carried out in the laboratory before being able to choose the biorestoration operation. Indeed, it is after having checked that i) the pollutants are biodegradable, ii) the microorganisms of the site have the capacity to degrade these compounds, iii) the nature of the medium and the environmental conditions are favourable to the development of the microflora, that bioremedation could be set up. It is thus very interesting to characterize the bacterial communities, such as those degrading oil, because the knowledge of the relationship between the microorganisms and their environment is an essential element to understand an ecosystem, and therefore to consider a strategy of bioremediation.

Bioremediation processes are considered interesting alternatives to the traditional techniques of depollution. Indeed, these processes make possible to carry on the decontamination of environments polluted by decreasing the impacts on the treated ecosystems. In order to improve these processes of depollution, it is however necessary to include/understand the operating mode of these microbial communities with respect to the pollutant. Moreover, once natural depollution is initiated; the continuous monitoring of the microorganisms is necessary to make sure that the "biosystème" does not deviate from its initial function. Due to the microorganism's organization in consortia and their specific needs, they are not easily cultivable and their study by traditional microbiological techniques remains consequently limited. Thus, the identification and follow-up of the bacterial stocks being of interest in bioremediation requires the implementation of molecular techniques (Amann, 1995). Different techniques were used with certain success (Margesin & Schinner, 2001; Romantschuk et al., 2000); however, the effective application of these bioremediation techniques still requires considerable research and optimization.

4.3 Biodegradation of oil hydrocarbons in marine environment

Among the microorganisms, bacteria are qualitatively and quantitatively dominant in hydrocarbon degradation. Indeed their capacity to develop in hydrocarbons is not limited to some microbial species. Multiple studies have demonstrated the properties of hydrocarbonoclastic bacteria (Zobell since 1946; Reisfled et al., 1972; Walker & Colwell, 1974, Bartha & Atlas, 1977). The diversity of the species was also established by studies of numerical taxonomy (Bertrand & Thousand, 1989), which showed that the use of hydrocarbons by *Enterobacteriacea* would be acquired by transfer of plasmids. In addition, the composition and the effectiveness of the microflora will be a function of the oil origin, climatic conditions, and seasonal variations (Walker & Colwell, 1974). The biological degradation of a hydrocarbon mixture is not the sum of the biological degradation of each compound taken individually. Olson et al. (1999) evaluated the individual degradation of the principal chemical families of oil (n-alkanes, iso and cyclo-alkanes, and aromatics) compared with the degradation of a fraction made up (mixture) of these three families. This study demonstrated that the susceptibility to biodegradation seems to arise in the following order: n-alkanes > aromatic > iso and cyclo-alkanes.

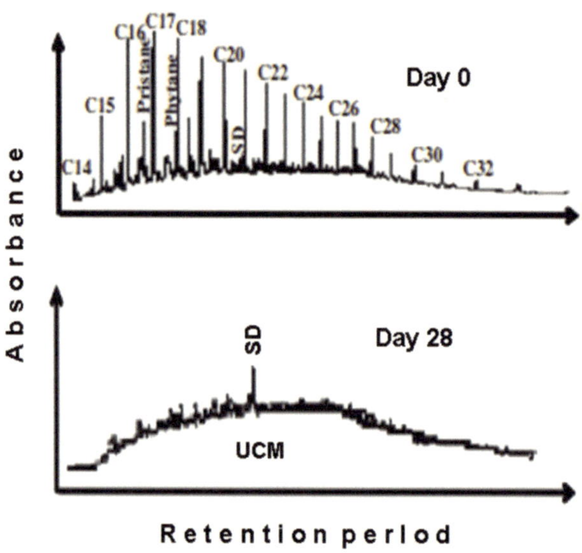

Fig. 4. Chromatogram spectra showing the biodegradation of non aromatic hydrocarbon (NAH) fractions, at different incubation periods (Zrafi-Nouira et al., 2009); UCM: Unresolved complex mixture; SD: standard.

The individual degradation of hydrocarbon compounds has been extensively studied for years. However, few studies have focused on the evaluation of the biodegradation of mixtures of complex products as crude oils. The aliphatic fraction is degraded more quickly than the aromatic fraction. However, the degradation is slower for mixtures of compounds than on the individual compounds. Data concerning the degradation of oil result often from observations carried out on polluted sites, i.e. open ecosystems, in which a complete assessment of the biodegradation process is not possible. Several studies concerning petroleum biodegradation, on site or in the laboratory, were carried out by Salanitro, (2001); who proved that the rates of degradation were often over-estimated, thus stressing the importance to work in a control environment (laboratory) to effectively carry out the assessment of the carbon degradation tests. Research has shown that the degradation of complex mixtures as oil depends on various types of bacteria whose catabolic capacities are complementary can thus cooperate in the degradation. Richard & Vogel (1999) isolated a consortium from the ground containing seven different bacteria (five *Pseudomonas*, an *Achromobacter*, and another bacterium that was not identified) of which only three were able to use hydrocarbons as growing substrate. It seems that the rest of the bacteria present in the consortium used the metabolic intermediaries as growing substrate. Head et al., (2006) observed that in marine environment following an oil spill, there is a correlation between hydrocarbon composition and bacteria dynamics. Bacteria take a significant share in natural decontamination following an oil discharge because of their capacities to quickly adapt to changing conditions. Figure 4 shows chromatograms showing non aromatic hydrocarbon (NAH) fractions, at different incubation periods during crude oil (Zarzatine) degradation by microflora in polluted seawater from Tunisia.

Table 1 summarizes the biodegradation of different carbon sources and briefly describes the microorganisms involved in the process.

Carbon Sources	Incubation period	Biodegradation %	Microorganisms description	Reference
Zarzatine Crude oil (ZCO)	28 days		Adapted microflora from polluted seawater adjacent to oil refinery (Tunisia)	Zrafi-Nouira et al.(2009)
Saturated fraction		92.6		
Aromatic fraction		68.7		
Oil refinery tank bottom sludge (OTBS)	10 days		Adapted microbial consortia from polluted soil	Gallego et al. (2007)
n-alkanes		100		
Branched alkanes		44		
cyclo-alkanes		85		
aromatics		31-55		
Crude oil	10 weeks		Microbial consortium from seawater near a oil refinery (UK)	McKew et al. (2007)
n-alkanes (C10-C18)		99		
n-alkanes (C20-C32)+Pr		41-84		
PAHs		32-88		
Crude oil	7 days		Seawater adjacent to oil refinery (Tunisia)	Ozaki and Fujita (2006)
Saturated fraction		56		
Aromatic fraction		46		
Saturated + aromatics		20		
Crude petroleum	200 days		Eubacterial from soil contaminated with wood treatment plant (near Barcelone)	Vinas et al. (2005)
TPH		72-79		
PAHs		83-87		

Pr: Pristane; TPH: total petroleum hydrocarbons; PAHs : Polycyclic aromatic hydrocarbon

Table 1. Biodegradation of crude oil compounds.

5. Place of metagenomics in petroleum bioremediation

5.1 From culture-dependent approaches to culture-independent approaches

Investigations of microbial species that are present in petroleum polluted environments and during petroleum spill bioremediation have been traditionally conducted using samples to grow bacterial cultures in the laboratory. However, laboratory growth medium does not reproduce the actual diversity of the complex system (polluted environment). The current culture-dependent methods used can only account for a small subset of the total microbial diversity. This causes the underestimation of the microbial communities present in petroleum polluted environments. The challenge of isolating even a small fraction of these organisms seems overwhelming (Tyson & Banfield, 2005). Traditional microbiological methods of cultivation recover less than 1% of the total bacterial species present in the polluted site sample, and the cultivable portion of bacteria is not representative of the total phylogenetic diversity. Indeed, most of the cultivated microorganisms are those that grow quickly and are capable to growth in nutrient-rich media (Leadbetter, 2003) and those that are dominant and key players in the environmental system.

Thus, an improvement of cultivation methods has recently been addressed and non-traditional culture methods have been developed (Green & Keller, 2006; Tyson & Banfield, 2005). These methods are based on modified traditional approaches to isolate previously uncultured and phylogenetically distinct microorganisms that grow in diluted nutrient media or simulated natural environments (Green & Keller, 2006; Tyson & Banfield, 2005). Non-traditional approaches include single cell manipulation techniques such as optical tweezers and laser microdissection, targeted isolation using 16S rRNA-direct probes, and the development of high-throughput methods to grow encapsulated single cells (Tyson & Banfield, 2005). Despite these advances, cultivation attempts of petroleum degrading-bacteria and their isolation from complex polluted site samples fail. Genetic methods can be used to overcome the difficulties associated with the laboratory cultivation of petroleum-degrading-bacteria, which depend upon environmental conditions, unsuspected growth factors, and multiples relationships within a complex community. The use of genetic techniques to detect, identify, and quantify bacteria has largely replaced microbial growth tests. These modern biotechnology methods have been recently employed in petroleum polluted environments. Genetic techniques will be the methods of choice for monitoring bioremediation bacteria players in the future. Initial efforts to introduce the use of genetic techniques for monitoring petroleum-degrading-bacteria in environmental samples have involved a type of biotechnology called metagenomic. There are molecular techniques available that can be used to study genomic material obtained directly from the environment, instead of cultured samples, which provides the opportunity to extract sequence data from microbial communities as they exist in nature. Extraction of genome sequences from metagenomic data is crucial for reconstructing the metabolism of microbial communities that cannot be mimicked in the laboratory.

Crude oil metagenomics could be used to examine the bacteria dynamics during petroleum biodegradation and *in situ* bioremediation in hydrocarbon polluted sites.

The comprehension of petroleum polluted environments and the identification of the microorganisms playing key roles during biological degradation remain uncertain so that the optimal development of bioremediation strategies is still unknown. The advantages of these molecular techniques are that they allow accessing the microbial diversity of complex systems (Theron & Cloete, 2000) and allow supervising and evaluating the bioremediation process (Bachoon et al., 2001; Brockman, 1995; Widada et al., 2002). Most of these studies using molecular techniches were based on the construction of 16S rRNA clone libraries and subsequent sequencing of individual 16S rRNA clones. The resulting nucleotide sequences were then taxonomically and phylogenetically classified to deduce the structure of the underlying community.

5.2 Molecular tools to monitor bacterial composition in petroleum polluted sites

Assessing environmental microbial community structures through molecular techniques requires a satisfactory sampling strategy that takes into account the high microbial diversity and the heterogeneous distribution of microorganisms in polluted sites. So that, microbial ecologists have invested substantial effort in optimizing DNA and RNA extraction procedures from various environmental samples (water, sediments, soil and sludge).

The key molecular player in microbial classification has been the RNA component of the small subunit of ribosomes (SSU rRNA, or 16S/18S rRNA), which Carl Woese in the early 1970s considered to be a convenient and reliable « universal molecular chronometer ». These nucleic acids contain preserved regions allowing an easy isolation and variable regions allowing the differentiation of taxonomic units. These biological molecules are the targets of choice for molecular ecology studies (Olsen et al., 1986).

The screening of metagenomic libraries has traditionally followed two paths: sequence-based and function-based screening. Some of the metagenomic libraries have been screened by hybridization or by PCR to detect genes with homology to known genes. The analysis of genome sequence data that has been recovered from the environment is motivated by many objectives, which include the establishment of gene inventories and natural product discovery (Handelsman, 2004). Genomics can resolve the phyogenetic composition and metabolic potential of communities during bioremediation processes and establish how functions are played among populations, reveal the dynamics of population diversity in response to changes in hydrocarbon changes during biodegradation of petroleum compounds. The evaluation of bioremediation can be studied by molecular fingerprinting techniques applied to degradation genes. The polymorphism of genes considered is detected by enzymatic restriction and examination of the profiles of restriction after genic amplification. Watanabe et al. (1998) used temperature gradient gel electrophoresis (TGGE) of gene coding a phenol hydroxylase (LmPH) and gene coding 16S rRNA in order to detect and to characterize the prevalent bacteria degrading phenol in activated sludge of a station of purification. MacNaughton et al. (1999) used denaturing gradient gel electrophoresis DGGE to identify the populations responsible for decontamination and to evaluate two techniques of bioremediation after oil spill (biostimulation and bioaugmentation). Two sequences affiliated to *Flexibacter-Cytophaga-Bacteroides* were detected in the biostimulated samples but the participation of these microorganisms in the hydrocarbon degradation was not proven.

The combination of several methods offers interesting information in microbial diversity of the studied ecosystems. Watts et al. (2001) recently compared three methods using 16S rDNA: ARDRA (amplified ribosomal DNA restriction analysis), T-RFLP (Terminal-restriction fragment length polymorphism), and DGGE that were applied to study the communities involved in the degradation of chlorinated compounds. They conclude that the three methods render different results, namely the relative frequency of the clones for ARDRA, intensity of the bands for the DGGE (resolution of freezing) and the peaks height for the T-RFLP. However, the use of these three different methods seems to allow an appropriate analysis of the studied communities and the information obtained from the different methods employed was complementary.

It is also interesting to couple the analysis of microbial diversity with a degradation activity study. Kanaly et al. (2000) used the DGGE technicque to access the dynamics of a bacterial consortium during the degradation of benzo[α]pyrene. They confirmed the presence, in their consortium, of microorganisms known to degrade the aromatic compounds, such as *Sphingomonas paucimobilis EPA505, Mycobacterium str.PYR-1,* and *Alcaligenes denitrificans WW1.* In addition, it is interesting in bioremediation evaluation to detect the functional degradation genes. The expression of these genes can be detected based on ARNm analysis (Van Hamme et al., 2003). The extraction of ARNm followed by RT-PCR (Reverse transcription polymerase chain reaction) coupled to an analysis of 16S rDNA makes possible to get evidence of the active microorganisms in a microflore. These methods offer an interesting alternative to detect the activities of non- cultivated bacteria.

In fact, certain degradation pathways are well characterized and some degradation genes are well known, which allowed the development of a large variety of specific primers. Currently, the detection of the *alk*B gene, implied in the metabolism of alkanes, and others genes implied in the metabolism of aromatic hydrocarbons is well documented (Whyte et al., 1997; Whyte et al., 1998; Churchill et al., 1999; Smits et al., 1999; Widada et al., 2002). Margesin et al. (2003) showed that the genotypes of alkB of *P. putida* are more frequently found in high-altitude polluted grounds than in the not polluted grounds. Wilson et al. (1999) developed a method to isolate and characterize the ARNm of microorganisms degrading naphthalene in a hydrocarbon contaminated aquifer. Recently Debruyn et al. (2011) examined the homologous genomic regions in four PAH-degrading Mycobacterium (strains JLS, KMS, MCS, and M. gilvum PYR-GCK) isolated from two PAH-contaminated sediments. These isolates had nidA (and some, nidA3) genes that were homologous to Mycobacterial ARHDO (aromatic ring hydroxylated dioxygenase) genes, suggesting that horizontal gene transfer events had occurred. Militon et al. (2007) proposed a probe design algorithm that is able to select microarray probes targeting SSU rRNA at any phylogenetic level. Application of the combined array strategy may help identifying unknown bacterial species. Neufeld et al. (2006) used microarray technology to characterize and compare hexachlorocyclohexane (HCH) contaminated soils from Spain and their results highlighted the power of habitat-specific microarrays for comparing complex microbial communities. Gomes et al. (2010) used the combination of culture enrichments and molecular tools to identify bacterial guilds, plasmids, and functional genes potentially important in the process of petroleum hydrocarbon decontamination in mangrove microniches (rhizospheres and bulk sediment), and to recover degrading consortia for future use in remediation strategies. Fingerprinting could be obtained with proteomics and metabolomics approaches but results are difficult to relate to a precise identification at the phylogenetic level. Databases must be constructed and related genomics, proteomics, and metabolomics information to obtain easily interpretable results for microorganism identification. In fact, shotgun sequencing approaches have been used to reconstruct genome fragments and near complete genomes from uncultivated species and strain from natural consortia (Tyson & Banfield, 2005). Sequence data from uncultivated consortia are more useful for metabolic profiling if the genome fragment can be assigned to an organism type. Sequence similarity to previously characterized organisms and the presence of phylogenetically informative gens would assist genomes reconstruction. Wang et al. (2011) generated a complete *Methanococcus maripaludis* genome from metagenomic data derived from a thermophilic subsurface oil reservoir. Comparison of the genome from the thermophilic, subsurface environment with the genome of the species type provided insight into the adaptation of a methanogenic genome to an oil reservoir environment. Thus a robust database of genomic information from cultivated microorganisms can greatly enhance the ability to link genome fragments to a particular organism (Figure 5). Currently, both the cost of sequencing and the challenges that are associated with the management of vast datasets precludes comprehensive genomic studies of highly complex communities. It would be optimal if the amount of sequencing is appropriate with the diversity of the sample to be analyzed.

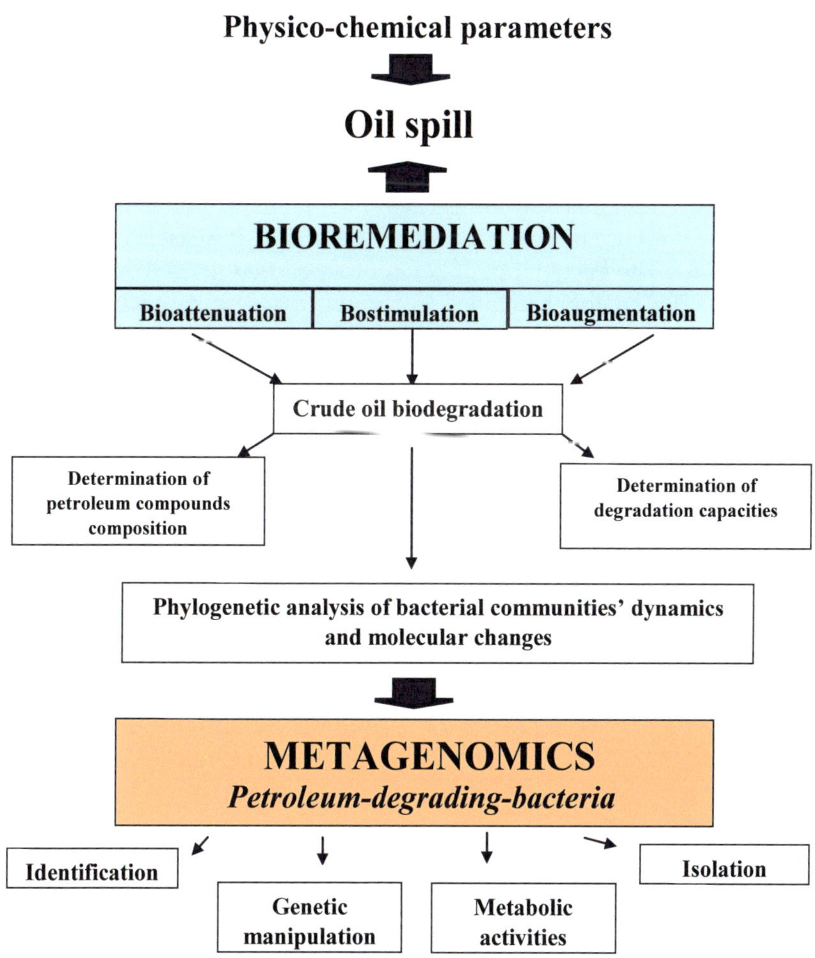

Fig. 5. Schematic representation of the bioremediation process of an oil spill and its possible improvement by metagenomics.

5.3 An overview of phylogenetic analysis reports in characterising bacteria composition in polluted environments

The analysis of bacterial diversity in different polluted environments helped to discover the black box of adapted consortia dynamics in response to the presence of petroleum compounds. Phylogenetic analysis of different consortia originated from petroleum polluted environment shows that the abundant phylotypes distributed within the group of *Proteobacteria* are essentially *alph-, beta-,* and *gamma-proteobacteria*. The predominance of *Proteobacteria* in polluted hydrocarbons environments has been largely documented (seawater, ground, ice, estuary, and river). Thus the prevalence of *Proteobacteria* in polluted seawater is well known (Brakstad & Lødeng, 2004; Brakstad et al., 2004; Kasai et al., 2005; Marc et al, 2005). Using hydrocarbon polluted sand, Mac-Naughton et al. (1999) showed the predominance of the *Alphaproteobacteria* compared to the *Gammaproteobacteria*. They also reported that *Gammaproteobacteria* are detected after 8 weeks following contamination by oil. Popp et al. (2006) concluded that the predominance of *Gammaproteobacteria* can be correlated to high levels of contamination. McKew et al. (2007) reported that the indigenous microflora from an estuary in Brittany (France) adapted to crude oil degradation were dominated by *Thalassolitucus Oleivorans* and were composed of various other phylum such as *Oceanospirillum, Roseobacter,* and *Arcobacter*. In grounds contaminated by crude oil in the north of Canada, Juck et al. (2000) showed that the bacterial population was composed mainly by *Nocardioides, Arthrobater,* and *Xanthomonas*. In Germany (the North Sea) Brakstad & Lødeng (2004) reported that the indigenous microflora of polluted environment was characterized by the presence of *Sphingobacteria Flavobacteria, Pseudoalteromonas, Alteromonas, Vibrio,* and *Roseobacter*. The bacterial population of a polluted ground zone of Shizuoka, in Japan (Kasai et al., 2005) mainly contained *Variovorax, Acidovorax, Burkholderia, Thiobacillus, Alcaligens,* and other microorganisms. In the clone library obtained from asphalts of the tar wells in California, the prevalent bacteria were affiliated to *Chromatiales, Xanthomonadaceae, Pseudomonadaceae,* and *Rhodobacteraceae* (Kim & Crowley, 2007). *Moraxellaceae* within the *Acinetobacter* are often present in polluted sites with shown capacities to degrade linear alkanes in C16 (Bogan et al., 2003). The presence of *the Xanthomonas, Stenotrophomonas,* and *Hydrocarboniphaga* has been reported (Popp et al., 2006). Similarly, the abundance of other dominant groups in polluted environments has also been reported; in fact, *Actinobacteria, Firmicutes,* and *Bacteroidetes* were described to be also dominant (Chang et al., 2000; Popp et al., 2006). Recently, Dos Santos et al. (2011) studied the effect of the oil spill on the existing microbial community using direct pyrosequencing, which confirmed that the phylum *Proteobacteria,* in particular the classes *Gammaproteobacteria* and *Deltaproteobacteria,* were prevalent before and after the simulated oil spill. The authors claim that the order *Chromatiales* and the genus *Haliea,* and three other genera, *Marinobacterium, Marinobacter,* and *Cycloclasticus* increased, which make them possible targets for the biomonitoring of the impact of oil in mangrove settings.

The difference in bacteria group dominance in petroleum polluted environment indicates that indigenous population is specific to this setting. These differences can be allotted to the environmental conditions (temperature, pH, salinity, O_2, etc.) and the nature of the petroleum compounds present in the site (crude oil, refined product, asphalt, gasoil, etc).

The dominant described taxa represent specific bacteria that are strongly resistant to the presence of petroleum hydrocarbon. General literature (Zrafi-Nouira et al., 2009; Head et al.,

2006; Popp et al., 2006) describes a range of phylogenetic bacteria groups in which the most abundant taxa are closely related to: *Pseudoaminobacter, Alcaligenes Nitrate redactor* and *Halomonas, Alcanivorax, Pseudomonas , Mesorizobuim , Shewanella,* and *Marinobacter.*

Indeed various indigenous total colony counts were worldwide described, and each microflora was specific to its environment. These studies described various microbial communities which can live on the complex mixtures of oil hydrocarbons. Nevertheless, little is known on the changes of the consortia after a specific enrichment by crude oil (Zrafi-Nouira et al., 2009).

5.4 An overview of bacterial dynamics during *in vitro* biodegradation and *in situ* bioremediation

There is a general interest to study the microflora diversity that is able to degrade crude oils (Abed et al., 2002; MacNaughton et al., 1999; Orphan et al., 2000).

The phylogenetic analyses carried out in various stages of the biodegradation incubation during crude oil degradation showed differences in the chemical composition of the oil and the microbial composition of the communities (Table 2). Brakstad & Lødeng (2004) indicated that the reduction of microbial diversity is observed from day 7 to day 21 during the degradation of crude oil. Indeed, the presence of oil reduces clearly bacteria diversity (Röling et al., 2002). During biodegradation bacterial divisions persist dominant but with a variation of their proportions and their compositions.

Library	Affiliation group	Affiliation sub-group	Number of clones	Number of OTUs	% of OTUs	% of cultivated OTUs	% of uncultivated OTUs	% of novel OTUs
Adapted microflora	Proteobacteria	Alpha	27	8	36.4	87.7	12.5	0
		Beta	36	3	13.6	66.7	0	33.3
		Gamma	14	6	27.3	83.3	0	16.7
	Firmicute	Clostridia	13	3	13.6	33.3	0	66.7
	Actinobacteria	Actinobacteria	3	1	4.55	0	0	100
	Bacteroidetes	Flavobacteria	1	1	4.55	100	0	1
Day 7	Proteobacteria	Alpha	43	8	29.6	50	0	50
		Beta	3	1	3.7	100	0	0
		Gamma	49	16	59.3	62.5	0	37.5
	Firmicute	Clostridia	2	1	3.7	100	0	0
	Bacteroidetes	Flavobacteria	2	1	3.7	100	0	0
Day 14	Proteobacteria	Alpha	68	7	58.4	42.8	0	57.2
		Beta	5	1	8.3	0	100	0
		Gamma	13	3	25	0	33.3	66.7
	Actinobacteria	Actinobacteria	1	1	8.3	0	0	100
Day 21	Proteobacteria	Alpha	8	2	22.2	100	0	0
		Beta	2	1	11.1	0	100	0
		Gamma	45	5	55.6	80	0	20
Day 28	Actinobacteria	Actinobacteria	1	1	11.1	0	100	0

[a]: The number of OTUs was calculated with a threshold value of 97% 16S rRNA sequence similarity

Table 2. Distribution of clone sequences and OTUs analyzed during *in vitro* crude oil biodegradation using an adapted bacterial microflora from polluted seawater (Zrafi-Nouira et al., 2009).

Several studies suggest a balance between Alpha- and Gamma-Proteobacteria, which depend on the nature, level, and composition of the oil sources. In our previous study (Zrafi-Nouira et al., 2009); we noticed that after day 7 of crude oil degradation the microbial flora from polluted seawater was significantly diversified. We also noted the appearance of a novel Operational Taxonomic Unit (OTUs) with less than 97% of similarity with the nearest parent in the GenBank database. This study made possible to identify bacterial phylotypes potentially implied in the processes of decomposition of aliphatic and aromatic hydrocarbons during the degradation of crude oil.

We identified a potential implication of *Gamma-Proteobacteria: Alcanivorax, Halomonas, Enterobacter, and of Uncultred bacteruim*. We noticed a phenomenon named "gamma-shift", which consists of a deviation of diversity towards the γ-Proteobacteria during an increase in the nutrients of the contamination source (Popp et al., 2006). According to Evans et al. (2004), this tendency could be caused by a high number of bacteria degrading hydrocarbons belonging to this class (γ-Proteobacteria). The abundance of bacteria from the *Alpha-Proteobacteria,* primarily of the *Pseudaminobacter* also seems to suggest their adaptation to the presence of hydrocarbons. These bacteria are even found during the first phases of degradation. Indeed, the detected species *Pseudaminobacter sp. W11-4* A was described as pyrene degrading bacteria. However, after the 28 day of degradation, only the *Actinobacteria* were present in the consortium. We hypothesized that at this stage the prevalent species are implied in the decomposition of recalcitrant aromatic and cyclic hydrocarbons with the possibility of the anaerobic decomposition of the residual hydrocarbons. Similarly, another study showed the association of *Actinobacteria* in the anaerobic decomposition of hydrocarbons oil (Rehmann et al., 2001; Pinda-flora et al., 2004). *In vitro* studies of the crude oil degradation process provides insights on the composition of the bacterial communities and their evolution during degradation. It is thus essential to be able to identify the purifying bacteria composing this microflora and to understand their adaptive mechanisms in the presence of petroleum. Thus, information obtained in this regard could guide the installation of processes stimulating the activity of the bacteria implied in the degradation of the pollutant or allowing their massive establishment in an environment to be treated *in situ*. The bacterial composition is then of primary importance for the comprehension of the process of *in situ* bioremediation after oil spills. The study of bacterial dynamics and the diversity during the natural attenuation, biostimulation or bioaugmentation of the oil discharge can help identifying bacterial phylotypes potentially implied in the biodegradation processes. Indeed during bioremediation, we can observe an evolution of the bacterial communities which seems to be correlated with the reduction of oil charge in the sites. This could be due to the decomposition of hydrocarbons by specialized microorganisms that generates the intermediate compounds, which produce new pressures of selection on the microbial communities involving their evolution. The adapted bacteria persist and even grow and those not having the adaptive mechanisms decline because the conditions in this environment do not promote their development. Observations conducted during bioremediation indicated that *Proteobacteria* decreased while *Actinobacteria* increased. Although bacterial communities show evolution during bioremediation, the presence of some phylotypes seems to be constant, which is named bacterial "core-set". This "core-set" is composed of phylotypes belonging primarily to *Actinobacteria α-Proteobacteria, β-Proteobacteria, γ-Proteobacteria, and Flavobacteria.*

Biotreatability of contaminated sites must be carried out first in the laboratory setting to gain knowledge on the proper biorestoration operation; nevertheless the *in situ* study is needed to access real information of the bioremediation process in nature. *In situ* characterization of bacterial communities degrading oil provides information on the relationship between microorganisms and the environment. In addition, the composition and the effectiveness of the microflora will be a function of the origin of oil and the environmental conditions (Harayama et al., 1999). Several studies concerning biodegradation of petroleum products, on site or at laboratory scale, were carried out (Salanitro, 2001). Some studies have proved that the rates of degradation were often over-estimated in laboratory evaluations. An explanation for this might be the fact that each medium has a potential of degradation and a specific indigenous microflora, so that similar contamination does not cause the development of similar microbial communities (Bundy et al., 2002). In consequence, high concentration of pollutants decreases the density of the microbial population, whereas a lower content of pollutants enriches the microflora able to degrade hydrocarbons (Long et al., 1995). According to Smit et al. (2001), the distribution of bacteria would be rather correlated with the availability of nutrients. Indeed, availability of the substrates would select species supporting a fast and massive division, such *Proteobacteria* whereas a limited availability of nutriments would support the species having slower but more effective division like the *Acidobacterium* and *Actinobacteria.* These two theories are however not conflicting if we take into account that in marine environment the presence of unstable niches from the nutritive point of view will support microorganisms with fast adaptation.

In addition, the interactions between various species (phenomena of co-operation), with variable generation times from one species to another allows the relative abundance of each species within the population (a less abundant population can replace abundant population but less adapted to the selected culture medium) that can act in the selection of the microflores on site (Amann et al., 1995; Ward et al., 1997).

Comparison between *in vitro* and *in situ,* based on the bacterial composition analysis changes and its dynamics during bioremediation, is valid for a better comprehension of these processes and their applications at various levels (decontamination of the polluted sites, industrial and domestic effluent, and waste processing). There is a difference in the fine composition of the bacterial phylotypes. The microflora acting on site has a wealth of bacterial diversity, whereas the microflora obtained during various incubation periods at laboratory scale are less diverse and more specific. This is the result of multiple factors: the effects of the conditions of the sites, the interference of others organic compounds present on site, the interference of the microorganisms present in other compartments of the environment (sediments and rocks). For the microflora acting *in vitro,* multiple factors can intervene on the bacterial composition. The major one is the effect of acclimation to crude oil. Indeed, acclimation supports a specific selection to the type of oil used. As well as the effect of the selection that is imposed by the contribution of continuous pollution sources on the indigenous microflora. Under laboratory conditions, the adapted microflora is subjected to the effect of the metabolites and their accumulation in a closed microcosm that can be toxic and limit their growth. In addition to these differences within the phylotypes, a bacterial "core-set" was highlighted in terms of bacterial groups. Indeed, the bacterial inventories (Zrafi-Nouira et al., 2009) showed the prevalence of *Proteobacteria* in the majority of the cases with a swinging between *Gamma* and *Alpha Proteobacteria* in terms of abundance and diversity according to the nature and the degree of the contamination. A proliferation of *Gamma proteobacteria* phenomenon called "Gamma-shift" could be related to an enrichment of polluting compounds. During *in vitro* degradation and *in situ* bioremediation the proliferation of *Gamma proteobacteria* was observed in the early phases of degradation, whereas an exclusive abundance of *Actinobacteria* was observed in the late phases of bioremediation. We suggested an implication of these *Actinobacteria* in the decomposition of recalcitrant branched hydrocarbons or aromatics present at the end of the degradation process. The adaptation of the bacterial populations following petroleum pollution can highlight the degrading capacities of certain microorganisms. Thus, in order to identify the bacteria having these abilities, comparison of population dynamics in different polluted matrices is useful to monitor active spices.

6. Conclusions

In nature and especially in marine environments crude oil is considered to be an important source of pollution. After oil spills accidents, petroleum products are subjected to environmental changes resulting in their degradation. However, human intervention through the application of bioremediation technology is necessary to accelerate the bioremediation process. Advances in bioremediation technologies are based in the improvement of our comprehension of the composition of the microflora present in the polluted site and the activity of the microorganisms implied in petroleum degradation, which help to select bioremediation strategies for polluted sites. It is thus necessary to develop a multi-field approach taking into account all the parameters intervening in the natural bioremediation process. The use of methods based on the molecular approach could be helpful to improve the evaluation of the diversity, identification, isolation, and functional properties of microorganisms involved in the degradation of different petroleum compounds. The molecular analysis of the bacterial composition present in polluted sites represents an essential stage to improve our understanding of the microorganisms involve in the degradation of hydrocarbons. Similarly, metagenomics represent the technology of choice to enhance the application of bioremediation in the future. The key point of this review is that nature seems to offer a diversity of bacteria capable of degrading and neutralize crude oil pollution. It is up to us to optimize the application of what nature provides.

7. References

Aas E., Baussant T., Balk L., Liewenborg B., Andersen O.K. (2000). PAH metabolites in bile, cytochrome P4501A and DNA adducts as environmental risk parameters for chronic oil exposure: a laboratory experiment with Atlantic cod. *Aquat. Toxicol.* 51: 241-258.

Abed R.M., Safi N.M., Köster J., De Beer D., El-Nahhal Y., Rullkötter J., Garcia-Pichel F. (2002). Microbial diversity of a heavily polluted microbial mat and its community changes following degradation of petroleum compounds. *Appl Environ Microbiol* 68 : 1674-1683.

Amann R., Ludwig W., Schleifer K.H. (1995). Phylogenetic identification and *in situ* detection of individual microbial cells without cultivation. *Microb Rev.* 59: 143-169.

Atkinson R., Arey J. (1994). Atmospheric chemistry of gas-phase polycyclic aromatic hydrocarbons : formation of atmospheric mutagens. *Environ. Health Prespect.* 102 (4): 117-126.

Bachoon D.S., Araujo R., Molina M., Hodson R.E. (2001). Microbial community dynamics and evaluation of bioremediation strategies in oil -impacted salt marsh sediment microcosms. *J Ind Microbiol Biotechnol.* 27: 72-79.Bartha and Atlas, 1977

Bartha R. & Athlas R.M. (1977). The microbiology of aquatic oil spills. *Adv.appl. Microbiol.* 22: 225-266.

Bathe S., Schwarzenbeck N., Hausner M. (2005). Plasmid-mediated bioaugmentation of activated sludge bacteria in a sequencing batch moving bed reactor using pNB2. *Lett Appl Microbiol.* 41: 242-247.

Baumard P., Budzinski H., Garrigues P. (1998b). Polycyclic aromatic hydrocarbons in sediments and mussels of the western Mediterranean sea. *Environ Toxicoland Chem.* 17: 765-776.

Baumard P., Budzinski H., Michon Q., Garrigues P., Burgeot T., Bellocq J. (1998a). Origin and bioavailability of PAHs in the Mediterranean Sea from mussel and sediment records. *Estuar Coast Shelf S.* 47: 77-90.

Beckles D.M., Ward C.H., Hughes J.B. (1998). Effect of mixture of polycyclic aromatic hydrocarbons and sediments on fluoranthene biodegradation patterns. *Environ Toxicol Chem.* 17 (7): 1245-1251.

Bertrand J.C., Mille G. (1989). *Devenir de la matière organique exogène. Un modèle: les hydrocarbures.* In: Bianchi M, Marty D, Bertrand JC, Caumette P, Gauthier MJ (eds) Microorganismes dans les écosystèmes océaniques, Chapitre 13. Masson, Paris, pp 85-343.

Bingham A. (1992). Gulf War Pollution leaves long-term legacy. *Pollution Prevention,* 2 (4): 19-23.

Bogan B.W., Sullivan W.R., Kayser K.J., Derr K.D., Aldrich H.C., Paterek J.R. (2003). Alkanindiges illinoisensis gen. nov., sp. nov., an obligately hydrocarbonoclastic, aerobic squalane-degrading bacterium isolated from oilfield soils. *Int J Syst Evol Microbiol,* 53: 1389-1395.

Bouchez M., Blanchet D., Haeseler F., Vandecasteele J.P. (1996). Les hydrocarbures aromatiques polycycliques dans l'environnement : Propriétés, origines, devenir. *Revue de l'Institut Français du Pétrole.* 51(3) : 407-419.

Bouloubassi I., Saliot A. (1991). Sources and transport of hydrocarbons in the Rhone delta sediments (Northwestern Mediterranean). *Fresenius J Anal Chem.* 339: 765-771.

Bragg J.R Prince R.C, Harer E.J, Atlas R.M. (1994). Effectiveness of bioremedation for the Exxon Valdez oil spill. *Nature.* 368:413-418.

Brakstad O.G., Lødeng A.G.G. (2004). Microbial diversity during biodegradation of crude oil in seawater from the North Sea. *Microb. Ecol.* 49(1):94-103.

Brakstad O.G., Bonaunet K., Nordtug T., Johansen O. (2004). Biotransformation and dissolution of petroleum hydrocarbons in natural flowing seawater at low temperature. *Biodegradation.* 15(5):337-346.

Brockman F.J. (1995). Nucleic-acid-based methods for monitoring the performance of *in situ* bioremediation. *Mol. Ecol.* 4: 567-578.

Budzinski H., Jones I., Bellocq C., Pierard P., Garrigues P. (1997). Evaluation of sediment contamination by polycyclic aromatic hydrocarbons in the Gironde estuary. *Mar. Chem.* 58 : 85-97.

Bundy, J.G., Paton, G.I., Campbell, C.D. (2002). Microbial communities in different soil types do not converge after diesel contamination. *J Appl Microbio.* 92: 276-288.

Burns K.A., Teal J.M. (1979). The West Falmouth oil spill: Hydrocarbons in the salt marsh ecosystem. *Estuar. Costal. Mar. Sci.* 8: 349-360.

Chang Y.J., Stephen J.R., Richter A.P., Venosa A.D., Brüggemann J., Macnaughton S.J., Kowalchuk G.A., Haines J.R., Kline E., White D.C. (2000). Phylogenetic analysis of aerobic freshwater and marine enrichment cultures efficient in hydrocarbon degradation: effect of profiling method. *J Microbiol Methods*. 40(1):19-31.

Churchill S.A., Harper J.P., Churchill P.F. (1999). Isolation and characterization of a Mycobacterium species capable of degradin threeand four-ring aromatic and aliphatic hydrocarbons. *Appl Environ Microbiol.* 65: 549-552.

Clark R.C.J., Blumer M. (1967). Distribution of n- paraffins in marine organisms and sediment. *Limnol Oceangr.* 12: 79-87.

Curtis T.P., Sloan W.T. (2004). Prokaryotic diversity and its limits: microbial community structure in nature and implications for microbial ecology. *Curr Opin Microbiol.*7: 221-226.

Da Silva M.L., Alvarez P.J. (2004). Enhanced anaerobic biodegradation of benzenetoluene-ethylbenzene-xylene-ethanol mixtures in bioaugmented aquifer columns. *Appl Environ Microbiol* 70: 4720-4726.

Debruyn Jm, Mead Tj, Sayler GS. (2011). Horizontal Transfer of PAH Catabolism Genes in Mycobacterium: Evidence from Comparative Genomics and Isolated Pyrene-Degrading Bacteria. *Environ Sci Technol.* [Epub ahead of print]

Djomo J.E., Ferrier V., Gauhier L., Zoll-Moreux C., Marty J. (1995). Amphibian micronucleus test in vivo: Evaluation of the genotoxicity of some major polycyclic aromatic hydrocarbons found in a crude oil. *Mutagenesis.* 10: 223-226

Dos Santos H.F., Cury J.C., Do Carmo F.L., Dos Santos A.L., Tiedje J. (2011). Mangrove Bacterial Diversity and the Impact of Oil Contamination Revealed by Pyrosequencing: Bacterial Proxies for Oil Pollution. *PLoSone* 6(3): e16943. doi:10.1371/journal.pone.0016943

El Fantroussi S., Agathos S.N. (2005). Is bioaugmentation a feasible strategy for pollutant removal and site remediation?. *Curr Opin Microbiol.* 8: 268-275.

El Fantroussi S., Urakawa H., Bernhard A.E., Kelly J.J., Noble P.A., Smidt H., Yershov G.M., Stahl D.A. (2003). Direct profiling of environmental microbial populations by thermal dissociation analysis of native rRNAs hybridized to oligonucleotide microarrays. *Appl Environ Microbiol.* 69: 2377-2382.

Elloumi J., Guermazi W., Ayadi H., Bouaïn A., Aleya L. (2008). Detection of Water and Sediments Pollution of an Arid Saltern (Sfax, Tunisia) by Coupling the Distribution of Microorganisms with Hydrocarbons. *Water Air Soil Poll.* 187:157-171.

Ericson G., Lindesjö E., Balk L. (1998). DNA adducts and histopathological lesions in perch (Perca fluviatilis) and northern pike (Esox lucitus) along a polycyclic aromatic hydrocarbon gradient on the Swedish coastline of the Baltic Sea. *Can J Fish Aquat Sci.* 55: 815-824.

Evans F.F., Seldin L., Sebastian G.V., Kjelleberg S., Holmstrom C., Rosado, A.S. (2004). Influence of petroleum contamination and biostimulation treatment on the diversity of Pseudomonas spp. in soil microcosms as evaluated by 16S rRNA based-PCR and DGGE. *Lett Appl Microbiol.* 38: 93-98.

Fernandes M.B., Sicre M.A., Boireau A., Tronczynski J. (1997). Polyaromatic hydrocarbon (PAH) distributions in the Seine River and its estuary. *Mar Pollut Bull.* 34: 857-867.

Gallego J.L., García-Martínez M.J., Llamas J.F., Belloch C., Peláez A.I., Sánchez J. (2007). Biodegradation of oil tank bottom sludge using microbial consortia. *Biodegradation.* 18(3):269-281.

Gao J.P., Maguhn J., Spitzauer P., Kettrup A. (1998). Distribution of polycyclic aromatic hydrocarbons (PAHs) in pore water and sediment of a small aquatic ecosystem. Int *J Environ Anal Chem.* 69(3): 227-242.

Gomes Nc, Flocco CG, Costa R, Junca H, Vilchez R, Pieper Dh, Krögerrecklenfort E, Paranhos R, Mendonça-Hagler Lc, Smalla K. (2010). Mangrove microniches determine the structural and functional diversity of enriched petroleum hydrocarbon-degrading consortia. *FEMS Microbiol Ecol.* 74(2):276-90. doi: 10.1111/j.1574-6941.2010.00962.x. Epub.

González J.J., Viñas L., Franco M.A., Fumega J., Soriano J.A., Grueiro G. (2006). Spatial and temporal distribution of dissolved/dispersed aromatic hydrocarbons in seawater in the area affected by the Prestige oil spill. *Mar Pollut Bull.* 53: 250-259.

Green, B.D. And Keller, M. (2006). Capturing the uncultivated majority. *Current opinion in biotechnology*, 17, 236-240.

Handelsman, J. (2004). Metagenomics: application of genomics to uncultured microorganisms, *Microbiol Mol Biol Rev*, 68, 669-685.

Harayama S., Kishira H., Kasai Y., Shutsubo K. (1999). Petroleum biodegradation in marine environments. *J Mol Microbiol Biotechnol.* 1(1): 63-70.

Head I.M., Martin Jones D., Röling Wf.M. (2006). Marine microorganisms make a meal of oil. *Nature Rev Microbiol.* 4: 173-182.

Hinchee R.E., Arthur M. (1991). Bench scale studies of petroleum hydrocarbons. *App Biochem Biotechnol.* 901(6):28-29.

http://www.independent.co.uk/environment/bp-oil-spill-disaster-by-numbers-2078396.html.

Jaffrennou, C., Stephan, L., Giamarchi, P., Cabon, J. Y., Burel-Deschamps, L., Bautin, F. (2007). Direct fluorescence monitoring of coal organic matter released in seawater. *Journal of Fluorescence,* 17, 564–572.

Juck D., Charles T., Whyte L.G., Greer C.W. (2000). Polyphasic microbial community analysis of petroleum hydrocarboncontaminated soils from two northern Canadian communities. *FEMS Microbiol Ecol.* 33(3):241-249.

Kanaly R A., Bartha R., Watanabe K., Harayama S. (2000). Rapid mineralization of Benzo[a]pyrene by a microbial consortium growing on diesel fuel. *Appl Environ Microbiol.* 66: 4205-4211.

Kasai Y., Takahata Y., Hoaki T., Watanabe K. (2005). Physiological and molecular characterization of a microbial community established in unsaturated petroleum-contaminated soil. *Environ Microbiol.* 7(6):806-818.

Killops S.D., Killops V.J. (2005). *Introduction to organic geochemistry,* 2nd Edition. Blackwell, Oxford, 30-70, pp.

Kim J.S., Crowley D.E. (2007). Microbial diversity in natural asphalts of the Rancho La Brea Tar Pits. *Appl Environ Microbiol.* 73(14):4579-4591.

Laflamme R.E., Hites R.A. (1978). The global distribution of polycyclic aromatic hydrocarbons in recent sediments. *Geochim Cosmochim Acta.* 42: 289-3003.

Leadbetter J.R. (2003). Cultivation of recalcitrant microbes: cells are alive, well and revealing their secrets in the 21st century laboratory. *Curr Opin Microbiol.* 6: 274-281.

Long S.C., Aelion C.M., Dobbins D.C., Pfaender F.K. (1995). A comparison of microbial community characteristics among petroleum-contaminated and uncontaminated subsurface soil samples. *Microb Ecol* 30: 297-307.

Louati A., Elleuch B., Kallel M., Saliot A., Dagaut J., Oudot J. (2001). Hydrocarbon contamination of coastal sediments from the Sfax area (Tunisia), Mediterranean sea. *Mar Pollut Bull* 42. 445-452.

Macnaughton S.J., Stephen J.R., Venosa A.D., Davis G.A., Chang Y.J., White D.C. (1999). Microbial population changes during bioremediation of an experimental oil spill. *Appl Environ Microbiol.* 65: 3566-3574.

Marc V., Sabate J., Espuny M.J., Solanas A.M. (2005). Bacterial community dynamics and polycyclic aromatic hydrocarbon degradation during bioremediation of heavily creosote contaminated soil. *Appl Environ Microbiol.* 71(11):7008-7018.

Margesin R., Schinner, F. (1997). Efficiency of indigenous and inoculated cold-adapted soil microorganisms for biodegradation of diesel oil in alpine soils. *Appl Environ Microbiol* 63: 2660-2664.

Margesin R., Schinner F. (2001). Bioremediation (natural attenuation and biostimulation) of diesel-oil-contaminated soil in an alpine glacier skiing area. *Appl Environ Microbiol.* 67: 3127-3133.

Margesin R., Labbé D., Schinner F., Greer C.W., Whyte L.G. (2003). Characterization of Hydrocarbon-Degrading Microbial Populations in Contaminated and Pristine Alpine Soils. *Appl Environ Microbiol.* 69: 3085-3092.

Marteil L. (1974). La conchyliculture francaise. Premiere partie : Le milieu naturel et ses variations. *Revue de l'Institut des Pêches Maritimes.* 38: 217-337.

Mckew B.A., Coulon F., Osborn A.M., Timmis K.N., Mcgenity T.J. (2007). Determining the identity and roles of oil-metabolizing marine bacteria from the Thames estuary, UK. *Environ Microbiol.* 9(1):165-176.

Militon, C., Rimour, S., Missaoui, M., Biderre, C., Barra, V., Hill, D., Mone, A., Gagne, G., Meier, H., Peyretaillade, E. And Peyret, P. (2007). PhylArray: Phylogenetic Probe Design Algorithm for MicroArray, *Bioinformatics.* doi: 10.1093

Moslemy P., Neufeld R.J., Guiot S.R. (2002). Biodegradation of gasoline by gellan gumencapsulated bacterial cells. *Biotechnol Bioeng.* 80: 175-184.

Mulligan C.N., Yong R.N. (2004). Natural attenuation of contaminated soils. Environ Int. 30: 587-601.

Neufeld J.D, Mohn W.W., De Lorenzo V. (2006). Composition of microbial communities in hexachlorocyclohexane (HCH) contaminated soils from Spain revealed with a habitat-specific microarray. *Environ Microbiol.* 8(1):126-40.

Oda J, Maeda I., Mori T., Yasuhara A., Saito Y. (1998). The relative proportions of polycyclic aromatic hydrocarbons and oxygenated derivatives in accumulated organic particulates as affected by air pollution sources. *Environ Technol.* 19: 961-976

Ollivon D., Garboan B., Chesterikoff A. (1993). Analysis of the distribution of some polycyclic aromatic hydrocarbons in sediments and suspended matter in the river Seine (France). *Water Air Soil Poll.* 81: 135-152.

Olsen G.J., Lane D.J., Giovannoni S.J., Pace N.R., Stahl D.A. (1986). Microbial ecology and evolution: a ribosomal RNA approach. *Annu Rev Microbiol.* 40: 337-365.

Olson J.J., Mills G., Herbert B.E., Morris P.J. (1999). Biodegradation rates of separated diesel components. *Environ Toxicol Chem.* 18: 2448-2453.

OPEC. (2008). Organization of the Petroleum Exporting Countries. Monthly Oil Market Report.

Orphan V.J., Taylor L.T., Hafenbradl D., Delong E.F. (2000). Culture-dependent and culture-independent characterization of microbial assemblages associated with high-temperature petroleum reservoirs. *Appl Environ Microbiol.* 66: 700-711

Ozaki S., Fujita Kishimoto T. (2006). Isolation and phylogenetic characterization of microbial consortia able to degrade aromatic hydrocarbons at high rates. *Microbes Environ.* 21(1):44-52.

Palayer J. (1997). Le point sur les hydrocarbures aromatiques polycycliques. Agence de l'eau Seine-Normandie. 63 pp.

Pérez-Carrera E., León,V.M., Parra,A.G.,, González-Mazo, E. (2007). Simultaneous determination of pesticides, polycyclic aromatic hydrocarbons and polychlorinated biphenyls in seawater and interstitial marine water samples, using stir bar sorptive extraction-thermal desorption-gas chromatographymass spectrometry. *Journal of Chromatography A*, 1170(1–2), 82–90.

Perfumo A., Banat I.M., Marchant R., Vezzulli, L. (2007). Thermally enhanced approaches for bioremediation of hydrocarbon-contaminated soils. *Chemosphere.* 66: 179-184.

Pineda-Flores G., Ball-Arguello G., Lira Caleana C., Mesta-Howard M.A. (2004). A microbial consorptium isolated from a crude oil sample that uses asphaltenes as a carbon and energy source. *Biodegradation* 15:145-151.

Popp N., Schlomann M., Mau M. (2006). Bacterial diversity in the active stage of a bioremediation system for mineral oil hydrocarbon-contaminated soils. *Microbiology.* 152: 3291-3304.

Prince R.C. (1993). Petroleum Spill Bioremediation in Marine Environment. *Crit. Rev. Microbial* 19, 217–242.

Ramade F. (1992). Précis d'écotoxicologie. Collection d'écologie (22). Masson (Eds), 300pp.

Rehmann K., Hertkorn N., Kettrup A.A. (2001). Fluoranthene metabolism in Mycobacterium sp. strain KR20: identity of pathway intermediates during degradation and growth. *Microbiology.* 147(10):2783-2794.

Reisfeld A., Rosenberg E., Gutnick D. (1972). Microbiol degradation of crude oil: factors affecting the dispersion in sea water mixed and pure cultures. *Appl. Microbiol.* 24: 363-369.

Rice, S. D., Spies, R. B., Wolfe, D. H., Wright, B. A. (1996). Proceeding of the Exxon Valdez Oil Spill Symposium. American Fisheries Society Publication, vol. 18, Bethesda, Maryland.

Richard J.Y., Vogel T.M. (1999). Characterization of soil bacterial consortium capable of degrading diesel fuel. *Int Biodeter Biodeg*. 44: 93-100.

Röling W.F., Van Verseveld, H.W. (2002). Natural attenuation: what does the subsurface have in store? *Biodegradation*. 13: 53-64.

Röling W.F., Milner M.G., Jones D.M., Lee K., Daniel F., Swannell Rj, Head I.M. (2002). Robust hydrocarbon degradation and dynamics of bacterial communities during nutrient enhanced oil spill bioremediation. *Appl Environ Microbiol*. 68(11):5537-5548.

Romantschuk M., Sarand I., Petanen T., Peltola R., Jonsson-Vinanne T., Koivula K., Rjala Y. , Haahtela K. (2000). Means to improve the effect of *in situ* bioremedation of contaminated soil: an overview of novel approches. *Environ Poll*. 107(2): 245-254.

Salanitro J.P. (2001). Bioremediation of petroleum hydrocarbons in soil. *Adv in Agron*. 72: 53-105.

Schilderman A.A.E.L., Moonen E.J.C., Maas L.M., Welle I., Kleinjans J.C.S. (1999). Use of crayfish in biomonitoring studies of environmental pollution of the river Meuse. *Ecotoxicol Environ Safe*. 44: 241-252.

Serrano A., Gallego M., Gonzalez J.L., Tejada M. (2007). Natural attenuation of diesel aliphatic hydrocarbons in contaminated agricultural soil. *Environ Pollut*. 151(3): 494-502.

Shaw G.R., Connell D.W. (2001). DNA adducts as a biomarker of polyaromatic hydrocarbon exposure in aquatic organisms: relationship to carcinogenicity. *Biomarkers*. 6 (1): 64-71.

Shwannell R.P., Lee K., Mc Dongh M. (1996). Field evaluations of marine oil spill bioremedation. *Microbiol Rev*. 60 (2): 342-365.

Simpson C.D., Harrington C.F., Cullen W.R. (1998). Polycyclic aromatic hydrocarbon contamination in marine sediments near Kitimat, British Colombia. *Environ Sci Technol*. 32: 3266-3272.

Smit E., Leeflang P., Gommans S., Van Den Broek J., Van Mil S., Wernars K. (2001). Diversity and seasonal fluctuations of the dominant members of the bacterial soil community in a wheat field as determined by cultivation and molecular methods. *Appl Environ Microbiol*. 67: 2284-2291.

Smits T.H., Röthlisberger M., Witholt B., Van Beilen J.B. (1999). Molecular screening for alkane hydroxylase genes in gram-negative and gram-positive strains. *Environ Microbiol*. 1: 307-317.

Theron J., Cloete T.E. (2000). Molecular techniques for determining microbial diversity and community structure in natural environments. *Crit Rev Microbiol*. 26: 37-57.

Tuvikène A. (1995). Responses of fish to polycyclic aromatic hydrocarbons (PAHs). *Ann Zool Fenn*. 32: 295-307.

Tyson, G.W. And Banfield, J.F. (2005) Cultivating the uncultivated: a community genomics perspective, *Trends in microbiology*, 13, 411-415.

Van Hamme J.D., Singh A., Ward D.M. (2003). Recent Advances in Petroleum Microbiology. Microb Mol Biol Rev. 67: 503-549.

Van Veen J. A., Van Overbeek L. S., Van Elsas J.D. (1997). Fate and activity of microorganisms introduced into soil. *Microbiol Mol Biol Rev* 61: 121-135.

Viñas M., Sabaté J., Espuny M.J., Solanas A.M. (2005) .Bacterial community dynamics and polycyclic aromatic hydrocarbon degradation during bioremediation of heavily creosotecontaminated soil. *Appl Environ Microbiol.* 71(11):7008-7018.

Wakeham, S. G. (1996). Aliphatic and polycyclic aromatic hydrocarbon in Black Sea sediments. *Marine Chemistry,* 53, 187–205.

Walker J.D., Colwell R.R. (1974). Microbiol degradation of model petroleum at low temperature. *Microbiol. Ecol.* 1: 63-95.

Wang X, Greenfield P, Li D, Hendry P, Volk H, Sutherland T.D. (2011). Complete Genome Sequence of a Nonculturable Methanococcus maripaludis Strain Extracted in a Metagenomic Survey of Petroleum Reservoir Fluids. *J Bacteriol.* 193(19):5595.

Ward D. M., Santegoeds C.M., Nold S.C., Ramsing M.J., Ferris M. J., Bateson M.M. (1997). Biodiversity within hot spring microbial mat communities: molecular monitoring of enrichment cultures. *Ant van Leeuw* 71: 143-150.

Watanabe K., Teramoto M., Harayama S. (2002). Stable augmentation of activated sludge with foreign catabolic genes harboured by an indigenous dominant bacterium. *Environ Microbiol.* 4: 577-583.

Watanabe K., Teramoto M., Futamata H., Harayama S. (1998). Molecular detection, isolation and physiological characterization of functionally dominant phenol-degrading bacteria in activated sludge. *Appl Environ Microbiol* 6: 4396-4402.

Watts J.E.M., Wu Q., Schreier S.B., May H.D., Sowers K.R. (2001). Comparative analysis of polychlorinated biphenyl-dechlorinating communities in enrichment cultures using three different molecular screening techniques. *Environ Microbiol.* 3: 710-719.

Whyte L.G., Bourbonnière L., Greer C.W. (1997). Biodegradation of petroleum hydrocarbons by psychrotrophic Pseudomonas strains possessing both alkanes (alk) and naphthalene (nah) catabolic pathways. *Appl Environ Microbiol* 63: 3719-3723.

Whyte L.G., Hawari J., Zhou E., Bourbonniere L., Inniss W.E., Greer C.W. (1998). Biodegradation of variable-chain-length alkanes at low temperatures by a psychrotrophic Rhodococcus sp. *Appl Environ Microbiol* 64: 2578-2584.

Widada J., Nojiri H., Omori T. (2002). Recent developments in molecular techniques for identification and monitorin of xenobioticdegrading bacteria and their catabolic genes in bioremediation. *Appl Microbiol Biotechnol.* 60: 45-59.

Wilcok R.J., Corban G.A., Northcott G.L., Wilkins A.L., Langdon A.G. (1996). Persistence of polycyclic aromatic compounds of different molecular size and water solubility in surficial sediment of an interdial sandflat. *Environ Toxicol Chem.* 15 (5): 670-676.

Wilson M.S., Bakermans C., Madsen E.L. (1999). *In Situ*, Real-Time Catabolic Gene Expression: Extraction and Characterization of Naphtalene Dioxygenase mRNA Transcripts from Groundwater. *Appl Environ Microbiol* 65: 80-87.

Woese, C.R. And Fox, G.E. (1977). Phylogenetic structure of the prokaryotic domain: the primary kingdoms, Proceedings of the National Academy of Sciences of the United States of America, 74, 5088-5090.

Yunker M.B., Snowdon L.R., Macdonald R.W., Smith J.N., Fowler M.G., Skibo D.N., Mclaughlin F.A., Danyushevskaya A.I., Petrova V.I., Ivanov G.I. (1996). Polycyclic aromatic hydrocarbon composition and potential sources for sediment samples from the Beaufort and Barents seas. *Environ Sci Technol.* 30: 1310-1320.

Zobell C.E. (1946). Action of microorganisms on hydrocarbons. *Bacterial Rev.* 10:1-49.

Zrafi-Nouira I, Khedir-Ghenim Z, Zrafi F, Bahri R, Cheraeif I, Rouabhia M, Saidane-Mosbahi D (2008) Hydrocarbon pollution in the sediment from the Jarzouna-Bizerte coastal area of Tunisia (Mediterranean Sea). *Bull Environ Contam Toxicol* 80(6):566–572.

Zrafi-Nouira Ines; Zouhour Khedir-Ghenim; Raouf Bahri; Imed Cheraeif; Mahmoud Rouabhia; Dalila Saidane-Mosbahi. (2009). Hydrocarbons in Seawater and Interstitial Water of Jarzouna-Bizerte Coastal of Tunisia (Mediterranean Sea): Petroleum origin investigation around refinery rejection place. *Water Air and Soil Pollution*. Vol. 202 No.1: 19-31.

Zrafi-Nouira Ines; Sonda Guermazi; Rakia Chouari; Nimer M. D. Safi; Eric Pelletier; Amina Bakhrouf; Dalila Saidane-Mosbahi; Abdelghani Sghir (2009). Molecular diversity analysis and bacterial population dynamics of an adapted seawater microbiota during the degradation of Tunisian zarzatine oil. *Biodegradation*.Vol. 20 No.4: 467-486.

Zrafi-Nouira Ines; Safi Nimer; Bahri Raouf; Mzoughi Nadia; Aissi Ameur; Abdennebi Hassen; Saidane-Mosbahi (2010). Distribution and sources of polycyclic aromatic hydrocarbons around a petroleum refinery rejection area in Jarzouna-Bizerte (Coastal Tunisia). *Soil and Sediment Contamination*, Vol.19 No.3 : 292-306.

11

Oil-Spill Bioremediation, Using a Commercial Biopreparation "MicroBak" and a Consortium of Plasmid-Bearing Strains "V&O" with Associated Plants

Andrey Filonov[1], Anastasia Ovchinnikova[1], Anna Vetrova[1],
Irina Puntus[1], Irina Nechaeva[2], Kirill Petrikov[2], Elena Vlasova[2],
Lenar Akhmetov[1], Alexander Shestopalov[3],
Vladimir Zabelin[3] and Alexander Boronin[1]
[1]*G.K. Skryabin Institute of Biochemistry and Physiology of Microorganisms, Russian Academy of Sciences (IBPM RAS)*
[2]*Tula State University*
[3]*JSC "Biooil"*
Russia

1. Introduction

Oil is the most widely distributed source of energy in the world and the largest-scaled environmental pollutant. Oil, oil products, and oil-containing industrial wastes pollution is ranked second place after radioactive pollution on account of their harmful action to ecosystems (Chernyakhovsky et al., 2004). Oil deposits in Russia are mainly located in the northern regions of the European part and Western Siberia. Therefore oil spills in terrestrial and aquatic ecosystems occur primarily in cold climate regions. Approximately 8 to 9 billion tons per year of oil and oil products are discharged into the environment during extraction, transportation, processing and storing oil. Half of this amount goes into the soil and groundwater, and the other half contaminates surface water and air (Chernyakhovsky et al., 2004). In Western Siberia, in the areas of hydrocarbon accumulations, over 200 thousand hectares of land were polluted with oil reaching depths up to 10 cm in 1995. By 2003, the oil polluted area reached 700 thousand hectares.

The process of natural restoration of polluted environment requires long times. The autoremediation of an oil-spilled soil is a process that requires from 2 to 30 years or more to complete at a level of pollution of 5 g oil/kg soil. The rate of this process is slower in the northern regions. Thus, the aftermath of oil spills is observed over decades (Oborin et al., 1988).

In Russia, this ecological problem is of a special significance because of the scale of the oil spills when soil excavation and restoration *ex-situ* are impossible. Bioremediation, which consists on the microorganisms' ability of utilizing and transforming oil hydrocarbons, plays the main role during *in-situ* remediation. Bioremediation provides an economically beneficial and high-specific clean-up technology to remove pollutants concentration by targeting individual pollutants or their mixtures. The planning of bioremediation strategies for polluted land and water areas should consider the use of the existing indigenous oil-oxidizing microorganisms with different affinity to oil fractions and their activation by addition of fertilizers (nitrogen, phosphorus, and potassium). The inoculation of effective microorganisms is necessary in the northern regions where the warm season is short and the natural microflora has no time to adapt to the changing environmental conditions, which is the case for Russia located in a zone of cold and moderate climate.

Bacteria such as *Acinetobacter, Alcaligenes, Arthrobacter, Burkholderia, Brevibacterium, Corynebacterium, Micrococcus, Mycobacterium, Nocardia, Pseudomonas, Rhodococcus, Serratia,* and *Bacillus* among others possess significant pollutant degradation potential; it has been shown that these bacteria can use recalcitrant pollutant and xenobiotics as energy source. These microorganisms are often isolated from oil-polluted sites and used to clean-up (bioremediate) oil spills. The remediation of polluted soils requires the study of the microorganisms' diversity in the environment and the determination of the ability of different microbes and their consortia to degrade pollutants in the presence of high salt concentration (Van Hamme et al., 2003; Ventosa et al., 1998). This is important because it has been observed that the introduction of a single oil-oxidizing strain into the oil-spilled environment does not assure a complete clean-up.

The majority of the studies have focused on determining the effect of oil spill on natural populations of microorganisms from different ecological niches (Delille et al., 2002; Prince & Bragg, 1997). However, few studies have addressed the selection of microorganisms capable of degrading oil and oil products at low temperatures.

The research summarized in this chapter describes the screening and collection of microorganisms capable of degrading different hydrocarbons. Several active oil-degrading strains were selected with the following characteristics:

- Psychrotrophic that is the bacteria ability to grow in a wide temperature range (from 4 to 30°C), that is useful for bioremediation of oil-spilled territories in different regions of Russia;
- Halotolerant that refers to bacteria capable of degrading oil and oil products at high salt concentrations (3-7% NaCl). The application of halotolerant bacteria to degrade oil products is of a special importance when conducting bioremediation at seashore, salted water areas, and salted marshes;
- Capacity to produce biosurfactants. The strains must be capable to synthesize biosurfactants when cultivated in mineral media with oil products as carbon and energy sources. The persistence of oil hydrocarbon pollutants in the environment is due to their low water solubility. Therefore, the action of microbial emulsifiers on hydrocarbon pollutants will enhance the strains bioutilization efficiency of these contaminants.
- Capacity to target polycyclic aromatics (PAHs). Microorganisms of the genus *Pseudomonas* bear plasmids that may encode for biodegradation of PAHs. The presence of conjugative plasmids harbouring PAH catabolism genes promotes the increase the PAHs degradation potential because of genes dissemination among indigenous microorganisms.

2. Biopreparation of "MicroBak" and field testing

2.1 Strains selection

A total of 165 strains obtained from the Laboratory of Plasmid Biology of IBPM RAS were examined for their ability to utilize diesel fuel or crude oil as the sole carbon and energy source at temperatures ranging from 4 to 6°C and 24°C. The screening of the most active strains was performed according to the following criteria: their ability to grow on diesel fuel and/or crude oil both at moderate temperatures ranging from 24°C to 32°C and at low temperatures ranging from 2°C to 6°C in the presence of 3 to 10% NaCl and their capability to produce bioemulsifiers when cultivated on hydrophobic substrates. Based on these screening criteria, 9 psychrotrophic strains degrading oil hydrocarbons were selected as follows: *Rhodococcus* sp. S25, *Rhodococcus* sp. S26, *Rhodococcus* sp. S67, *Rhodococcus* sp. X5, *Rhodococcus* sp. X25, *Rhodococcus* sp. Ars38, *Microbacrerium* sp. Ars25, *Pseudomonas* sp. 142NF(pNF142), and *Pseudomonas putida* BS3701(pBS1141, pBS1142) were chosen. The selected microorganisms had different ability to grow on diesel fuel, crude oil, oil fuel, hexadecane, benzoate, benzene, toluene, and naphthalene as a sole carbon and energy source in temperatures ranging from 2°C to 32°C. All the strains were capable of growing in a mineral Evans medium (Evans et al., 1970) in the presence of 3% NaCl, however two of them grew well at NaCl concentrations of 5% and only one bacterium (*Rhodococcus* sp. X5) grew at NaCl concentrations in the range from 7 to 10%. All microorganisms were able to grow in a pH range from 5 to 8.

After determining the main physiological and morphological strait of the strains, the identification of the most active degrader microorganisms was performed. The strains were identified based on their physiological and morphological characteristics. Seven gram-positive strains X5, X25, S25, S26, S67, and Ars38 were identified as representatives of the *Rhodococcus* genus and the strain Ars25 was identified as a *Microbacterium* using nucleotide sequencing of 16S rRNA genes.

The ability of the strains to degrade crude oil was assessed according to the reduction of oil in a liquid medium determined by gravimetric analysis (Baryshnikova et al., 2001). Single strains were able to degrade from 26% to 66% of oil at 24°C, and from 28% to 47% of oil at 4-6°C in 20 days. Eight of the nine strains chosen were more effective in degrading oil at low temperatures (4-6°C) than at a temperature of 24°C (Fig. 1).

Thus, these eight strains, which showed potential for oil bioremediation in cold climates, were used to prepare a consortium of oil degrader strains.

The selection of a mixed consortium of microorganisms was carried out in a liquid mineral medium through batch cultivation with oil as the sole carbon and energy source in the presence of *Eriophorum vaginatum* grass based sorbent or without it. After cultivation, the dominant rhodococci and pseudomonads strains were determined by analysis of the selected population. The representative strains of the genus *Pseudomonas* were distinguished by their cultural and morphological traits. Dominant rhodococci species were determined by performing genotyping of the selected strains because these strains had identical cultural and morphological traits. Random Amplification of Polymorphic DNA (RAPD) PCR-analysis was applied (Fig. 2).

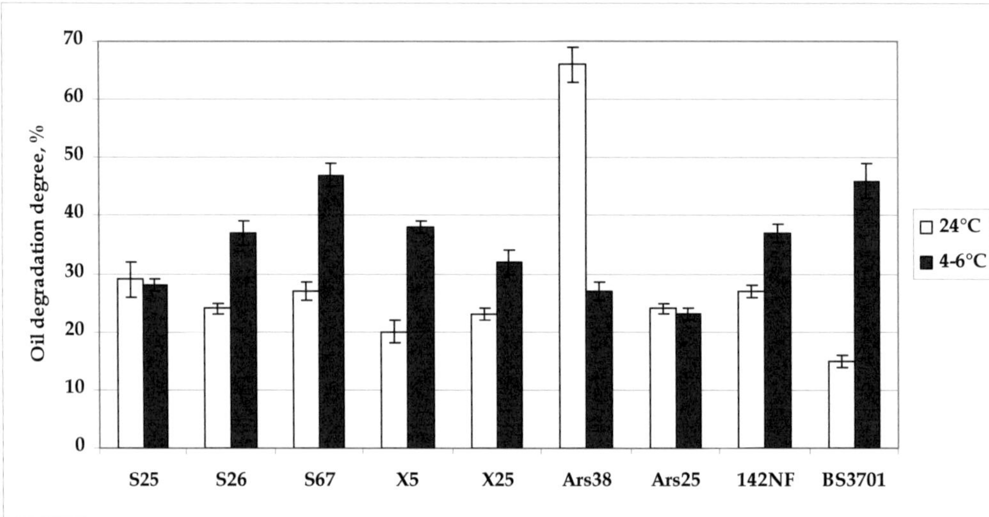

Fig. 1. Oil degradation by single microorganisms strains in Evans liquid medium at different temperatures for a period of 20 days

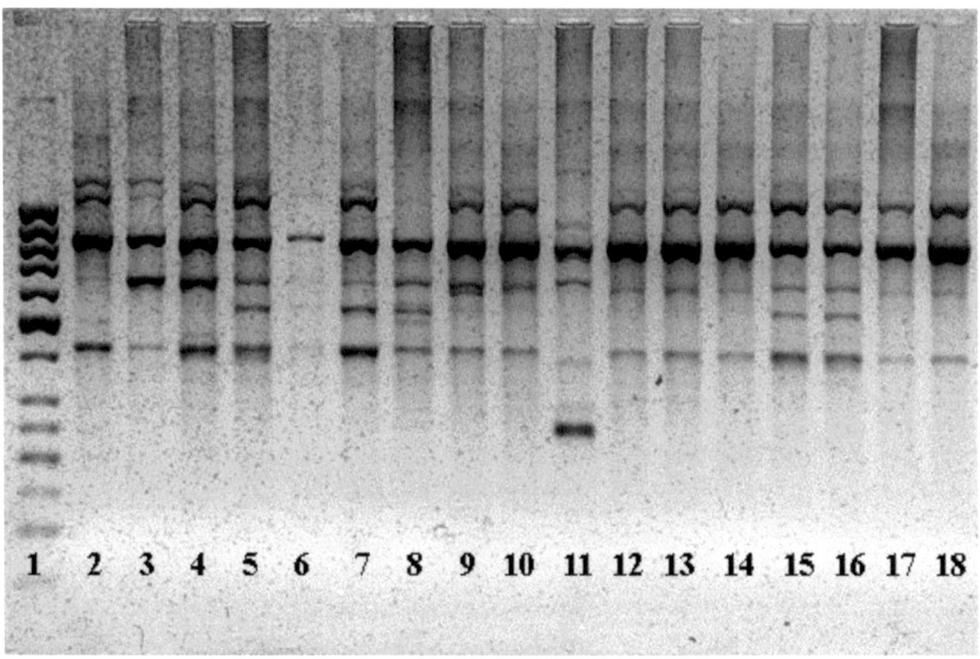

Fig. 2. RAPD PCR-analysis of rhodococci strains with OA20 primer (GTTGCGATCC):
1 - 50 bp Ladder (Fermentas);
2 - *Rhodococcus* sp. S25; 3 - *Rhodococcus* sp. S26; 4 - *Rhodococcus* sp. S67; 5 - *Rhodococcus* sp. X5; 6 - *Rhodococcus* sp. X25; 7 - *Microbacterium* sp. Ars25; 8 - *Rhodococcus equi* Ars38; 9-18 - clones presenting the largest population after cultivation in liquid mineral medium

Thus, a consortium of microorganisms including *Rhodococcus* sp. X5, *Rhodococcus* sp. S67, *Pseudomonas* sp. 142NF(pNF142), and *Pseudomonas putida* BS3701(pBS1141,pBS1142) was selected. This consortium was used as the basis for the biopreparation to target the bioremediation of soils polluted by crude oil and oil products. Characteristics of the microbial consortium are: psychrotrophy, halotolerant (3-5% NaCl), capability to synthesize bioemulsifiers, carrier of plasmids in pseudomonades capable of targeting the biodegradation of polycyclic aromatic hydrocarbons.

2.2 Role of catabolic plasmids in oil biodegradation

Many examples of plasmid genes encoding for hydrocarbons degradation, including short chain alkanes, substituted and non-substituted aromatic hydrocarbons, and other xenobiotics are known (Harayama et al., 1990; Wallace et al., 1992). The concentration of aromatic compounds in oil hydrocarbons depends on the crude oil and it could range from 10% to more than 50%. Aromatic compounds are the most toxic and recalcitrant components in the crude oil. The role of catabolic plasmids in oil biodegradation was determined by gravimetric analysis of the oil degradation efficiency of *Pseudomonas* strains bearing PAH biodegradation plasmids and their plasmid-free variants. It was revealed that the presence of naphthalene degradation plasmid pBS216 in the *Pseudomonas chlororaphis* PCL1391 strain promoted a significant increase (10-fold) of oil degradation in 7 days if compared with the plasmid-free strain. However, the presence of plasmids pOV17 or pNF142::Tc in the strain did not have the same effect (Fig. 3). Plasmids pBS216 and pOV17 are known to have similar structures but different activities of catechol-2, 3-dioxygenase (Volkova et al., 2005). Genes encoding for the enzyme are often localized in one of the plasmids from a bacterial host bearing different plasmids that influences the oil degradation capability. The results obtained show that the catabolic potential of microorganisms in oil degradation is given by a combination of "host bacterium - plasmid". The presence of pBS1141 plasmid in the strain *Pseudomonas putida* BS3701 induced the removal of oil hydrocarbons (up to 3-fold) in comparison with an eliminant BS3701E, which does not present the PAH biodegradation plasmid pBS1141.

Figure 4 indicates that the cell biomass growth in oil shows a higher increase for the plasmid-bearing bacteria than the cell biomass growth of plasmid-free microorganisms. These plasmid genes have the ability of using different aromatic hydrocarbons (naphthalene, phenanthrene etc.) in comparison with plasmid-free strains. For example, naphthalene dioxygenase encoded by genes of naphthalene biodegradation plasmid catalyzes about 76 reactions (deoxygenation, monooxygenation, dehydratation, O- and N-dealkylation and sulfooxidation).

Crude oil and oil products are complex multicomponent pollutants containing hundreds of chemical compounds. During oil degradation both quantitative (content decrease) and qualitative (transformation of fractions composition) changes take place, for instance, selective degradation or transformation of single oil components occurs. Fractionation of the residual oil on silica gel separate oil components in three fractions: the hexane fraction containing paraffin, naphthene, and aromatic hydrocarbons, the benzene fraction containing polycyclic aromatic hydrocarbons; and the alcohol-benzene fraction contains naphthenic acids and tars (Babaev et al., 2009). The plasmid-bearing strains evaluated in this study showed that the naphthalene biodegradation plasmids increased the degree of degradation of hexane, benzene, and benzene-alcohol fractions.

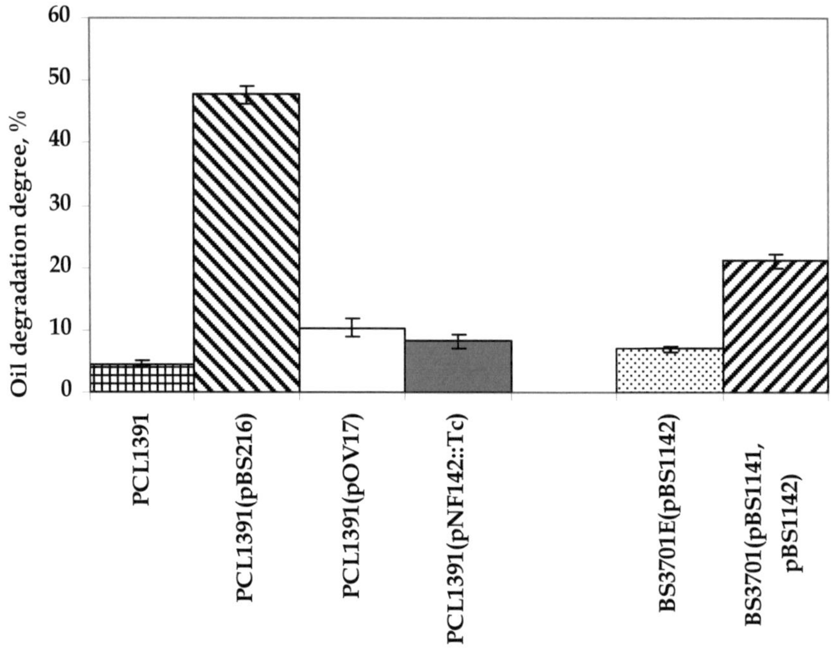

Fig. 3. The degree of oil destruction by native plasmid-bearing strains and their plasmid-free variants in Evans medium in 7 days at 24°C

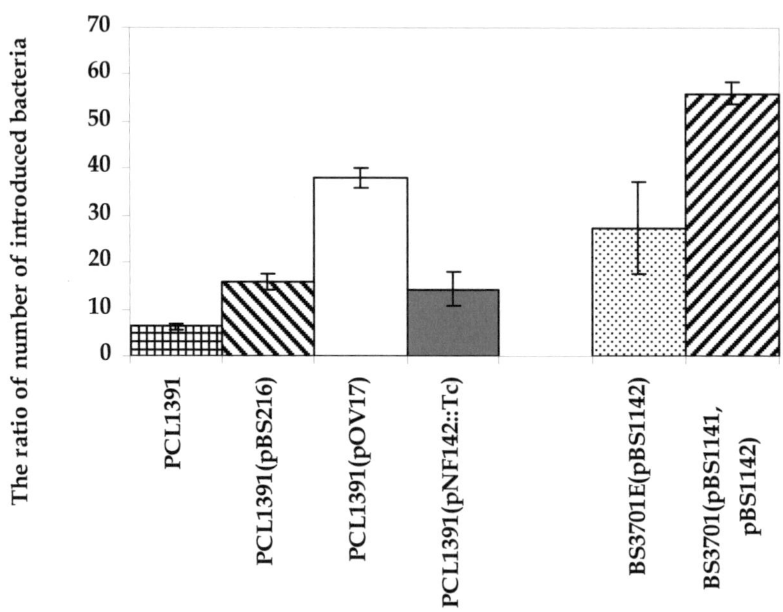

Fig. 4. The ratio of the final to the initial number of microorganisms for a testing period of 7 days

The highest degradation of paraffin-naphthene, mono- and polyaromatic hydrocarbons, asphalthenes and tars was detected in the sample containing the strain *P. chlororaphis* PCL1391(pBS216), in which the degradation degree for the three fractions was 38%, 31% and 26% respectively, causing a total oil degradation of 48%. Thus, the presence of PAH degradation plasmids in host strains promotes the increase of oil degradation.

2.3 Evaluation of biosurfactants

The capacity of bacteria to produce surface-active compounds (biosurfactants) during hydrocarbon degradation is one of the most important mechanisms allowing microorganisms to use oil components poorly soluble or insoluble in water (Desai & Banat, 1997). Biosurfactants contribute to the solubilization of hydrocarbons by forming emulsions that facilitates the contact of microbial cells with the hydrophobic substrate and its influx into the cell. Currently biosurfactants are a subject of research. Advantages of biosurfactants over synthetic surfactants are their effectiveness, low toxicity, and biodegradability. The appropriate knowledge of biosurfactant properties, structures, and their biosynthesis conditions will promote the formulation of biopreparations (Banat et al., 2010). Despite of the numerous reports on new strains-producers of surface-active compounds a unified methodological approach for the screening and identification of these organisms from native samples has not been developed yet. The general accepted criteria for the evaluation of the surface activity of microbial cultures are still unknown.

Different strains were screened to determine the most efficient producers of biological surface-active agents. The microorganisms chosen for this evaluation were the following: gram-negative *Pseudomonas fluorescens* 142NF (pNF142), *Pseudomonas putida* BS3701 (pBS1141,pBS1142) (which bear plasmids of biodegradation of mono- and polycyclic aromatic hydrocarbons); gram-positive *Rhodococcus* sp. S67, *Rhodococcus* sp. X5, and *Rhodococcus* sp. S26. These strains are highly effective oil destructors.

When carrying out the selection of microorganisms capable of producing surfactants, the surface tension of the culture broth is usually estimated (Satpute et al., 2010). Simple and rapid methods based on measuring the emulsification of hydrophobic compounds such as hexadecane are applied that allow measuring the index of emulsification and the degree of emulsification activity. The determination of glycolipid biosurfactants is performed using photocolorimetry reactions such as the Molisch reaction.

It is well known that the carbon source may have a significant impact on biosurfactants synthesis (Muthusamy et al., 2008). Biosurfactant synthesis is often observed in various microorganisms during their growth on hydrophobic substrates such as carbohydrates and vegetable fats. On the other hand, intensive biosurfactant synthesis is detected during the growth of microorganisms (representatives strains of *Pseudomonas aeruginosa*) on hydrophilic carbon sources (glucose, glycerol) (Abdel-Mawgoud et al, 2010). In this study, biosurfactants-producer microorganisms were compared by their cultivation on two different substrates: hexadecane as a hydrophobic substrate and glucose as a hydrophilic substrate.

It was found (Table 1) that hexadecane stimulates the intensive synthesis of biosurfactants. In this case, the content of biosurfactants in the culture broth for all the strains evaluated was high and a significant decrease of the surface tension was observed from 77 mN/m down to 34-31 mN/m (control). The largest values of the emulsification index and degree of emulsification activity were detected. For instance, rhodococci strains appeared to be more effective producers of biosurfactants compared to pseudomonades when cultivated on hexadecane.

Strain	Glycolipids content(4), mg/l	Surface tension(5), mN/m	Index of emulsification, %		Emulsifying activity(4) (λ=540 nm), units of optical density
			CFS(4)	CB(5)	
P. fluorescens 142NF	190±10	34±1	50±4	50±5	0,9±0,2
P. putida BS3701	250±20	34±1	53±6	53±5	0,6±0,1
Rhodococcus sp. S67	310±20	33±1	78±9	78±6	0,7±0,1
Rhodococcus sp. X5	400±30	31±1	75±6	75±7	1,0±0,2
Rhodococcus sp. S26	740±50	32±1	47±3	47±3	1,5±0,2

(4) The value measured for the cell-free supernatant (CFS)
(5) The value measured for the culture broth (CB)

Table 1. Surface-active properties of microorganisms' growth in the medium with hexadecane.

Biosurfactants can be extracellular (exo-type) and cell-bounded (endo-type). The index of emulsification and the emulsifying activity were determined for only the solubilized biosurfactants (exo-type) in the broth after cells precipitation. Nevertheless, there was some effect of the endo-type biosurfactants on the index of emulsification in the non-centrifuged culture broth.

The results show that the indices of emulsification in the culture broth and in the cell-free supernatant for the strains cultivated on hexadecane were the same (Table 1). Thus, the five strains producing extracellular biosurfactants were evaluated when using hexadecane as the sole carbon and energy source.

In the case of glucose as the source of carbon and energy, the content of glycolipids did not exceed 50 mg/l, and the values of surface tension obtained were much higher than for the case of hexadecane. This suggests that the growth of microorganisms on the hydrophilic substrate (glucose) was not accompanied by the intensive formation of exo- biosurfactants as in the case of hexadecane. The emulsification activity and emulsification index determined during the growth of *Pseudomonas* on glucose indicates the synthesis of exo-type biosurfactants (Table 2). At the same time, the emulsification index values appeared to be 0% for cell-free supernatants of rhodococci, and 7-29% - for the culture broth. Moreover, cell suspensions of *Rhodococcus* were able to stabilize hexadecane/water emulsions effectively (Table 3). Emulsification was not observed for the *Pseudomonas* cells at the same conditions. Thus, rhodococci produced endo-type biosurfactants, when glucose was used as a growth substrate.

Biosurfactants were extracted from the culture broth using a mixture of chloroform : methanol (3:1), followed by the evaporation and purification of the organic phase, using silica gel column chromatography. The identification of the glycolipid biosurfactants was performed by thin layer chromatography (TLC) on plates of silica gel through a specific reaction to sugar: treatment with phenol and sulfuric acid followed by heating. The presence of carbohydrate in the biosurfactant molecule resulted in a blue-violet coloration of the compound.

Strain	Glycolipids content, mg/l[4]	Surface tension, mN/m[5]	Index of emulsification, %		Emulsifying activity[4] (λ=540 nm), units of optical density
			CFS[4]	CB[5]	
P. fluorescens 142NF	10±2	54±1	33±4	33±4	1,1±0,2
P. putida BS3701	19±2	61±1	33±3	33±3	0,5±0,1
Rhodococcus sp. S67	49±9	53±1	0	7±2	0,2±0,1
Rhodococcus sp. X5	42±8	56±1	0	29±4	0,2±0,1
Rhodococcus sp. S26	19±5	49±1	0	29±3	0,2±0,1

[4] Emulsification index measured for the cell-free supernatant (CFS)
[5] Emulsification index measured for the culture broth (CB)

Table 2. Surface-active properties of microorganisms' growth in the medium with glucose.

Strain	Medium	Emulsifying activity, %
P. fluorescens 142NF	Evans medium + glucose	0
	Luria-Bertany medium[6]	0
P. putida BS3701	Evans medium + glucose	0
	Luria-Bertany medium	0
Rhodococcus sp. S67	Evans medium + glucose	35±4
	Luria-Bertany medium	40±5
Rhodococcus sp. X5	Evans medium + glucose	38±4
	Luria-Bertany medium	46±5
Rhodococcus sp. S26	Evans medium + glucose	46±6
	Luria-Bertany medium	48±6

[6] Carhart & Hegeman, 1975

Table 3. Emulsification properties of cell suspensions cultivated in agar media

According to TLC results it could be assumed that two strains of *Pseudomonas* produced the same rhamnolipids. The biosurfactants samples of *Pseudomonas* had only one spot with a flow rate (R_f) of 0.32 (Fig. 5). It is well known that, when performing TLC analysis under the conditions described, the glycolipids of pseudomonads are separated according to the number of rhamnose residues contained in the molecule. Mono-rhamnolipids have a greater mobility, with the known values for the retention reaching 0.7 (Zhang & Miller, 1994; Robert et al., 1989), and for di-rhamnolipids R_f reported values are 0.32 (Matsufuji et al., 1997) and 0.45 (Zhang & Miller, 1994).

If the results obtained in this work are compared to the published data, then it can be reasoned that the strains *P. fluorescens* 142NF and *P. putida* BS3701 produce di-rhamnolipids. Another common type of biosurfactants, which are produced by microorganisms of the genus *Pseudomonas*, are lipopeptides (Desai & Banat, 1997). Consequently, TLC analysis was performed by staining the chromatograms with ninhydrin to test the presence of lipopeptides in the extracts. The absence of a characteristic violet-pink color indicated the absence of lipopeptide biosurfactants in the extracts.

TLC of trehalose lipids showed the presence of four (strain *Rhodococcus* sp. S67) or five (strains *Rhodococcus* sp. X5 and S26) components with the following values of R_f: 0.32, 0.46, 0.51 (absent in the strain S67), 0.57, and 0.63. The identification of the trehalose lipids produced by the representatives of the genus *Rhodococcus* using the retention values was difficult because there is no enough published data available that could be used for comparison purposes. The TLC analysis of the succinoyl trehalose lipids that are produced by the strain *Rhodococcus* sp. MS11 show a single spot with R_f of 0.41 (Rapp & Gabriel-Jurgens, 2003). This study revealed distinctive retaining values for the compounds isolated; the spots with R_f of 0.32 and 0.46 are the only compounds There is no published data on R_f for trehalose lipids in the system chloroform : methanol : water (65:15:2). It should be noted that the separation of rhodococci biosurfactants on four or five components rather than trehalose lipids by TLC had not been previously described.

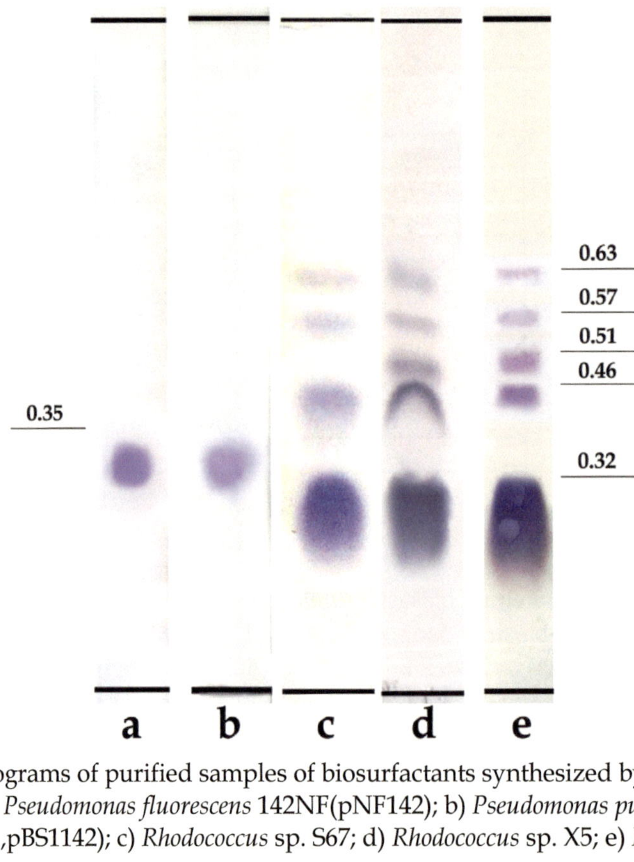

Fig. 5. Chromatograms of purified samples of biosurfactants synthesized by microorganisms (R_f.is shown): a) *Pseudomonas fluorescens* 142NF(pNF142); b) *Pseudomonas putida* BS3701(pBS1141,pBS1142); c) *Rhodococcus* sp. S67; d) *Rhodococcus* sp. X5; e) *Rhodococcus* sp. 26

The biosurfactants samples obtained were characterized using mass-spectrometry with electron spraying in a positive ionization mode. For glycolipids of the strains of *Pseudomonas fluorescens* 142NF and *Pseudomonas putida* BS3701 a series of signals was observed in a range of 700-900 Da (Fig. 6).

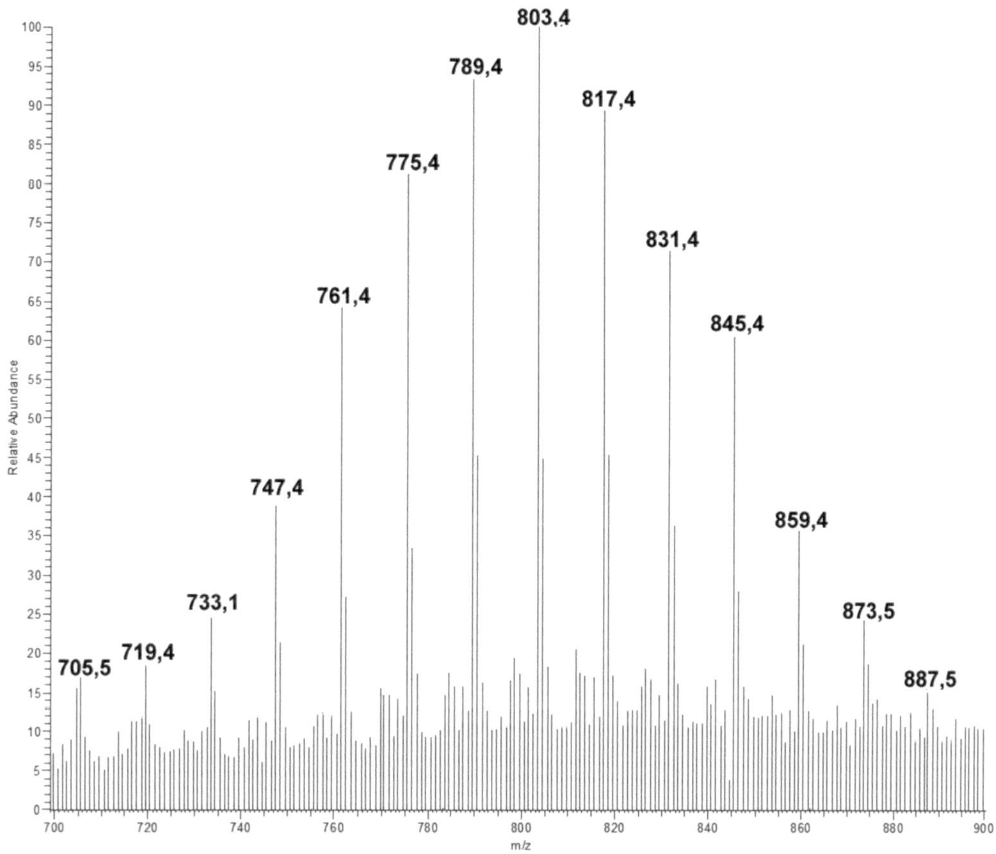

Fig. 6. The mass spectrum of glycolipid biosurfactants produced by the strain *Pseudomonas fluorescens* 142NF

One of the most intense peaks detected corresponded to the pseudo-molecular ion [M + H +] with a mass of 803 Da. It has been reported that ramnolipid B has a molar mass of 802 Da that contains two residua of hydroxydecanoic acid and one residuum of decenoic acid (Fig. 7a) (Abdel-Mawgoud et al., 2010). The difference in the masses for each pair of adjacent peaks was 14 Da that corresponds to the mass of a fragment of $-CH_2-$. It can be considered that the strains *Pseudomonas fluorescens* 142NF and *Pseudomonas putida* BS3701 produced a mixture of homologous rhamnolipids type B.

The same peaks were detected in the mass-spectrum of the biosurfactants produced by the strains *Rhodococcus* sp. X5 and *Rhodococcus* sp. S26: 866.4, 871.5, 877.2, 894.4, 899.9 (Fig. 8). Previous research on biosurfactants produced by rhodococci indicated that the main signals in the mass-spectra of electron spray were induced by pseudo-molecular ions [M + Na +] with masses of 871.5 and 899.6 (Rapp & Gabriel-Jurgens, 2003; Tuleva et al., 2008). That corresponded to the homologous succinoyl trehalose lipids: dioctanoyl-decanoyl (848 Da) and octanoyl-didecanoyl (876 Da), differing by 28 Da, i.e. on a double methylene fragment (-CH_2-) (Fig. 7). Thus, the strains *Rhodococcus* sp. X5 and *Rhodococcus* sp. S26 evaluated in this work produced trehalose lipids of the similar structure. The remaining three signals in Fig. 8 could not be identified.

Infrared spectroscopy showed the presence of functional groups typical for the proposed structures of glycolipids in biosurfactants isolated from pseudomonads and rhodococci. A broad absorption band of a hydroxyl group at 3450 cm^{-1} was distinguished on the spectra obtained. Bands of valent oscillations of carbonyl groups of esters and carboxylic acids were observed in the areas of 1745 cm^{-1} and 1630 cm^{-1}, respectively. The peak of 1047 cm^{-1} belonged to the asymmetric valent oscillations of C-O-C bonds. Spectrum peaks of valent oscillations for aliphatic C-H bonds in the area of 2924 and 2852 cm^{-1} were observed, the absorption of deformation vibrations of these bonds was present at 1380 cm^{-1}.

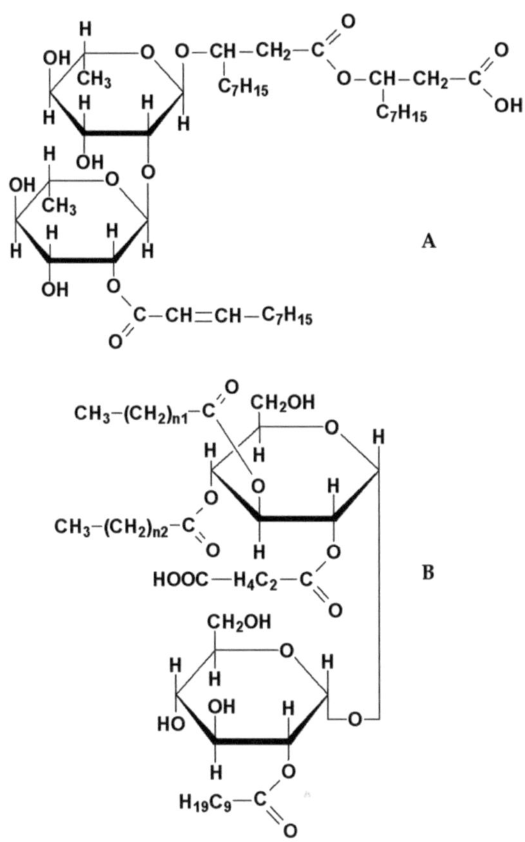

A

B

Fig. 7. The proposed structure of glycolipid biosurfactants: A - rhamnolipid B, produced by pseudomonads (Abdel-Mawgoud et al., 2010); B - succinoyl-dioctanoyl-decanoyl trehalose: $n_1 = n_2 = 6$; succinoyl-dioctanoyl-*di*decanoyl trehalose: $n_1 = n_2 = 6, 8$, produced by rhodococci (Rapp & Gabriel-Jurgens, 2003; Tuleva et al. , 2008)

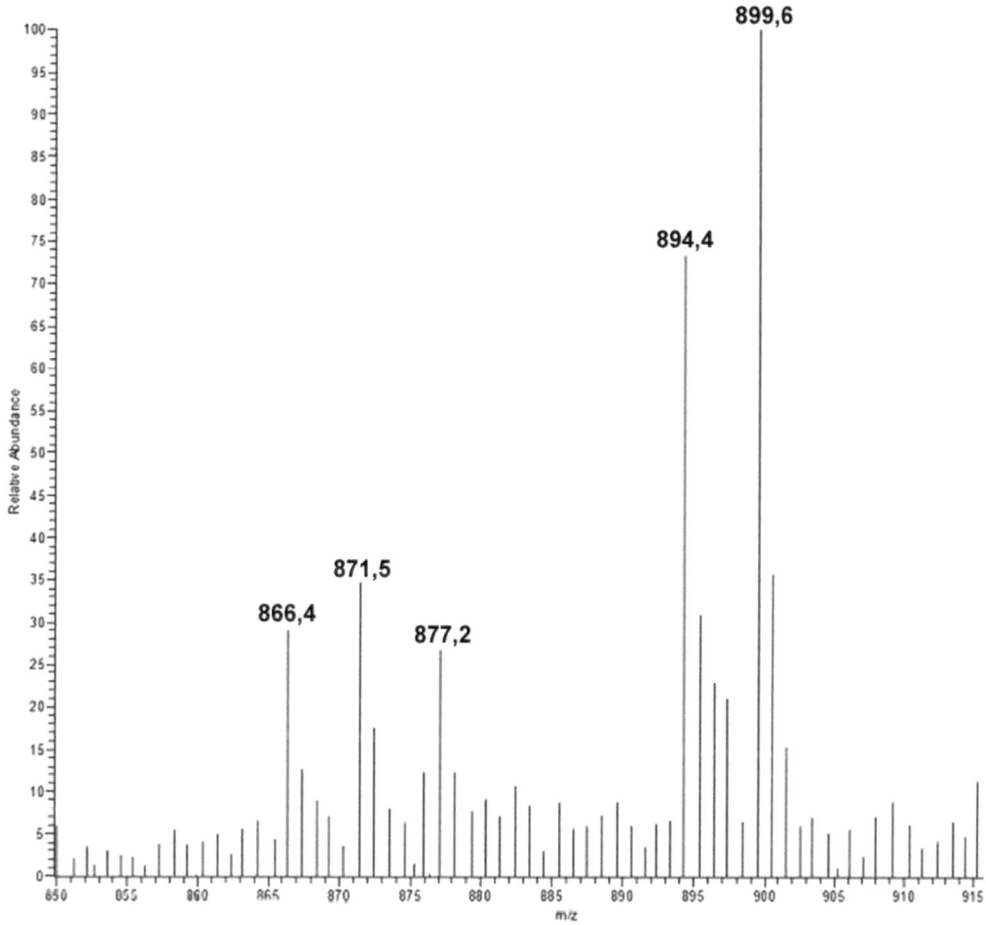

Fig. 8. The mass spectrum of glycolipid biosurfactants produced by the strain Rhodococcus sp. X5

Thus, all the microorganisms evaluated in this study were able to synthesize biosurfactants. The most efficient producers of exo-type biosurfactants were rodococci bacteria when cultivated on hexadecane and when cultivated in glucose they produced cell-bound biosurfactants (endo-type). Pseudomonads produced only extracellular biosurfactants. The formation of homologous di-rhamnolipids by the strain *Pseudomonas putida* BS3701 and the strain *Pseudomonas fluorescens* 142NF was demonstrated. The studied rhodococci spp. strains S67, X5 and S26 produced compounds of glycolipid nature, two of which were related to succinoyl-trehalose lipids.

2.4 Pilot testing of the "MicroBak" biopreparation for *in situ* clean-up of oil-spilled soil

Pilot testing was carried out in an open environment located in the municipal waste refinery of Pushchino from September 2006 to September 2007. The land tested included four plots of 1 m^2 each (Fig. 9). The composition of crude oil used as a model contaminant (25 g oil per 1 kg soil) is presented in Table 4.

Density, g/cm^3	Water content, %	Salts content, mg/ml	Contamination, %	Sulfur content, %	Composition		
					Hexane fraction, %	Benzene fraction, %	Alcohol-benzene fraction, %
0,868	0,06	45	0,0080	1,42	62,29	13,49	11,21

Table 4. Properties of the crude oil used

Plot 1 Control	**Plot 3** Consortium "MikroBak" fertilizer
Plot 2 Control with crude oil	**Plot 4** Consortium "MikroBak" fertilizer

Fig. 9. The scheme of the pilot field experiment

The degree of oil degradation was assessed in plot 2 (control) and plot 4 (where the consortium "MicroBak" (10^7 colony forming units (CFU)/g soil) and the "Nitroammophoska" fertilizer (3 g per 1 kg soil) were introduced) as indicated in Fig. 10. The pilot testing indicated that the oil degradation degree was higher in plot 4 where microorganisms and the fertilizer were introduced. In plot 4, 34% of the oil was degraded after two months of testing, while only 22% of the oil was degraded in the control plot (plot 2).

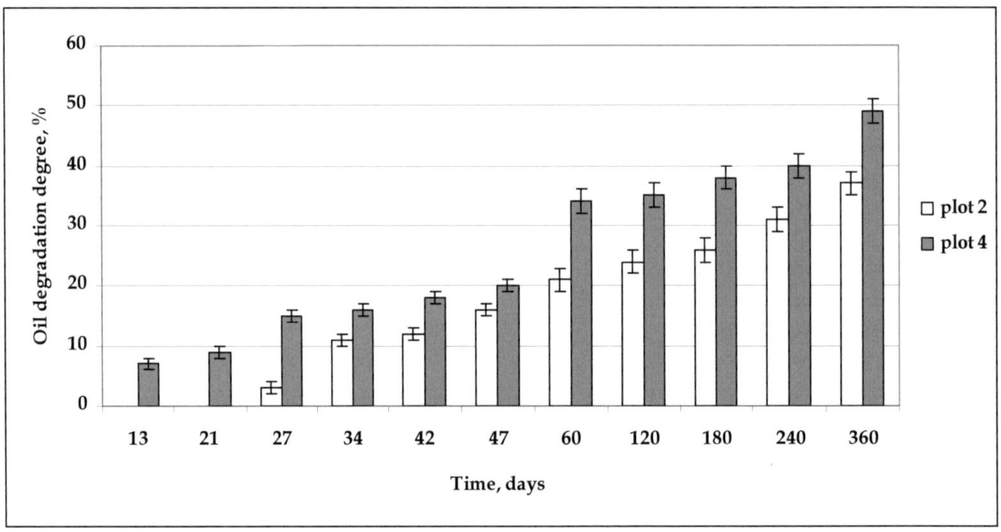

Fig. 10. Oil removal in the pilot field experiment

After 360 days of testing (end of field testing, Fig. 10) the total amount of oil removed was 49% and 38% for plots 4 and 2, respectively.

Soil toxicity evaluation in the previously bio-treated plot (plot 4) compared to the toxicity of the control plot (plot 2) was conducted using oat as bioindicator. The results obtained after a month of testing demonstrated that the soil toxicity decreased in plot 4 where the consortium "MicroBak" and a fertilizer were introduced (Fig. 11).

A B

Fig. 11. Pilot testing: soil phytotoxicity evaluation. June, 2007. A - Control plot (soil with oil), B - Plot 4 pre-treated with the consortium "MicroBak" and a fertilizer

2.5 Pilot testing of the "MicroBak" biopreparation in an industrial site polluted with hydrocarbons

Pilot testing of the biopreparation was performed in the territory of JSC "Tulskaya Toplivno-Energeticheskaya Compania" (Tula region, Kireevsky rajon, Rozhdestvenka) from September to November, 2007 at temperatures ranging from 0-22°C (Table 5). Soil was polluted by a mixture of oil products (diesel fuel, gasoline, gas condensate, oil fuel) due to the emergency flows during handling and pipelines breaking.

The polluted sites were sanded, and then the sand : soil mix (to 5 cm in depths, 1 : 2) was removed to an open plot where the soil clean-up process from took place. The initial level of oil pollutants in the soil was 14.2 g/kg soil (1.4 wt% oil residuals). The clean-up was performed by introducing "MicroBak" biopreparation (5×10^8 CFU/kg soil) and the mineral fertilizer "Nitroammophoska" (1.5 g/kg soil). After 14 days of testing, additional amounts of biopreparation (5×10^8 CFU/kg soil) and fertilizer (1.5 g/kg soil) were introduced into the polluted soil by ground loosening in depths of 10 to 20 cm.

The bioremediation efficiency of the polluted soil was estimated by determining the residual content of hydrocarbons (Table 5). After one month of testing, the content of hydrocarbons in the polluted soil decreased by 99.6% due to the simultaneous supplementing of the biopreparation and mineral fertilizer into the soil, while in the control site, the percentage of oil components removal was 90.7%.

Sampling date	Air temperature, °C	Soil temperature, °C	pH of soil water extract	Soil humidity, %
September 14, 2007	9	11	7.35	24
September 28, 2007	22	16	7.11	20
October 17, 2007	15	8	7.41	28
November 15, 2007	0	0	7.32	30

Table 5. Characteristics of soil and air during field tests

Sampling date	Residual content of oil products, mg/kg soil		Clean-up degree, %	
	Control plot	Experimental plot	Control plot	Experimental plot
September 14, 2007	13026±4559	11994±4198	-	-
October 17, 2007	1214±425	52.0±23.4	90.7	99.6
November 15, 2007	19.6±8.8	18.0±8.1	99.8	99.8

Table 6. Residual concentration of hydrocarbons in the control plot and in the bio-treated plot

After two months of field testing, the number of heterotrophic and hydrocarbon-oxidizing microorganisms was determined (Fig. 12). Figure 12 indicates that the addition of mineral fertilizers and soil tillage in depths from 10 to 20 cm aided the growth of microorganisms (by two orders of magnitude) compared to the control plot.

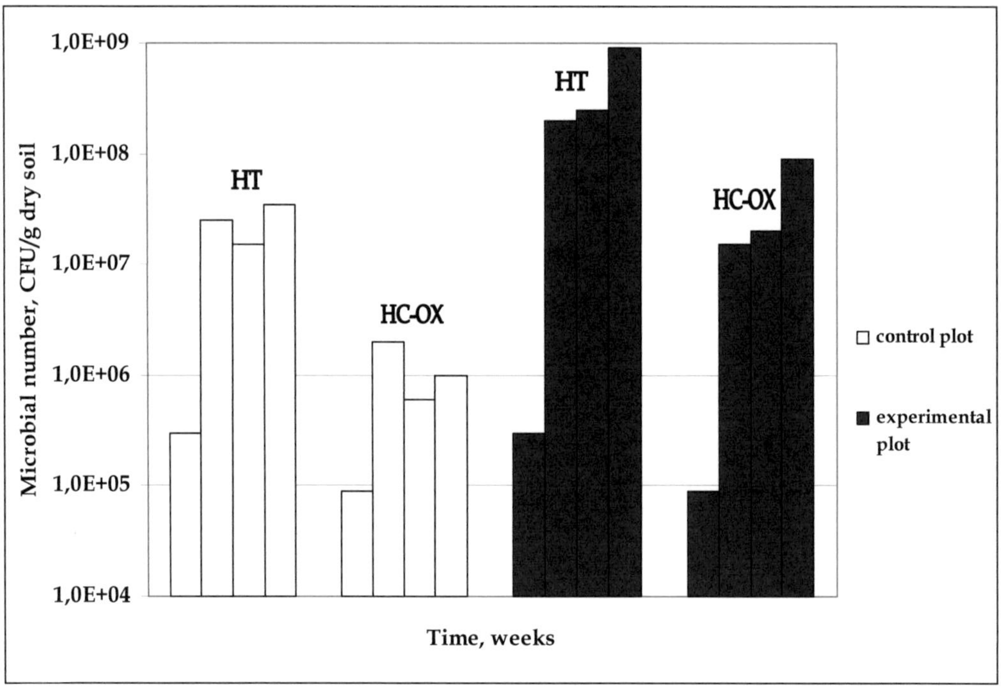

Fig. 12. Kinetics of microbial populations number (HT - heterotrophic microorganisms, HC-OX - hydrocarbon-oxidizing microorganisms)

These pilot test results demonstrate the bioremediation efficiency and the potential *in- situ* bioremediation of the biopreparation at low temperatures (0-15°C).

3. Oil biodegradation by the "V&O" consortium in a wide temperature and pH range at high levels of petro-pollution

Highly-polluted areas are common in sites of oil extraction, oil transportation, and oil storage such as tailings ponds, aged oil fuel spots, and oil storage pits. Thus, the application of specialized microorganisms' preparations capable of degrading high concentrations of oil in the polluted areas in a wide range of temperatures and pH values is of special interest.

3.1 Development of the consortium "V&O"

Seven bacterial strains obtained from the Laboratory of Plasmid Biology of IBPM RAS and ten bacteria obtained from JSC "Biooil" were chosen to formulate the specialized biopreparations.

All the microorganisms were capable of growing on crude oil, diesel fuel, and fuel oil. The strains 142NF and *Pseudomonas putida* F701 demonstrated the ability of using naphthalene or salicylate as the sole carbon and energy source. While the strains *Rhodococcus* sp. S25, *Rhodococcus* sp. S67, *Pseudomonas putida* F701, 5 and *Acinetobacter baumannii* 1B were able to use toluene as the sole carbon and energy source. Microorganisms 1A, 142NF and F701 were not capable of degrading decane, nonane, and hexadecane. The strains S67, *Rhodococcus* sp. X5, *Rhodococcus* sp. X25, 142NF and F701 showed capability of degrading benzoate.

All the strains evaluated (17 strains) showed tolerance to sea salt at a concentration of 5 wt% in a liquid mineral medium polluted with 2% of diesel fuel. Strains 142NF and 2B utilized diesel fuel in the presence of 7 wt% of salt concentration, while the bacterium X5 was capable of degrading diesel fuel at a salt concentration of 10 wt%. Moreover, the strains from IBPM RAS were capable of degrading diesel fuel in a pH range of 5 to 7, with the strains S25, S26, and F701 being resistant to a pH of 8. Most bacteria from the "Biooil" collection, excluding the strains 2C, 3, 4, and 5, were able to grow in the liquid mineral medium containing 2% of oil diesel fuel in a pH range from 5 to 9. The growth of strains 1B, 2A, *Serratia* sp. 6, and the *Acinetobacter baumannii* 7 was observed at acidic conditions (pH 4) and at basic (pH 10) conditions. Almost all the microorganisms studied were able to degrade oil and diesel fuel (2%) in a temperature range of 4°C to 30°C, while strains 4, 2B, and 2C were able to grow in a temperature range of 14 to 30°C. Strains F701, 1B, 6, 7, S26 and S25 demonstrated the capability of pollutant degradation (oil or diesel fuel) at 42°C.

Bacterial growth in Evans medium with different oil content was evaluated at temperatures of 24°C and 4°C. The highest growth values at 24°C in 10 days of testing in a medium containing 40% of oil were detected only for the strains S25, S26, 1B, 6, and F701. At 4°C under similar conditions, the best bacterial growth was only detected for the S26 strain. Bacteria S25, S26, 1B, 6, F701, X25, and 7 mineralized oil at a concentration of 30% at room temperature. Strains 7, 6, and S26 were capable of degrading similar oil content at 4°C. Under the conditions of microorganisms cultivation in the liquid mineral medium at concentrations of sea salt of 40% and 3% at 24°C, the best cultivation results were obtained for strains F701,1B, S26, 2C, and 6. At the same cultivation conditions but a temperature of 4°C, microorganisms 1B and S26 were capable of growing. Strains F701, 1B, S26, 2C, 7 and 6 were effective in mineralizing diesel fuel (30% concentration) in the presence of 3% of sea salt at 24°C and at lower temperatures (4°C).

Thus, microorganisms S26, S25, X25, 2C, 1B, 6, F701, and 7 mineralized crude oil and diesel fuel at high concentrations (up to 30%) in the presence of sea salt concentration of 3% in a temperature range of 4 to 42°C and pH values of 4 to 10. These observations indicate that these strains would be effective in degrading oil under natural salinity conditions, for instance in bioremediation applications involving oil-spilled in stratum containing salted water.

The clean-up of acid soils from oil spills is often required (Gemmell & Knowlas, 2000). For these types of applications, the strains evaluated in this work show potential for the successful bioremediation of oil-polluted soils at conditions of high acidity.

The strains *Rhodococcus erythropolis* S26, *Rhodococcus sp.* S25, *Serratia sp.* 6, and *Acinetobacter baumannii* 7 demonstrated the highest efficiency in removing hexane fractions at 24°C. Bacteria *Acinetobacter baumannii* 1B and *Pseudomonas putida* F701 utilized asphalthene-tar oil components effectively. At a temperature of 24°C, bacteria of the genus *Pseudomonas* F701 and 142NF degraded hydrocarbons of benzene fractions. At low temperature (4°C), the ability to degrade benzene-alcohol fraction was revealed in strains *Rhodococcus erythropolis* S26, *Rhodococcus sp.* S25, *Serratia sp.* 6 and *Acinetobacter baumannii* 1B.

An important criterion for choosing microorganisms to formulate a consortium was the presence of catabolic plasmids. From the 17 strains screened, plasmids were discovered in 8 of the strains as follows: *Rhodococcus* sp. X25, *Rhodococcus* sp. S25, *Pseudomonas putida* F701, *Acinetobacter baumannii* 7, *Serratia* sp. 6, *Acinetobacter baumannii* 1B, *Rhodococcus erythropolis* S26, and the strain 4 (Fig. 13).

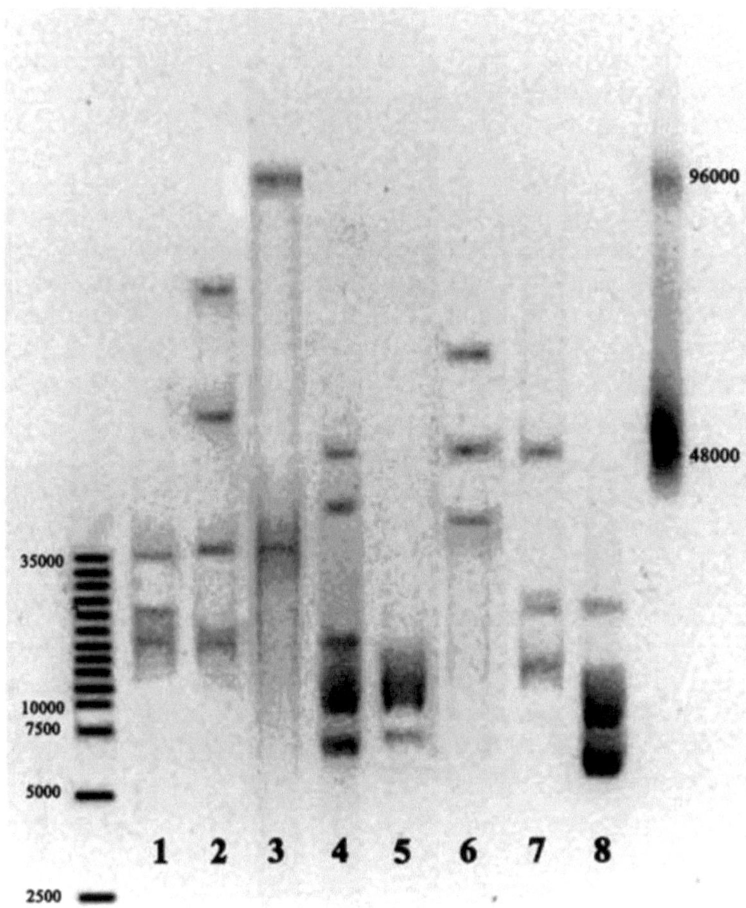

Fig. 13. Electrophoregram of plasmid DNA of degrader strains: *1 kb DNA-marker*, 1 – *Rhodococcus* sp. X25 (35, 21, 17 kb), 2 - *Rhodococcus* sp. S25 (77, 55, 36, 20 kb), 3 - *Pseudomonas putida* F701 (96, 36 kb), 4 - *Acinetobacter baumannii* 7 (50, 42, 20, 12, 6 kb), 5 - *Serratia* sp. 6 (14, 7 kb), 6 - *Acinetobacter baumannii* 1B (67, 48, 34 kb), 7 - *Rhodococcus erythropolis* S26 (48, 30, 15 kb), 8 - strain 4 (30, 8, 6 kb), *λ Ladder PFG*

The naphthalene-biodegradation plasmid pNF142 in the strain *Pseudomonas sp.* 142NF(pNF142) has been previously described (Gomes et al., 2005).

Determination of the localization of naphthalene-biodegradation genes (in the strain F701) or hexadecane-degradation genes (in the strains 1B, 4, 6, 7, S26, X25, S25) was performed through experiments on the spontaneous elimination of plasmids by strains cultivation under non-selective conditions (in Luria-Bertany medium) and through ethidium bromide treatment (10 µg/ml medium). Prolonged cultivation in the rich medium of *Pseudomonas putida* F701 and the strain 4 promoted the appearance of plasmid-free cell in a population. *P. putida* F701 and the strain 4 eliminants were not capable of growing on naphthalene and hexadecane, respectively. The plasmids discovered were stably inherited in strains 1B, 6, 7, S26, X25, and S25.

Ethidium bromide treatment induced the appearance of 1B, 6, 7, S26 and S25 plasmid-free eliminants loosing the ability to grow on hexadecane. Thus, a plasmid control for the hexadecane degradation was proposed.

Mating experiments were performed to discover possible conjugative plasmids. The naphthalene-degrader F701 and hexadecane-degraders 1B, 4, 6, 7, S26, and S25 were used as donors, and *Pseudomonas putida* KT2442 - as the recipient. Microorganisms 7 and F701 were shown to bear conjugative catabolic plasmids that are non-competitive in nature. These conjugative catabolic plasmids could enhance the degradative potential of soil microbial populations by disseminating genes among indigenous bacteria. The question about plasmid conjugativity of strains 1B, 4, S25, 6, and S26 is still open, however they could participate in the distribution of catabolic genes by transformation. The host cell dies and is lysed and the DNA with catabolic genes fluxes into the environment and may stay intact (Prozorov, 1999). Thus, applying microorganisms bearing conjugative plasmids of biodegradation seems to be promising when developing biopreparations for bioremediation technologies.

To create a consortium two steps were carried out: 1) analysis and further combination of physiological, metabolic and degradative traits of microbial properties and the presence of catabolic plasmids in strains as well; 2) selection of a mixed consortium during batch cultivation of the most active microorganisms in a liquid mineral medium with oil as a sole carbon and energy source (Table 7). Based on the traits mentioned above the following bacteria were chosen: *Rhodococcus erythropolis* S26, *Rhodococcus* sp. S25, *Acinetobacter baumannii* 1B, *Serratia* sp. 6, *Pseudomonas putida* F701, and *Acinetobacter baumannii* 7. The microorganisms were inoculated into a liquid Evans medium (pH 4) supplemented with 15% (v/v) of crude oil (Table 4) at temperatures of 4°C or 24°C.

The strains *Rhodococcus erytropolis* S26, *Acinetobacter baumannii* 1B, *Acinetobacter baumannii* 7, and *Pseudomonas putida* F701 are ahead in growth rate, which makes them dominant in the mixed population. This mixture of microorganisms was designated as "V&O" consortium. It should be noted that bacteria *Acinetobacter baumannii* 1B, *Rhodococcus erytropolis* S26, *Acinetobacter baumannii* 7 bear hexadecane degradation plasmids and the strain *Pseudomonas putida* F701 bears naphthalene biodegradation genes on a conjugative plasmid pF701a. Thus, this selected consortium of plasmid-bearing degrader strains is promising for cleaning-up soil and water sites polluted with high concentrations of oil in a wide temperature and pH range.

Criteria	Microorganisms
Medium pH 5 - 8	1A, **1B**, **2A**, 2B, **6**, **7**, S25, S26 and F701
5% NaCl	1A, 1B, 2A, **2B**, 2C, 3, 4, 5, 6, 7, **X5**, X25, S25, S26, S67, **142NF** and F701
Temperature 4 - 42°C	1B, 6, 7, S25, S26 and F701
Oil content 30% at 24°C	**1B**, **6**, 7, X25, **S25**, **S26** and **F701**
Oil content 20% at 2 - 4°C	**6**, **7**, S25, **S26** and F701
Diesel content 30% at 24°C	**1B**, **2C**, **6**, **7**, **S26** and **F701**
Diesel content 30% at 4°C	**1B**, 6, 7, **S26** and F701
Biosurfactants active producer	1B, 7, X5, S26, S67 and F701
The presence of catabolic plasmids	1B, 4, 6, 7, S26, S67, X25, S25 and F701

Table 7. Criteria for formulation of a consortium of microorganisms. Strains with the best criteria indicators are bold-typed.

3.2 "V&O" consortium: oil degradation efficiency

The oil degradation efficiency of the selected "V&O" consortium was assessed through crude oil degradation experiments carried out in a liquid Evans medium at temperatures of 4°C, 24°C, and 50°C using IR-spectrometry as the analytical technique. The total testing time was 30 days and crude oil concentration was 150 g/l. The maximum decrease in oil concentration was observed after 15 days of testing (Fig. 14). It is probable that during the first 15 days of testing, the microorganisms rapidly degraded *n*-alkanes (up to C_{12}) and cyclic compounds with one aromatic ring. A drop of degradation rate was observed in the second half of the experiment; in this period microorganisms targeted the destruction of heavy (asphalthene) fractions, which are the most recalcitrant oil components.

In terms of temperature, the highest degradation degree was observed in a system at a temperature of 4°C with an oil removal percentage of 44% higher that the oil removal observed in the control test without microorganisms. Oil degradation in the systems studied was assessed against a control test for abiotic oil removal. Since, of course, abiotic removal at 4°C was less than at 24°C, the residual oil content in the control at low temperature was higher that the one at 24°C.

The lowest oil removal (only 1%) was detected in a liquid mineral medium at 50°C (Fig. 14). At a temperature of 50°C, the consortium bacteria grew under critical conditions in the absence of light hydrocarbon fractions that are easily assimilated by the consortium as carbon and energy source. The lack of light hydrocarbon fraction is due to the rapid evaporation of these fractions, which in turn, increases the concentration of heavy compounds, which are more difficult to target during the initial growing stages of microorganisms.

Thus, these results demonstrate the high effectiveness of the "V&O" consortium for oil degradation at low and moderate temperature.

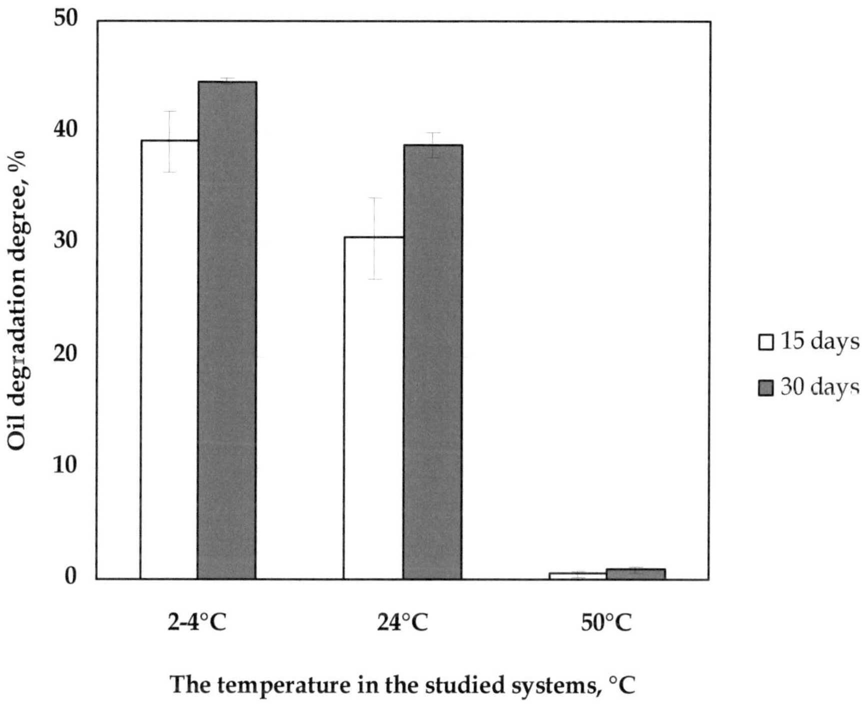

Fig. 14. "V&O" consortium: oil degradation as a function of temperature. Temperatures evaluated were 4°C, 24°C, and 50°C in a liquid mineral medium with 15% (v/v) of oil.

3.3 "V&O" consortium effectiveness versus "Microbak" and "Biooil"

The "Microbak" biopreparation and the "V&O" consortium, which were prepared at the Laboratory of Plasmid Biology (IBPM RAS), and the biopreparation "Biooil", which was provided by JSC "Biooil", were evaluated to determine their effectiveness as oil degraders.

The biopreparation "MicroBak" contains bacteria of the genera *Pseudomonas* and *Rhodococus;* while the biopreparation "Biooil" includes *Bacillus, Sacharomyces, Acinetobacter, Enterobacter.*

Assessment of consortia efficiency was performed in a liquid mineral medium containing crude oil (15% v/v) as a sole carbon and energy source (Table N). Infrared spectroscopy was used as the analytical technique to determine the concentration of residual oil in the samples. Figures 15 and 16 summarize the results of oil degradation during cultivation of the three consortia at temperatures of 4°C and 24°C in a period of 30 days. The "V&O" consortium at 4°C rendered the highest oil hydrocarbons removal (44%) after 30 days, while the oil degradation degree in the systems inoculated with the "Biooil" and the "MicroBak" biopreparations reached 36% and 40% after 30 days, respectively (Fig. 15).

At a temperature of 24°C, the V&O consortium induced the highest oil hydrocarbons removal both after 15 days (31%), and after 30 days (39%) (Fig. 16). The "Biooil" and "MicroBak" biopreparations rendered similar oil degradation after 30 days of cultivation with percentages of 36% and 34%, respectively.

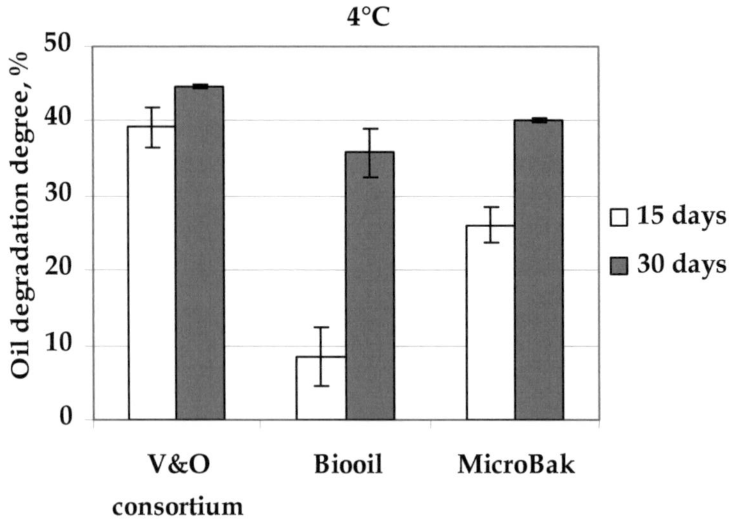

Fig. 15. Effectiveness of "V&O" consortium, biopreparations "Biooil" and "MicroBak" in degrading oil in a liquid mineral medium with 15% (v/v) oil at low temperature (4°C).

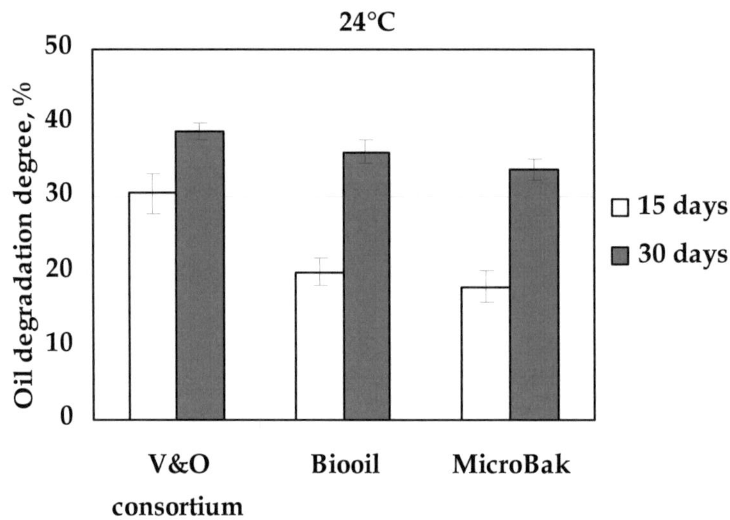

Fig. 16 Effectiveness of "V&O" consortium, biopreparations "Biooil" and "MicroBak" in degrading crude oil in a liquid mineral medium with 15% (v/v) oil at moderate temperature (24°C).

The oil consumption by bacteria in the "MicroBak" and in the "Biooil" biopreparations was slower in comparison with the "V&O" consortium strains, which showed a longer adaptation period to the pollution conditions that was confirmed by bacterial number change (data not shown).

4. Efficiency of plant-microbial consortia

Some plants apply a number of mechanisms to overcome the effect of toxic pollutants. These mechanisms include excretion, conjugation of toxic substances with intracellular compounds and further compartmentalization of conjugates, degradation of pollutants to the cellular metabolites and carbon dioxide. Thus, in the process of soil decontamination from oil and oil products, it seems to be promising to make use of the synergistic approach of microorganisms and plants associations.

Sterile soil experiments were conducted to study oil degradation degree, the interactions between microorganisms within the selected consortium, and to establish the influence of oil hydrocarbons - degrading microorganisms on plants growth.

4.1 Screening of plants to create an effective plant-microbial association

The screening of 20 different plants indicated that sunflower, corn, barley, lawn grass (a mix of grasses with the basic red fescue grass), and some kinds of beans (a string bean, peas) are resistant to petro-pollution (2% (v/w) of crude oil). During the screening process, besides the resistance to toxic oil hydrocarbons shown by the root system of the plants (spur or fibrous), it was also important to consider the branchiness and soil volume coverage of the plants. Thus, it was found that barley and lawn grass roots had a dense biomass and they occupied the highest soil volume. Thus, the associations "V&O - barley" and "V&O – lawn grass" were selected for evaluation in further experiments.

4.2 Removal of oil content by microbial-plant associations

Laboratory tests were carried out to evaluate the oil degradation performance of the "V&O - barley" microbial-plant association. Oil degradation was assessed using a "plant – single strain" association and a "plant – V&O consortium" association in sterile conditions for a testing period of 10 days.

The introduction of microorganisms into soil model systems polluted with oil and seeded with plants (barley) promoted the detoxification of the soil. The degree of detoxification was determined by measuring the shoot length of the plants in the presence of microbial-plant associations and without the microbial-plant associations (Control test: plants + crude oil without microorganisms) (Fig. 17). The detoxification induced by the plant-microbial associations was probably caused by the ability of bacteria to colonize plant roots and rhizosphere with the simultaneous degrading of crude oil, which minimizes the toxic effect of crude oil on plant development.

The number of bacteria in the roots (in water washouts from the roots) was higher (by 1-2 orders of magnitude) than in the rhizosphere (the number of bacteria was determined, using washouts from the near-rhizosphere soil). It has been previously demonstrated that during the simultaneous cultivation of *Rhodococcus* and *Pseudomonas* strains, the number of rhodococci strains decreases in more than one order of magnitude, while the number of pseudomonads strains increases (Baryshnikova et al., 2001). Our results demonstrate the absence of negative interactions between the "V&O - barley" microbial-plant association such as antagonism or allelopathy.

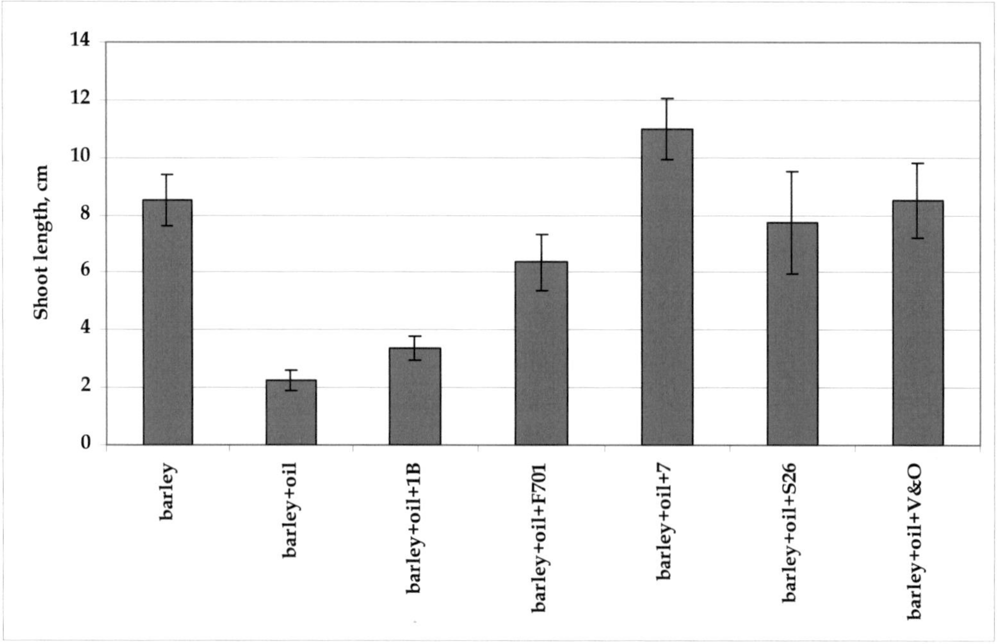

Fig. 17. The length of barley shoots in a sterile model experiment after 10 days of cultivation

In terms of percentage of oil degradation as a result of the application of microbial-plant associations, the experimental results indicate that the highest percentage of oil degradation (19%) was obtained by the application of the "V&O" consortium–barley association (Fig. 18). The other plant-microbial associations evaluated -"*Rhodococcus erythropolis* S26 – barley", "*Acinetobacter baumannii* 1B – barley", "*Acinetobacter baumannii* 7 – barley", and "*Pseudomonas putida* F701 – barley" – rendered a lower percentage of oil removal (< 15%) (Fig. 18).

During the microbial oil biodegradation within microbial-plant associations a cooperation process among bacteria and the plant accelerates the effective utilization of hydrocarbons. The contributions of the plant on the biodegradation of oil are related to enzymatic systems and pollutants detoxification mechanisms. Naumann et al. (1991) reported that the interaction of plant roots with organic compounds (including oil hydrocarbons) induces peroxidase activity that could function in the cell as a protection mechanism and/or influence on the degradation of pollutants present in the environment.

To establish if the microbial-plant association -"V&O – barley"- could be made more effective by the addition of mineral fertilizers non-sterile laboratory experiments were conducted in open systems with oil-spilled soil in the presence of a fertilizer ("Nitroammophoska") and without fertilizer for a period of 14 days. The experimental results demonstrated that supplementing the "V&O – barley "association with fertilizers did not influence on plant growth and did not change the number of the introduced microorganisms.

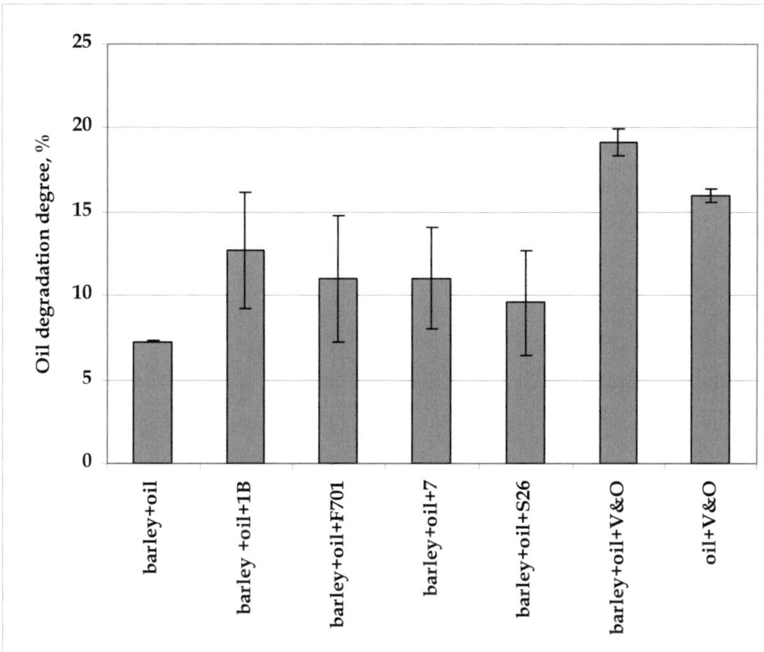

Fig. 18. Oil removal in model soil systems as a function of plant-microbial associations. Testing period: 10 days

5. Field tests of "V&O" consortium and JSC "Biooil" biopreparations

Field experiments on the Yamal-Nenetsky Autonomous Region in Western Siberia on sites of the "Pogranichnoe" oil deposit were conducted during July-August, 2008, with the objective of evaluating the efficiency of the microbial consortium "V&O" and the biopreparations "Biooil-SN" and "Biooil-Jugra" in degrading oil pollutants. The level of soil pollution in the sites ranged from 15 to 110 g/kg soil in an area of 8300 m^2. In this field testing, favourable conditions for the growth of microorganisms were created by adding mineral and organic fertilizers (calurea), tillaging, and water sprinkling.

After two months of the introduction of the "V&O" consortium the oil clean-up degree on the test site was 80%, while the JSC "Biooil" biopreparations rendered a clean-up ranging from 50 to 70%. The "V&O" consortium show high efficiency in degrading oil hydrocarbons already in a month after treatment (Fig. 19).

Before the experiment, the concentration of indigenous oil-degrading microorganisms was low (5.2×10^1 CFU per g soil) due to the high oil content (>10% of soil mass by weight). However, after the treatment with the "Biooil" preparations and the "V&O" microbial consortium the number of hydrocarbons-utilizing microorganisms increased by 3 orders of magnitude, this increased in microorganism population intensified the biodegradation of oil. These results are confirmed by the oil content removal data presented in Fig. 19.

Chemical toxicity of crude oil for living organisms is not always evident (Pikovsky, 2003). Often recovery of oil-spilled soils is traced by a pollutant change in the soil and sometimes pollutant removal does not show the decrease of its toxicity for living organisms. The soil is known to be a reservoir that can accumulate high concentration of pollutants. Thus, the soil can be highly polluted but non-toxic and vice versa, low polluted but highly toxic for living organisms, thus it is necessary to control the bioremediation process in terms of both residual oil content and phytotoxicity indicators (Knoke et al., 1999; Phillips et al., 2000). Therefore, phytotoxycity of the soil was visually evaluated by taking photographs of the site after 2 months of bioremediation. The photographs revealed a biomass increase on the plot treated with "V&O" consortium, which is an indication of low phytotoxicity.

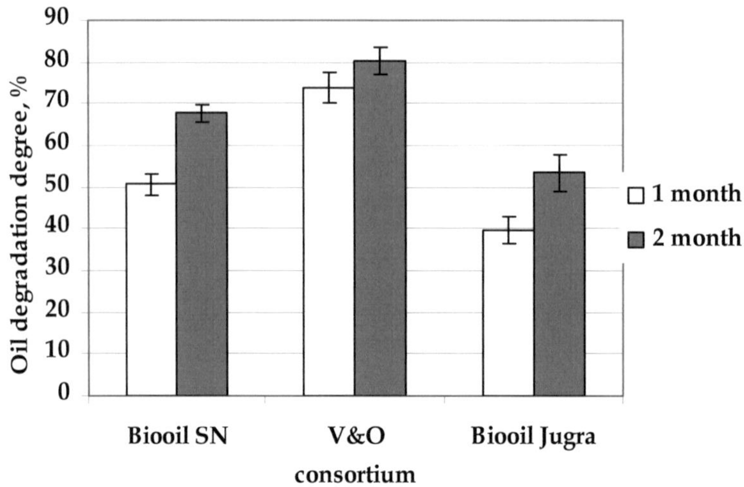

Fig. 19. Oil degradation in a field experiment in Western Siberia with biopreparations and a consortium

6. Conclusions

Results obtained from laboratory and field experiments demonstrated the efficiency of the "V&O" consortium in degrading oil. The percentage of oil removal obtained from the "V&O" consortium is higher (by 10 to 25%) if compared to the efficiency of the JSC "Biooil" biopreparations. Similarly, laboratory and field tests also demonstrated the high oil-oxidizing activity of the consortium "V&O" and the biopreparation "MicroBak". Thus, the consortium "V&O" and the biopreparation "MicroBak" are applicable to clean-up soil and water biotopes from crude oil and oil products spills at temperatures ranging from 4°C to 42°C.

7. Acknowledgment

The authors acknowledge the financial support provided by the Russian Foundation for Basic Research, Grants 11-04-97561-r_centre_a, 11-04-97562-r_centre_a.

8. References

Abdel-Mawgoud, A.M., Lépine, F. & Déziel, E. (2010). Rhamnolipids: Diversity of Structures, Microbial Origins and Roles. *Applied Microbiology and Biotechnology*, Vol.86, No.5, (May 2010), pp. 1323–1336, ISSN 0175-7598

Babaev, E.R., Mamedova, P.Sh., Quliyeva, D.M. & Movsumzade, M.E. (2009). The Choice of Active Hydrocarbon Destructing Microorganism for Clean-Up of Oil-Polluted Soils of Balakhanskoe Deposit. *Bashkirian Journal of Chemistry* (in Russian), Vol.16, No.1, (January 2009), pp. 103-106, ISSN 0869-8406

Banat, I., Franzetti, A., Gandolfi, I., Bestetti, G., Martinotti, M., Fracchia, L., Smyth, T. & Marchant, R. (2010). Microbial Biosurfactants Production, Applications and Future Potential. *Applied Microbiology and Biotechnology*, Vol.87, (June 2010), pp. 427–444, ISSN 0175-7598

Baryshnikova, L.M., Grishchenkov, V.G., Arinbasarov, M.U., Shkidchenko, A.N. & Boronin, A.M. (2001). Biodegradation of Oil Products by Degrader Strains and Their Associations in Liquid Media. *Applied Biochemistry and Microbiology*. Vol.37, No.5, (September-October 2001), pp. 542-548. ISSN 0555-1099

Carhart, G. & Hegeman, G. (1975). Improved method of selection for mutants of *Pseudomonas putida*. *Applied Microbiology*, Vol.30, No.6, (December 1975), pp. 1046, ISSN 0003-6919

Chernyakhovsky, E.R., Shkidchenko, A.N., Yuhmatova, O.A. & Chushkin, Z.Yu. (2004). Application of Various Technologies for Clean-Up of Oil, Petroleum Products and Oily Waste Processing Spills. *Problems of Security and Emergency* (in Russian), No.6, pp.34-40, ISSN 0869-4176

Delille, D., Delille, B. & Pelletier, E. (2002). Effectiveness of Bioremediation of Crude Oil Contaminated Subantarctic Intertidal Sediment: the Microbial Response. *Microbial Ecology*, Vol.44, No.2, (August 2002) pp. 118-126, ISSN 0095-3628

Desai, J. & Banat, I. (1997). Microbial Production of Surfactants and Their Commercial Potential. *Microbiology and Molecular Biology Reviews*, Vol.61, No.1, (March 1997), pp. 47–64, ISSN 1092-2172

Evans, C.G.T., Herbert, D. & Tempest, D.B. (1970). The continuous cultivation of microorganisms. 2. Construction of a chemostat. *Methods in Microbiology*, Vol.2, pp. 277–327, ISSN 0580-9517

Gemmell, R.T. & Knowles C.J. (2000). Utilisation of aliphatic compounds by acidophilic heterotrophic bacteria. The potential for bioremediation of acidic wastewaters contaminated with toxic organic compounds and heavy metals. *FEMS Microbiology Letters*, Vol.192, No.2, (September 2000), pp. 185-190, ISSN 0378 1097

Gomes, N.C.M., Kosheleva, I.A., Abraham, W.-R. & Smalla K. (2005). Effects of the Inoculant Strain *Pseudomonas putida* KT2442 (pNF142) and of Naphthalene Contamination on the Soil Bacterial Community. *FEMS Microbiology Ecology*, Vol.54, No.1, (September 2005), pp. 21–33, ISSN 0168-6496

Harayama, S. & Rekik, M. (1990). The *meta* Cleavage Operon of TOL Degradative Plasmid pWWO Comprises 13 Genes. *Molecular and General Genetics*, Vol.221, No.1, pp. 113-120, ISSN 0026-8925

Kireeva, N.A., Vodopyanov, V.V. & Miftakhova, A.M. (2001). Biological Activity of Oil-Spilled soils (in Russian), 376, Gilem, ISBN 5-7501-0225-4, Ufa, Russia

Knoke, K.L., Marwood, T.M., Cassidy, M.B., Liu, D., Seech, A.G., Lee, H. & Trevors, J.T. (1999). A comparison of five bioassay to monitor toxicity during bioremediation of pentachlorophenol-contaminated soil. *Water, Air and Soil Pollution*, Vol.110, No.1-2, (December 1997) pp. 157-169, ISSN 0049-6979

Matsufuji, M., Nakata, K. & Yoshimoto, A. (1997). High Production of Rhamnolipids by *Pseudomonas aeruginosa* Growing on Ethanol. *Biotechnology Letters*, Vol.19, No.12, (December 1997), pp. 1213–1215, ISSN 0141-5492

Muthusamy, K., Gopalakrishnan, S., Thiengungal, K.R. & Sivachidambaram, P. (2008) Biosurfactants: Properties, Commercial Production and Application. *Current Science*, Vol.94, No.6, (March 2008), pp. 736-747, ISSN 0011-3891

Naumann, D., Helm, D. & Labischnski, H. (1991). Microbiological Characterizations by FT-IR Spectroscopy. *Nature*, Vol.351, No.6321, (May 1991), pp. 81-82, ISSN 0028-0836

Oborin, A.A., Kalachnikova, I.G., Maslivets, T.A., Bazenkova, E.I., Plesheva, O.V. & Ogloblina, A.I. (1988). Autoremediation and Recultivation of Oil-Spilled Soils of Urals and Western Siberia, In: *Restoration of oil-spilled soil ecosystems* (in Russian), 140-159, Nauka, ISBN 5-02-003358-8, Moscow, Russia

Phillips, T.M., Liu, D., Seech, A.G., Lee, H., & Trevors J.T. (2000). Monitoring Bioremediation in Creosote-Contaminated Soil Using Chemical Analysis and Toxicity Tests. *Journal of Industrial Microbiology & Biotechnology*, Vol.24, No.2, (December 1999), pp. 132-139, ISSN 1367-5435

Pikovsky, Yu.I., Gennadiev, A.N., Chernyansky, S.S. & Sakharov, G.N. (2003). The Problem of Diagnostics and Valuation of Oil and Oil Products Spills in the Soil. *Eurasian Soil Science*, No.9, pp. 1132-1140 ISSN 0032-180X

Prince, R. & Bragg, J. (1997). Shoreline Bioremediation Following the Exxon Valdez Oil Spill in Alaska. *Bioremediation Journal*, Vol.1, No.2, pp. 97-104, ISSN 1088-9868

Prozorov, A.A. (1999). Horizontal Gene Transfer in Bacteria: Laboratory Modeling, Natural Populations, and Data from Genome Analysis. *Microbiology* (in Russian), Vol.68, No.5, (September-October 1999), pp. 635-646. ISSN 0026-3656

Rapp, P. & Gabriel-Jurgens, L.H.E. (2003). Degradation of Alkanes and Highly Chlorinated Benzenes, and Production of Biosurfactants, by a Psychrophilic *Rhodococcus* sp. and Genetic Characterization of Its Chlorobenzene Dioxygenase. *Microbiology*, Vol.149, No.10, (October 2003), pp. 2879–2890, ISSN 1350-0872

Robert, M., Mercade, M.E., Bosch, M.P., Parra, J.L., Espuny, M.J., Manresa, M.A. & Guinea, J. (1989). Effect of the Carbon Source on Biosurfactant Production by *Pseudomonas aeruginosa* 44T1. *Biotechnology Letters*, Vol.11, No.12, pp. 871-874, (December 1989), ISSN 0141-5492

Satpute, S.K., Banpurkar, A.G., Dhakephalkar, P.K., Banat, I.M. & Chopade, B.A. (2010). Methods for investigating biosurfactants and bioemulsifers: a review. *Critical Reviews in Biotechnology*, Vol.30, No.1, (October 2009), pp. 127–144, ISSN 0738-8551

Tuleva, B.K., Ivanov, G.R. & Christova, N.E. (2008). Biosurfactant Production by a New *Pseudomonas putida* Strain. *Zeitschrift für Naturforschung C*, Vol.57, No.3-4, (March-April 2008), pp. 356-360, ISSN 0939-5075

Van Hamme, J.D., Singh, A. & Ward, P. (2003). Recent Advances in Petroleum Microbiology. *Microbiology and Molecular Biology Reviews*, Vol.67, No.4, (December 2003), pp. 503–549, ISSN 1092-2172

Ventosa, A., Joaquhn, J. & Oren, A. (1998). Degradation of Phenol in Synthetic and Industrial Wastewater by *Rhodococcus erythropolis* UPV-1 Immobilized in an Air-Stirred Reactor with Clarifier. *Microbiology and Molecular Biology Reviews*, Vol.62, No.2, (March 2002), pp. 504-544, ISSN 1092-2172

Volkova, O.V., Anokhina, T.O., Puntus, I.F., Kochetkov, V.V., Filonov, A.E. & Boronin, A.M. (2005). Effect of Naphthalene Biodegradation Plasmids on Physiological Characteristics of Rhizospheric Bacteria of the Genus *Pseudomonas*. *Applied Biochemistry and Microbiology* (in Russian), Vol.41, No.5, (September-October 2005), pp. 525-529, ISSN 0555-1099

Wallace, W.H. & Sayler, G.S. (1992). Catabolic Plasmids in the Environment. *Encyclopedia of Microbiology*, Vol.1, pp. 417 – 430, ISBN 0122268911, 9780122268915, 012226892X, 9780122268922, 0122268938, 9780122268939, 0122268946, 9780122268946

Zhang, Y. & Miller, R.M. (1994). Effect of a Pseudomonas Rhamnolipid Biosurfactant on Cell Hydrophobicity and Biodegradation of Octadecane. *Applied and Environmental Microbiology*, Vol.60, No.6, (June 1994), pp. 2101-2106, ISSN 0099-2240